A Natural History of Vision

Nicholas J. Wade

A Natural History of Vision

A Bradford Book
The MIT Press
Cambridge, Massachusetts
London, England

This book was set in Times New Roman on the Monotype "Prism Plus" PostScript Imagesetter by Asco Trade Typesetting Ltd., Hong Kong, and was printed and bound in the United States of America.

Library of Congress Cataloging-in-Publication Data

Wade, Nicholas.
  A natural history of vision / Nicholas J. Wade.
     p.   cm.
  "A Bradford book."
  Includes bibliographical references and index.
  ISBN 0-262-23194-8 (alk. paper)
  1. Vision—History. 2. Physiological optics—History. I. Title.
  [DNLM: 1. Vision—physiology. 2. Visual Perception—physiology.
WW 103 W121n 1998]
QP475.W24   1998
612.8′4—dc21
DNLM/DLC
for Library of Congress   97-22697
                                                                CIP

To Christine

Contents

All that we know of nature, or of existence, may be compared to a tree, which hath its root, trunk, and branches. In this tree of knowledge, perception is the root, common understanding is the trunk, and the sciences are the branches.
—Reid (1764, p. 424)

Preface

It is with more than a little trepidation that one who is trained in neither history nor languages embarks upon a history of any branch of science. My chosen topic is vision, and its history mirrors, in many ways, the history of science. Interest in the history of vision developed from my research as a visual scientist. The experiments I conduct touch upon issues that have fascinated natural philosophers since time immemorial. To be sure, they can now be examined with an experimental sophistication that has not previously been matched, but it is arrogant in the extreme to consider that technological advance is a substitute for theoretical insight. Many of the theoretical developments in vision have been made by using the system under study—vision. Hence, the impetus for the present work has been a desire to follow the ways in which visual phenomena themselves have been described over the ages. However, the ages over which the descriptions are assembled do not reach to the present. They cease at about 160 years ago, when a variety of instruments were devised specifically to investigate visual phenomena. It could be said that from about 1840 vision became an experimental science, and its study was displaced from the natural environment to the laboratory. This argument can be more readily sustained for spatial than for color vision. Experimental advances in both departments were made when spatial and color phenomena could be studied independently of their object base. In the case of color, the prismatic spectrum enabled different components of white light to be isolated and combined with others, without reference to colored objects. From the seventeenth century onwards these methods of manipulating colored light have provided the foundations for color science. The case was somewhat more complicated for spatial vision, because it deals with the three-dimensional nature of objects. About 160 years ago an instrument (the stereoscope) was invented that enabled depth perception to be studied independently of viewing solid objects. This opened avenues of experimental enquiry that had not been possible previously, and it is here taken as a turning point in the history of vision.

This is an unconventional history of vision because it is based almost entirely on descriptions of visual phenomena in the (usually translated) words of the natural philosophers who reported them. The book is a history because it charts

the descriptions in chronological sequence, and it is a natural history because
it is principally concerned with observations of naturally occurring events.
Natural histories, like those of Pliny and Buffon, are catalogues of phenomena,
classified according to some scheme. The scheme adopted here reflects the
segments into which the study of vision has been divided in the past. Optics
in antiquity addressed the analysis of light and color. Descriptions of visual
phenomena covered a much wider spectrum, incorporating subjective states
(like afterimages), motion phenomena (like aftereffects), space perception (as
in apparent size and distance), and illusions (like the changes in apparent size
of the moon at different stages of its nightward path). Another constant source
of interest has been vision with two eyes. As a broad generalization, the aspects
of vision that warranted recording were those that were unusual. In normal
vision, this could follow from visual experiences that did not correspond to the
other senses. Light and color are themselves quintessentially visual, but so are
afterimages, or the double appearances of objects. Another source derived from
disease or operational intervention, like the effects that could be seen during
disease, particularly of the eye, or the beneficial effects of surgery, like those
of cataract removal. The nature of light and the structure of the eye are not
the primary concerns of this history, but they need to be surveyed briefly in
order to provide a context for the visual phenomena.

Thus, my intention is to treat vision as an observational discipline in its own
right. Writers have remarked on their own visual experiences since writing was
invented, so that a large body of observations has accumulated. This body is
dissected in the present volume. Descriptions of visual experience are likely to
be anchored in a more solid environment than the theories proposed to account
for them. The function of vision is to guide our behavior, and insofar as this
guidance is successful, there might seem to be little in the subject to warrant
enquiry. Indeed, the eternally entertained theory of naive realism speaks to this
issue—the world is as it is perceived. Nonetheless, there were circumstances in
which the phenomena of vision were remarkable and remarked upon. Visual
experiences in darkness (as a consequence of pressure or a blow applied to the
eyeball) were not only remarked on around 500 B.C., but they could have
provided the phenomenal source for emission theories of vision—that light
issues from the eye itself. Such theories might seem fanciful to us now, but the
phenomena upon which they are based are as readily experienced today as
they were two and a half thousand years ago, and descriptions of them have
been repeatedly refined throughout that period. Afterimages provide a similar
example; they can be seen following exposure to bright light, they take on the
shape of the intense stimulus, and they linger for an appreciable time. They
could have acted, together with the reflections seen in water and in the eye, as
a basis for the belief that vision was mediated by images or copies of external
objects. These phenomena and many others require an interpretation by any
adequate theory of vision; the theories might be supplanted but the phenomena

remain. This book is a catalogue of phenomena that were reported in what I call the observational era of vision research.

Some of the phenomena were illustrated as well as described, and the appropriate figures are included. After the fifteenth century illustrations printed in scientific treatises were woodcuts and engravings, often of exquisite quality, and the same applied to frontispiece illustrations, which often included portraits of the authors. This book follows a convention adopted in many histories of science and medicine, namely, of presenting images as well as words of most of the natural philosophers whose works are cited. It is not clear to me what the portraits add to understanding the material presented, but it is equally evident that my motivation to find portraits was immense. Once the search for a portrayal of a particular person was set in train the momentum seemed irresistible. It was usually dissipated when a portrait was located, but often the search continued for a better example. Better, in this context, implies a superior quality engraving rather than a more striking likeness—the criteria for similitude in portrayals are notoriously difficult to define. My main regret is that, despite my best efforts, several of the major figures in this history remain unrepresented, other than in words. The portraits themselves are presented in an unconventional way, since the face alone is featured and all background textures have been removed. It is my hope that this will lend some degree of uniformity to the portrayals.

It would not have been possible to compile the quotations and figures at all without access to the original volumes. In this regard I am indebted to the University Libraries of Cambridge, Dundee, Edinburgh, Glasgow, and St. Andrews, as well as to the British Library, the National Library of Scotland, the Bibliothèque nationale de France, and the Wellcome Institute of the History of Medicine. My particular thanks go to the librarians at Dundee University, to Christine Gascoigne of the Rare Books and Manuscripts Department at the University of St. Andrews, and to Jean Archibald of the Special Collections Department, Edinburgh University Library, both for their helpfulness in searching for sources and for providing many of the illustrations that have been adapted for use in the book.

Many people have offered advise and assistance in compiling the material presented in this book, and I would like to express my thanks to them all. I have enjoyed many discussions about the history of vision with Dieter Heller, Helen Ross, Mike Swanston, and Alan Wilkes, who have freely given of their single malti-faceted perspectives. I am particularly grateful to Véronique Pardieu for translating large sections of Ptolemy's *Optics*, and to Ian Howard for discussions regarding the binocular researches of both Ptolemy and Ibn al-Haytham. My appreciation of Purkinje's vision has been further enhanced by correspondence with Josef Brožek, who also generously gave me his collection of material on color vision by Matthias Klotz. Maurice Dorikens provided material about Plateau, most notably a copy of Plateau's doctoral dissertation.

Several people kindly made available translations of important articles or extracts: Du Tour's experiments on binocular rivalry (Robert O'Shea); La Hire's and Monge's descriptions of color constancy (Jacques Ninio and John Mollon, respectively); and Scheiner's account of the anatomy of the eye (Paola Feresin). Many of the translations are my own, and so the burden of their accuracy or otherwise rests with me alone, as it does for the interpretations presented in the commentaries preceding all the sections.

A Natural History of Vision

1 Introduction

Aristotle (ca. 384–322 B.C.) after an engraving in Baumeister (1885).

Aristotle (ca. 330 B.C.): *If the visual organ proper really were fire, which is the doctrine of Empedocles, a doctrine taught also in the* Timaeus *[of Plato], and if vision were the result of light issuing from the eye as from a lantern, why should the eye not have had the power of seeing even in the dark? It is totally idle to say, as the* Timaeus *does, that the visual ray coming forth in the darkness is quenched. What is the meaning of this "quenching" of light? That which, like a fire of coals or an ordinary flame, is hot and dry is, indeed, quenched by the moist or cold; but heat and dryness are evidently not attributes of light. . . . True, then, the visual organ proper is composed of water, yet vision appertains to it not because it is so composed, but because it is translucent—a property common alike to water and to air. (Ross, 1931, pp. 437b–438a)*

Science involves recording and interpreting natural phenomena. Nowadays, the records are the results of experiments and the many and varied phenomena are posited in well-defined compartments, like physics and physiology. These compartments are a relatively recent convention, as are the specialists who labor under titles such as physicists and physiologists. Neither the phenomena nor the practitioners were so clearly defined in the distant past. In antiquity, science, if such it should be called, was based on describing and classifying observations of naturally occurring events. The predominant sense of science was the eye, and vision itself was an integral part of the endeavor. What the eye saw could be described, cataloged, and even subjected to mathematical analysis. Plato's awe of the eye and vision was stated thus: "Vision, in my view, is the cause of the greatest benefit to us, inasmuch as none of the accounts now given concerning the Universe would ever have been given if men had not seen the stars or the sun or the heaven. But as it is, the vision of day and night and of months and circling years has created the art of number and has given us not only the notion of Time but also a means of research into the nature of the Universe" (1946, p. 107). His approach to natural phenomena, however, did not encourage the observational analysis of vision, because of his distrust of the evidence of the senses: the world of appearances was considered to be a world of illusions, and the essence of thought was to be sought in mathematics and ideal forms. Plato's idealism remained a dominant force in both science and philosophy. His preference for mathematics over matter influenced Euclid, who formulated a theory of vision in geometrical terms, with little concern for perceptual experience itself. These approaches to vision contrasted sharply with that of Aristotle, who placed more reliance on the evidence of his senses than on philosophical speculations.

The phenomena most intimately involved with vision were those of light. Indeed, the distinction between light and vision was not seriously entertained until the optical properties of the eye were described by Kepler early in the seventeenth century. Before Kepler, vision was essential to optics, and disorders of vision provided materials for medicine. Accordingly, histories of optics and ophthalmology have focused on vision during their early phases, but have tended to subordinate it when either the physical nature of light was established or the dioptrics, anatomy, and physiology of the eye was better understood. The present account of vision is neither a history of optics nor one of ophthalmology. It is a much more humble history, aimed at charting the course of descriptions of visual phenomena for the period up to 1840. Optics have been comprehensively chronicled by Smith (1738), Priestley (1772), Young (1807), Brewster (1830b), Whewell (1837), Wilde (1838), Helmholtz (1867, 1896), Mach (1926), Crombie (1964, 1967), Ronchi (1970), Lindberg (1976, 1983a), and Schmitz (1981, 1982). Ophthalmology has an even longer recorded history: a surviving Egyptian papyrus (referred to as the Ebers papyrus), dating from around 1650 B.C., describes many disorders of the eye, including dimness of sight and strabismus. A millenium later, there were specialists in diseases of the eye practicing in Egypt. However, as with optics, the starting point of this natural history will reside in Greece, and for medicine that is taken to refer to Hippocrates in the fifth century B.C. Ophthalmology has benefited from the analyses of many scholars, including Hirsch (1877), Hirschberg (1899, 1908,

1911a, 1911b, 1912), Magnus (1901), Shastid (1917), Polyak (1957), Duke-Elder (1961, 1968), Duke-Elder and Abrams (1970), and Duke-Elder and Wybar (1973).

Vision, as opposed to optics and ophthalmology, became an experimental discipline rather late. To be sure, what would be considered as experiments were conducted intermittently over the period under consideration, but an accepted armory of methods and machines was not available until the 1830s. In that decade: Weber in 1834 demonstrated that the nuances of visual discrimination could be measured by applying what became called psychophysical methods; Plateau (1833) devised a contrivance for synthesizing visual motion from a series of static pictures; Wheatstone (1838) demonstrated, by means of his invention, the stereoscope, that depth perception was influenced by retinal disparity; Treviranus (1837) described the cellular structure of the retina, heralding a new era for visual science, in which function could be related to microscopic structure. These instruments, and many others that were invented in the latter half of the nineteenth century, greatly expanded the range of visual phenomena and the ways in which they could be investigated (see Wade and Heller, 1997). The systematist and experimentalist Johannes Müller summarized the then contemporary state of vision in the two volumes of his *Handbook of Human Physiology*, published in German in 1834 and 1838, and translated into English a few years later. Paradoxically, although he had worked extensively on issues of binocular vision, he had overlooked the fact of disparity leading to depth perception, described by Wheatstone in 1838. Many quotations from the works of both Müller and Wheatstone appear in this book. They can be taken to reflect the purview of the past and the prospects for the future. On the one hand, Müller provided a compendium of the extant knowledge at the time this survey ends. On the other hand, Wheatstone initiated the new experimental era of vision, by enlisting novel instruments to study phenomena and by applying the methods of physics to their examination. The integration of the physics and physiology of vision with its psychology was undertaken by Helmholtz (1867) in the three volumes of his *Handbook of Physiological Optics*. The centenary of Helmholtz's birth (1921) was marked by their translation into English (see Cahan, 1994; Wade, 1994a).

For these reasons I refer to the period prior to 1840 as the observational era of vision, and it is this period to which the term "natural history of vision" is applied. Earlier starting points have been proposed; Boring (1929) suggested that 1796 was of significance because systematic studies of the personal equation (reaction time) were undertaken thereafter (see Mollon and Perkins, 1996). In the case of color, Newton's experiments could be taken as heralding the onset of experimental studies of vision. While color vison is chronicled in this book, more emphasis is focused on spatial vision, and it is in this area that 1840 serves as a suitable end point.

The story is here presented through the descriptions of visual phenomena given by the observers themselves. There are several excellent histories of vision (e.g., Priestley, 1772; Helmholtz, 1896; Boring, 1942; Pastore, 1971; van Hoorn, 1972; Lindberg, 1976; Meyering, 1989), which tend to focus on theory rather than providing the observations in the words of the observers. I am not aware of a history of the type presented here, in which extracts of the works of many scholars are presented under various headings in chronological sequence. The style demands much more of the reader, as a pattern is not explicitly imposed on the extracts. The selections I have made might introduce their own biases, and I am sure that others would not have chosen the same extracts. In many cases the descriptions of phenomena are brief, and frequently in a language different from that in which they were originally made. Of course there is always a danger of reading too much into such brief extracts, and that might exist in the translations that are included here, but an attempt has been made to use early translations. The survey is necessarily selective; it will start before Aristotle and proceed no further than Wheatstone's description of the stereoscope at the end of the 1830s. Both the starting and finishing points are to some extent arbitrary. On the one hand, the natural history of vision did not start with Greek philosophers; Aristotle drew on the works of his predecessors, but his writings represent the fullest account of early vision. On the other hand, phenomenal descriptions did not end with Weber and Wheatstone. Indeed, the phenomenological tradition of Goethe (1810) and Purkinje (1823a, 1825) was then in its heyday, but it can be argued that the 1830s represented the onset of the experimental era in visual science. For the first time, instruments for the manipulation of space and time (like the stereoscope and the stroboscopic disc) enabled students to apply the methods of physics to visual phenomena.

Vision was also the province of the artist, and styles for representing external space changed markedly over this period. These changes were determined in large measure by the developing theories of image formation, culminating in the rules for linear perspective established early in the fifteenth century (see Kemp, 1990). The principles of linear perspective were closely related to image formation in the eye, but the art and science of vision did not come together until the early seventeenth century, when the dioptrics of the eye were first accurately described. Accordingly, this natural history will draw sparingly on the statements of visual artists where they are relevant to a particular topic. Art was transformed in the nineteenth century by the invention of photography. The results of the experiments by Daguerre and Talbot were announced the year after Wheatstone's description of the stereoscope, providing an additional reason for terminating this natural history at the end of the 1830s. Daguerre's metal-based positives and Talbot's paper negatives were each presented to the public in 1839. When the French Academy was deliberating over the significance of Daguerre's discovery, its chairman proclaimed "From today painting is dead" (Vaizey, 1982, p. 9). This statement was as inaccurately prophetic for painting as the invention of a variety of scopes was for phenomenological descriptions of vision. Nonetheless, the date suitably severs natural from experimental inquiries into vision.

There is, however, an artistic dimension to this book, and it concerns the portraits included in it. The illustrations are either representations relating to the visual phenomena described or portraits of the scholars who gave them. The natural history here surveyed is terminated at around 1840. As was noted above, photography was invented at about the same time, and this was to have a dramatic impact on both scientific illustration and portraiture. Prior to that time, printed figures were mostly engravings, the details of which were far superior to the photomechanical reproductions that supplanted them. Accordingly, the illustrations presented in this book are derived, wherever possible, from engravings. Those relating to the descriptions themselves have generally been taken from the same source as the text. Sometimes, the figures have been derived from earlier or later printed sources than the text used, in which case the source will be cited. This is particularly the case for the illustrations from Descartes's *Dioptrics* and his *Treatise of Man*. Indeed, the illustrations used for the latter have their own complex history, largely because Descartes's manuscripts described but did not depict the form of the figures (see Hall, 1972). *Dioptrics* was published during Descartes's life (in 1637), but the *Treatise* first appeared over a decade after his death, and two separate versions of it were printed. The first, in 1662, was translated into Latin (*De homine*) and illustrated by Schuyl, who is said to have worked from a defective manuscript copy of the French. The French version (*Traité de l'Homme*) appeared in 1664: the text was given to two illustrators (van Gutschoven and La Forge), who each made a complete set of drawings independently of the other; van Gutschoven's were the ones most generally printed, though some of La Forge's were included, too. The whole set of La Forge's illustrations can be found in the Latin edition of 1677. In order to reflect the degree of variation that exists among the figures, all those taken from the *Treatise of Man* are shown in duplicate, one from the Latin version of 1662 and the other from the 1664 French version. The latter are both more widely known and more readily available, because they are reproduced in Adam and Tannery's *Oeuvres de Descartes* published early in this century. It is from this source that the illustrations from *Dioptrics* and *Treatise of Man* are taken.

The technique of engraving was invented in the fifteenth century and, when married with the printing press, it provided an ideal medium for making pictorial images that otherwise existed in isolation widely available. Most of the portraits included here were originally copied from paintings. This did not, of course, apply to the representations of the figures from antiquity. In some cases the engravings have been copied from statues or busts, many of which still survive (see Bernoulli, 1882, 1901; Delbrück, 1912; Hekler, 1912; Richter, 1965); in other cases their similarities to their subjects are likely to be more fanciful. Most of the observers whose works are cited are portrayed; some portraits are derived from relatively simple woodcuts, whereas others are after exquisitely executed line or stipple engravings. Each portrait has been placed alongside one of the most important quotations from that author, together with dates of birth and death, and the source of the original illustration. I have endeavored to present the portraits in a consistent manner, with the emphasis on the facial features of the subjects, by removing the background textures almost always engraved in the originals; hence all the portrayals are in a form not previously seen. I regret that not all the prominent contributors to visual science are represented; to my mind, the most unfortunate omissions are those of William Porterfield (ca. 1696–1771) and William Charles Wells (1757–1817), neither of whom were in any of the many sources I searched.

The descriptions are grouped under various headings and subheadings. The headings or chapters represent principal divisions of investigations into visual phenomena, and the subheadings or sections will be recognized as topics in which studies continue to be conducted. Light and color are obvious divisions, because they can be related to the physical characteristics of the stimulus. The other topics are less clearly defined in historical terms. One of the most penetrating accounts of vision in antiquity was that of Ptolemy in the second century A.D. (see Lejeune, 1956, 1989; Smith, 1996). Book II of his *Optics* classified visual phenomena into those of color, body, position, size, form, movement, and rest, and he went on to consider illusions that occur under each of these headings. Binocular vision was addressed in the context of the perception of position in Book II, and it was returned to at length in Book III. Books I–III of Ibn al-Haytham's *Optics*, written in the eleventh century, employed similar but more extensive divisions to those of Ptolemy, and also assessed the errors of sight in each of them (Sabra, 1966, 1989). Descartes, in his *Dioptrics*, considered that all the qualities of sight could be reduced to light, color, location, distance, size, and shape. The chapter headings that are employed here reflect these historical distinctions, but rely perhaps most strongly on Ptolemy. Light, color, motion, binocularity, and illusions were all categories employed by Ptolemy; his classes of size and form are incorporated here into space, and he also drew a distinction between subjective and objective aspects of visual phenomena.

Light, color, motion, and space would be included in most surveys of vision, and they are headings used here. In addition, I have included subjective visual phenomena, binocularity, and illusions. Illusions are always problematic in both definitional and categorical terms. They have been assembled after the section on space, because most of them do relate to aspects of spatial vision. There remain problems, however, concerning what constitutes an illusion, and some subheadings might be seen by some readers as misplaced. Binocularity has received abundant attention because I consider that it encompasses the most clearly psychological set of phenomena, which relate to all the other categories. The aspect of binocular vision of concern throughout the period under survey was almost always singleness of vision; it was Wheatstone's demonstration that singleness and stereopsis could coexist that is taken as defining the end of the observational era of the study of vision.

In many cases, books could (and often have) been written about the historical work sampled under a single subheading or section, and so the material presented here cannot in any way be considered to cover the topics comprehensively. The quotations cited are not necessarily the earliest descriptions of the various phenomena, but the sequence under a particular subheading can give some flavor of the development of detail in the observations and in their interpretations. This is not to suggest that the later authors drew on the works of earlier ones sampled here; often they wrote in isolation or ignorance of earlier deliberations on vision. Indeed, it can be very difficult to determine the earliest recorded reports of particular phenomena; it is only recently that citing earlier sources has become commonplace. In antiquity, phenomena were described and redescribed without attribution; many of Aristotle's observations were derived from earlier reports, and they have resurfaced constantly since his time as if they were novel.

As was mentioned above, it is impossible to do justice to an article or book by taking brief extracts from it. However, I hope that part of this problem has been avoided by concentrating on phenomena rather than theory, and by using several extracts, under different headings, from particularly important writers. Wherever possible, published translations into English have been used, but in some cases those given are, to the best of my knowledge, original translations. This applies most particularly to the writings of Purkinje (1823a, 1825), and many extracts from Ptolemy had not previously been translated into English; now, however, there is an English translation available (Smith, 1996). Any words or sections that appear in *italics* were so written in the texts from which they are taken; words in square brackets have been added to clarify the meaning of some short extracts; paragraph breaks in the original works have been omitted; and original spellings and capitalizations have been retained except for those cases where particular characters are no longer in use. For some citations, the date given to an author differs from that at the end of the quotation; this is usually because the work in question has been reprinted or translated, as will be evident from the reference listed by the later date in the bibliography.

In one sense, the quotations have been assembled to obviate the need for any commentary. The descriptions themselves provide the thread of history, and it is hoped that following them will weave an impression of the pattern, regular or random, that existed in particular domains. For some subheadings the threads appear connected, whereas for others they may seem to be chaotically independent, usually reflecting the different theoretical positions imposed on the phenomena. Nonetheless, a brief commentary precedes the descriptions in each section. The aim of the commentary is to provide a context into which the quotations can be placed, and to cite sources of research that have examined these topics in much more detail, so that the interested reader can pursue them. Cross-references will be made, so that certain themes can be followed in tandem. References in the commentary will only include dates if the work cited is not given in the text under discussion. Since the commentary will survey the works cited in a similar sequence to the quotations given, it seemed unnecessary to clutter it with dates. Accordingly, most of the names will not be accompanied by dates; these can be found by consulting the appropriate quotations in the text. It is not my intention to attempt in the commentary to bring the topics up to date. Rather, the reader will be left with an impression of the state of knowledge existing at the time that this survey ends.

Each chapter will commence with an apposite quotation by a major contributor to that area and with his portrait. The first section in each chapter is generally addressed to the broad issues involved in it, with more detailed approaches assembled in the following sections. Chapter 2 is concerned with light and the eye; here the traditional topics of optics and ocular anatomy are surveyed. They do not constitute the principal theme of this natural history, but they have been so important to the interpretations provided that a brief history is warranted. Color constitutes the substance of chapter 3. It is perhaps the topic in vision that yielded most readily to a mechanistic interpretation that was supported by experimental evidence. Newton appreciated the subjectivity of color vision, and chapter 4 focuses on subjective visual phenomena; some of these, like afterimages, have a long history, whereas others were charted only toward the end of the period under examination. Chapter 5 is addressed to motion perception and eye movements. These were early recognized to be of importance, but the precise measurement of eye movements did not emerge until the late nineteenth century. Analyses of binocularity (chapter 6) have been dominated by the singleness of vision that is experienced with two eyes, but many other phenomena were described before stereoscopic depth perception was demonstrated by Wheatstone. These phenomena were generally placed in the context of space, which is the subject of chapter 7. The limits of spatial vision are considered, but the central topics are visual direction and distance. Two critical debates were based on these phenomena—erect vision on the basis of an inverted retinal image, and the recovery of sight following congenital blindness. Chapter 8 covers a number of visual illusions, the most intensively investigated of which is the moon illusion. The brief final chapter gives some pointers to the ways in which the various topics were to be developed during the experimental period that followed the observational one surveyed here.

Goethe (1810): *If we may at all hope that natural history will gradually be modified by the principle of deducing the ordinary appearances of nature from higher phenomena, the author believes he may have given some hints and introductory views bearing on this object. (1840, p. 292)*

Johann Wolfgang von Goethe (1749–1832) after an engraving in Knight (1835a).

2 Light and the Eye

Johannes Kepler (1571–1630) after an engraving in Knight (1834).

Kepler (1604): *Thus vision is brought about by a picture of the thing seen being formed on the concave surface of the retina. That which is to the right outside is depicted on the left on the retina, that to the left on the right, that above below, and that below above. Green is depicted green, and in general things are depicted by whatever colour they have. . . . the greater the acuity of vision of a given person, the finer will be the picture formed in his eye. (Crombie, 1964, p. 150)*

The basic aspects of sight—variations in illumination from night to day, reflections from polished surfaces and water, the illumination of otherwise invisible objects by fire, the variety of colors experienced in objects and in rainbows, etc.—would have been observed since the dawn of human history. A perplexing variety of interpretations has been applied to them, ranging from the mystical and magical to the material, and these were often interwoven in early accounts. The starting point for the present history is in Greek science, where optics was taken to be the study of light and vision. Ideas about the nature of light itself were inseparable from those of the system responding to it. That is, Greek theories of light incorporated the visual aparatus to varying degrees, thus confounding physical optics with the visual perception. In fact, the major advances in optics have involved differentiating physical from psychological phenomena. For dioptrics (the science of refracted light) it was achieved in 1604 by Kepler, who portrayed the manner in which images are formed in the eye; for color it was Newton who, in 1672, published the results of his prismatic experiments, which indicated that the spectrum is a property of light rather than glass. Light and vision were conflated in a variety of ways by Greek thinkers, and their ideas were transmitted and extended by Arabic writers such as Ibn al-Haytham (ca. 1040), to be reabsorbed into European thought from the thirteenth century onward to form the medieval *Perspectiva* (see Crombie, 1952; Lindberg, 1976).

This book is concerned principally with the psychological rather than the physical dimensions of vision. The emphasis is placed on visual phenomena, their descriptions and interpretations, rather than the use of vision to enlighten understanding of physical phenomena. Nonetheless, the early history of light and the eye cannot sustain such a distinction, nor can the basic steps in elucidating the anatomy and physiology of the eye be neglected. Optics in general, and physical optics in particular, has benefited from some excellent historical surveys, as indicated above, and we are fortunate now to have available a translation of the first three books of Ibn al-Haytham's *Optics*, since they deal with his observations on vision (Sabra, 1989). In like manner, disorders of the eye have drawn upon vision to classify and clarify the under-lying causes, with the result that ophthalmology has a long and distinguished history. Records of operations on the eye are older than treatises on optics, and the Greek doctors absorbed a well-established set of procedures to which they could add.

Thus, vision aided optics and ophthalmology in the early stages of their developments, but it has not generally been accorded the same attention for the periods following the separation into physical, physiological, and psychological domains. For example, it has been said that Kepler's dioptrical analysis of the retinal image represented a "successful solution of the problem of vision" (Lindberg, 1976, p. x). From the psychological point of view, this is at best an oversimplification. Kepler formulated the problem that generations of students of vision have since attempted to resolve: how do we perceive the world as three-dimensional on the basis of a two-dimensional retinal image? Indeed, this has been seen by some as a pseudoproblem (e.g., Gibson, 1966), and others have referred to the "legacy of Kepler" as having defined the problem in terms of single, static retinal images rather than considering the starting point as binocular and dynamic (Wade, 1990). Kepler himself was cautious regarding the conclusions that could be deduced from the inverted and reversed retinal image (see section 7.2): "I leave it to the natural philosophers to discuss the way in which this image or picture is put together by the spiritual principles of vision" (Crombie, 1964, p. 147). Philosophers have not been united in their opinions, but they have appreciated that physical optics was not the solution to vision. The policy I will adopt to the various divisions of optics is to restrict consideration of the physical dimensions of light mainly to the period in which it was confounded with the psychological. Following the revelations of Kepler and Newton, attention will be directed more to the characteristics of vision and the observations that were made by those who sought to understand it. Hence, the disputes between corpuscular and wave theorists will only be touched upon here.

2.1   Optics

Of the Greek philosophers and mathematicians, Euclid (ca. 300 B.C.) assembled and systematized the phenomena of optics most lucidly; he followed Plato's lead and defined optics mathematically, thus equating light and vision. Euclid based his optics on the then well-known fact that light travels in straight lines, and pursued the consequences of this with commendable persistence. Vision was restricted to the cone of rays emanating from the eye and meeting the objects within it. The geometrical projections to these objects were lawful, and this lawfulness was applied to vision, too. Those objects subtending a larger angle were perceived as larger. Thus, Euclid provided not only an account of optical transmission through space but also a geometrical theory of space perception itself. The perceived dimensions of objects corresponded precisely to the angles they subtended at the eye, and illumination of those objects had its source in the eye. The theory neither mentioned nor could account for any aspects of vision that involved color.

   Euclid's was not the only theory that was entertained in Greek science, but it was an elegant one. Vision was generally considered to involve some process of contact between the eye and objects (see Beare, 1906; Lindberg, 1978), and other means of achieving this contact were advanced. Democritus (ca. 400 B.C.) proposed that all nature was composed of atoms in motion; these atoms were continually emitted from objects to compress the air and carry impressions to the eye. These impressions were like a copy or image of the object that could be received by the eye, and this theory was amplified by Epicurus (ca. 300 B.C.). Democritus set in train a materialist philosophy that was to resurface with the scientific revolution of the seventeenth century, though its impact on Greek science was more limited. The concept of some copy of objects, carried through the air to the eye, was to have widespread and long-lasting appeal, and it was itself transformed into "eidola," "simulacra," "species," "images," etc. Indeed, by the end of the thirteenth century, Roger Bacon was able to list the terms "image," "species," "idol," "simulacrum," "phantasm," "form," "intention," "passion," "impression," "similitude of the agent," and "shadow of the philosophers," used by authors of works on vision (Lindberg, 1983b), to which could be added the "effigies," "figures," and "membranes" of Lucretius (ca. 56 B.C.). According to Epicurus, the copies were received by the eye, and so this theory was one of intromission or reception, in contrast to Euclid's projection or extramission theory.

A third factor was introduced by Plato (ca. 350 B.C.), who suggested that light was emitted from both the eye and objects, and vision took place externally where these two streams united. According to Theophrastus: "His view, consequently, may be said to lie midway between the theories of those who say that vision falls upon [its object] and of those who hold that something is borne from visible objects to the [organ of sight]" (Stratton, 1917, p. 71), although Lindberg (1976) disputed this interpretation. For Plato, as for Aristotle, it was not light but color that was the principal source of interest in vision, and their descriptions and interpretations of color phenomena will be presented in the next chapter.

In Aristotle (ca. 330 B.C.) we find a theory more in line with modern conceptions of light and vision. His interests were in observation, and the phenomena he experienced directed the interpretations he proposed. Thus, he queried extramission theory by the simple expedient of testing a prediction that would follow from it: if light was emitted from the eye, then vision should be possible at any time the eyes were open. The fact that the prediction was not supported led him to suggest an alternative theory of the nature of light, which was extended by his pupil Theophrastus (ca. 300 B.C.). Light, generated by the sun, was reflected from objects but required a medium, air, through which to travel before it could be received by the eye. The emphasis on the medium, variously called the transparent or the diaphanous, reflected Aristotle's distinction between light as a substance and light as a motion of the medium. Such motions could be instantaneous, and they could be perceived by many observers simultaneously. Aristotle's conception of light was not, however, widely adopted.

Greek theories of light were transmitted through the Roman period mostly by Greco-Roman writers, although the transmission was tempered by a growing desire to relate the optical theories to the facts of observation. Lucretius (ca. 56 B.C.) made many references to vision in his poem *De Rerum Natura* (On the Nature of Things); he believed that light (lumen) was emitted from the sun, and when it struck objects it carried images (eidola) of them to the perceiver. Lucretius appreciated that images in themselves would not be useful to perception unless they carried with them some index of the distance the objects were away from the observer, so that their dimensions could be determined. The mechanism that he proposed for this—of the image brushing aside the intervening air—was exceedingly vague, but he was addressing a general problem that exists in all accounts of spatial vision (see Ronchi, 1970). Lucretius followed in the line of the Epicureans, but the relative merits of such reception theories were still in conflict with emission theories, as supported by Hero of Alexandria and Heliodorus. The emergence of experimental studies of light is also evident in the quotation from Hero; the phenomena are divided into those of vision, refraction, and reflection. On a more pragmatic level, Pliny (ca. 108) drew attention to the fact that individuals differed from one another in what they could see, some having their keenest sight for near objects, others for distant ones.

Ptolemy (ca. 150) is usually cast in the theoretical mold of Euclid (see Crombie, 1967), but he leavened Euclid's geometrical optics with some facts of both physical optics and visual perception (Delambre, 1812). In particular, he appreciated: that light should be thought of as continuous rather than discrete; that color was an integral component of light; that visual size cannot be equated with visual angle; that vision is not equal throughout the visual pyramid (rather than cone); that two pyramids of vision (one for each eye) need to be combined; and that experiments could be performed to study this binocular combination (see Lejeune, 1948, 1956, 1989). He set in train a reconciliation between physical and psychological analyses of vision which was amplified by Ibn al-Haytham. We know relatively little about Ptolemy's theory of light, because the first book of his *Optics* has not survived. What is clear is that his approach was more experimental, and that he introduced measurements of both reflected and refracted light. In accordance with Aristotle, Ptolemy's primary concern was with color, and light served the function of rendering the color of objects visible.

Galen (ca. 175) addressed matters of vision in the context of anatomy, though he made many astute observations, particularly in the context of binocular vision. He also ventured, with some misgivings, into the arena of optics (May, 1968; Siegel, 1970). In his book *De Usu Partium Corporis Humani* (On the Usefulness of the Parts of the Body) he apologized to his readers for introducing optical concepts, since they were at that time deeply unfashionable in the context of ocular anatomy. Nonetheless, we find that he made an explicit equation between rays of light, as in a sunbeam, and visual rays. More important, his theory of vision was physiological: visual spirits, called pneuma, passed along the hollow tubes of the optic nerves to interact with returning images of external objects in the crystalline lens. Here we find another enduring notion, that the "seat of vision" resides in the lens of the eye. Indeed, Galen himself supported this proposal by virtue of the blindness that results from cataracts and the sight that is restored when they are surgically removed. By adopting an anatomical and physiological analysis of vision, Galen was confronted with the existence of two eyes and the observation by them of a single visual world. He was able to benefit from Ptolemy's analysis of certain aspects of binocular single vision (see Howard and Wade, 1996; Smith, 1996), and to suggest his own physiological theory for its occurrence. The pneuma was unified from a single site in the anatomical process—the optic chiasm—where the two optic nerves were thought to be united.

Little was added to optical theory in the late Roman period, and the Greek texts were retained and copied initially in Byzantium and later in Persia and North Africa (see Crombie, 1952; Lindberg, 1976, 1983a; O'Leary, 1949).

Translations of Greek works into Arabic reached their peak in the ninth and tenth centuries, and they in turn were translated into Latin from the twelfth century. Because of strictures against dissection, Galen's anatomy and physiology of the eye was generally accepted by Islamic scholars, but they did extend knowledge of optics. Al-Kindi (ca. 860) summarized the principal theories of optics proposed by Greek philosophers. Vision could follow from intromission, as the atomists like Democritus had argued, by extramission after the manner of Euclid's theory, by some form of Platonic interaction, or via some medium. Al-Kindi rejected three of the four possibilities, adopting a Euclidean extramission theory. His rejection of the others was largely a negation of any form of intromission in the process of vision (see Lindberg, 1976). Al-Kindi considered that only an extramission theory could account for perceptual constancy—the experience of perception is in accord with the physical characteristics of objects rather than their projected dimensions. Regular shapes like squares or circles rarely project their regularities to the eye. Only when they are viewed in the frontoparallel plane would they be projected as squares or circles. Nonetheless, they are perceived as square or circular from many directions. This is the problem of perceptual constancy, and it had been addressed by Ptolemy (Ross and Plug, 1998; Wade and Swanston, 1996). This apparent conflict between perspective and perception was to influence medieval scholars, too.

Both Avicenna (ca. 1020) and Ibn al-Haytham accepted that the crystalline lens was the receptive organ for vision, although Ibn al-Haytham did hint at times that the retina was involved, too. He was very familiar with the writings of Euclid and Ptolemy, as he supplemented his income by making translations of them (Sabra, 1989). However, he adopted a theory of light similar to that of Aristotle, in which the medium is of prime importance. As it was with Aristotle, Ibn al-Haytham's theory had virtually no impact on his contemporaries, but it was rediscovered almost two centuries later, and translated into Latin around the beginning of the thirteenth century. His name was latinized to Alhazen and his *Optics* was translated as either *Perspectiva* or *De Aspectibus*. It was this book which awakened Western scholars like Roger Bacon, John Pecham, and Witelo (Vitellonis) to the physics of light, its mathematical treatment, and its application to vision (Lindberg, 1967; 1974). Later still, in 1572, Alhazen's *Opticae Thesaurus* was published, together with Witelo's *Perspectiva*, in a single volume. It was in Kepler's (1604) reaction to the latter, that among the things omitted by Witelo was the optical analysis of the retinal image. The medieval *Perspectiva* shared a common assumption that vision should be analyzed in terms of a pyramid with its base on external objects and its apex located on the surface of or in the eye. This perspective pyramid carried with it the problems posed by Al-Kindi, namely the conflict between optical projection and visual perception. One consequence of this was to treat perception with great suspicion, while accepting the validity of perspective projections. Thus, through much of the late medieval period considerably more attention was directed to physical than to psychological dimensions of optics (Meyering, 1989; Ronchi, 1970). Contrary to Al-Kindi, both Al-Farabi (ca. 950) and Avicenna (ca. 1020) argued that vision was not mediated by some emission from the eyes, but that it was an immaterial process. Thus, perception could be studied independently of physical and mathematical concerns.

Grosseteste (ca. 1250) is not considered to have had access to Alhazen's *De Aspectibus*, and his analysis of light and vision was Platonic, with light emitted from the eye interacting with that reflected from objects. Somewhat earlier, Plato's distinction between the material, sensual body and the rational soul had been incorporated into Christian theology by St. Augustine, and it even permeated the nature of light: spiritual light was the internal illuminant of ideal forms, and physical light was considered to be analogous to this (Crombie, 1953). In a similar manner to Avicenna's problem, the ideal forms were rarely encountered in perspective projections. For the Scholastics, they were present in the mind and could be illuminated by divine light. Hence we find the emergence of distinctions between different forms of light, lux and lumen, which were maintained from the time of Albertus Magnus (ca. 1250) to Reisch (1486). Lumen was external light, as from the sun or fire, whereas lux was perceived light.

The impact of absorbing the optics of Ptolemy (which had been translated into Latin in the twelfth century) and of Ibn al-Haytham is clear in the contrast between Grosseteste and Roger Bacon: pyramids of light strike the eye, but the physiological dimension remained Galenic. The crystalline lens is still taken to be the "seat of vision" and "species" remain a part of the process.

Relatively little was added to the science of optics in the late medieval period. In the sixteenth century both Maurolico and Porta continued the tradition of the early medieval perspectivists, and also described the refraction of light through lenses. Porta likened the camera obscura to the eye in the second edition of his popular treatise *Magiae Naturalis* (1589), and wrote a more serious book *De Refractione* four years later. The work of Maurolico contains strands that were to be amplified by Kepler, although his work was unlikely to have been available to the latter (see Ronchi, 1970). It was written in manuscript form between about 1520 and 1555, but it was not published until 1611, after Kepler's (1604) critique of Witelo's *Perspectiva*. Witelo's work was widely circulated toward the end of the century: as noted above, it had been edited and published by Friedrich Risner in 1572, together with Alhazen's *Optics*, and it was these analyses of optics that stimulated Kepler's interests.

Physical optics came of age in the seventeenth century (see Mach, 1926; Ronchi, 1970; Sabra, 1967). In addition to his *Ad Vitellionem Paralipomena* of 1604, Kepler wrote a text on dioptrics in 1611. In the first of these he added many things to Witelo's perspective, both experimentally and theoretically. Among them was the formulation of the basic principle of photometry that the intensity of light diminishes with the square of the distance from the source. He devoted considerable attention to refraction in *Dioptrice*, but he did not determine the general sine law. Willebrord Snel, in an unpublished manuscript written around 1621, described the relationship between angles of incidence and refraction, upon which the subsequent technical advances in optical instrument manufacture were based. He did not use sines in his formulation, but the dimensions that he described are equivalent (see Vollgraff, 1936). Snell's law, as it became known, was elaborated by Descartes (1637). Had Huygens not been aware of Snel's manuscript and made reference to it in his *Dioptrique* (1653), the relation between sines of the angles of incidence and refraction might have been called Descartes's law. Montucla (1758) in the survey of optics in his *Histoire des Mathematiques* also reinforced the priority of Snel in discovering the sine law of refraction. Light, according to Descartes, acted like a mechanical force which is transmitted through transparent media. He made analogies between mechanical events like projectiles bouncing from surfaces, and applied these to reflections and refractions of light. His theory of light attracted much criticism in his day because of the inconsistencies it embraced. On the one hand he argued that light was propagated instantly, and on the other that it varied its velocity according to the density of the medium through which it traveled.

The phenomenon of diffraction was demonstrated by Grimaldi (1665), who suggested that light might act like a liquid, flowing in waves. Wave theory was supported and extended by Huygens: he proposed and illustrated the wave fronts that could be produced by points on luminous sources, and he made an analogy between light and sound; diffraction was analyzed in terms of the wave fronts originating at the aperture. The theoretical contrast was between Huygens's wave theory and Newton's corpuscular theory of light.

With the appreciation that light could be considered as a physical property, and that its reflections and refractions followed physical principles, its study became the province of natural philosophers (or what we would now call physicists), whereas the examination of vision was pursued by physiologists and philosophers. The separation of the physics of light from the philosophy of sight was to reflect the ancient schism between materialists and idealists: light was an external, material phenomenon whereas sight was internal and subjective. In the seventeenth and eighteenth centuries, philosophers cast ever more caustic eyes on optics, as is evident in the scepticism enunciated by Berkeley, Malebranche, and Condillac. Nonetheless, there were those, like Hartley, who did strive to apply Newtonian principles to the study of vision.

Newton's corpuscular theory of light was dominant throughout the eighteenth century, so much so that it became dogma, and those who were to challenge it (like Young) were roundly condemned for doubting the genius of Newton (see Ronchi, 1970; Cantor, 1977; Schmitz, 1982). Despite its seeming dominance, the storm clouds for the corpuscular theory of light were gathering. Euler (1769) gave a mathematical description of chromatic aberration, and supported Huygens's wave theory. The technical advances in correcting for this aberration in telescopes led to rapid advances in astronomy during the eighteenth century. Euler believed that the eye was achromatic, but Young described and measured the chromatic aberration of his own eyes. By the beginning of the nineteenth century, Malus accepted corpuscular theory as a conceptual convenience, but one which found increasing difficulty in accounting for facts of diffraction and interference phenomena; it was the latter that proved so convincing to Young in his advocacy of wave theory. Despite his reluctant adoption of Newton's theory, Malus analyzed the phenomenon of polarization (a concept that derived from Newton), which was to accelerate its demise (see Whewell, 1837). The situation with regard to the nature of light that obtained by 1815 was suitably summarized by Fresnel, who was particularly influential in establishing wave theory. By 1840, as a consequence of his experimental endeavors, allied to the mathematical analyses of Airy and John Herschel, the majority of physicists adopted the wave theory of light (Cantor, 1984; Ziggelaar, 1993).

Democritus (ca. 400 B.C.): ... the air between the object of sight is compressed by the object and the visual organ, and thus becomes imprinted; since there is always an effluence of some kind arising from everything. Thereupon this imprinted air, because it is solid and is of a hue contrasting [with the pupil], is reflected in the eyes, which are moist. (Stratton, 1917, p. 111)

Democritus (ca. 460–370 B.C.) after an engraving in *The Historic Gallery of Portraits and Paintings*, Vol. 1. London: Vernor, Hood, and Sharpe, 1807.

Plato (ca. 350 B.C.): Though vision may be in the eyes and its possessor may try to use it, and though colour be present, yet without the presence of a third thing specifically and naturally adapted to this purpose, you are aware that vision will see nothing and the colours will remain invisible. . . . when the eyes are no longer turned upon objects upon whose colours the light of day falls but that of the dim luminaries of night, their edge is blunted and they appear almost blind, as if pure vision did not dwell in them. (1935, pp. 99 and 103)

Aristotle (ca. 330 B.C.): Democritus misrepresents the facts when he expresses the opinion that if the interspace were empty one could distinctly see an ant on the vault of the sky; that is an impossibility. Seeing is due to an affection or change of what has the perceptive faculty, and it cannot be affected by the seen colour itself; it remains that it must be affected by what comes between. Hence it is indispensable that there be *something* in between—if there were nothing, so far from seeing with greater distinctness, we should see nothing at all. (Ross, 1931, p. 419a)

Euclid (ca. 300 B.C.): 1. Let it be assumed that the lines drawn directly from the eye pass through a space of great extent; 2. and that the form of the space included within our vision is a cone, with its apex in the eye and its base at the limits of our vision; 3. and that those things upon which the vision falls are seen, and those things upon which vision does not fall are not seen; 4. and that those things seen within a larger angle appear larger, and those seen within a smaller angle appear smaller, and those seen within equal angles appear to be of the same size; 5. and that things seen within the higher visual range appear higher, while those within the lower range appear lower; 6. and, similarly, that those seen within the visual range on the right appear on the right, while those within that on the left appear on the left; 7. but that things seen within several angles appear to be more clear. (Burton, 1945, p. 357)

Theophrastus (ca. 300 B.C.): Possibly, however, the reflection in the eye is caused by the sun, in sending light in upon the visual sense in the form of rays. (Stratton, 1917, p. 113)

Epicurus (ca. 300 B.C.): We must also consider that it is by the entrance of something coming from external objects that we see their shapes and think of them. For external things would not stamp on us their own nature of colour and form through the medium of the air which is between them and us, or by means of rays of light or currents of any sort going from us to them, so well as by the entrance into our eyes or minds, to whichever their size is suitable, of certain films coming from the things themselves, these films or outlines being of the same colour and shape as the external things themselves. (Lindberg, 1983a, pp. 339–340)

Epicurus (ca. 342–270 B.C.) after an engraving in Baumeister (1885).

Lucretius (ca. 56 B.C.): Moreover, to say that the eyes can discern nothing, but that the mind looks out through them as through open portals, is difficult, when their own feeling leads us to the opposite conclusion; for it is their feeling that draws us and pushes us on to the very eyeballs; especially since we are often unable to perceive glaring objects because our bright eyes are hindered by the brightness, which never happens with portals; for an open door through which we look out ourselves never receives any annoyance. Besides, if our eyes act as portals, why then take the eyes away, and it is obvious that the mind should perceive things all the better with doors, posts and all, removed. (1975, pp. 215–217)

Hero of Alexandria (ca. 60): The science of vision is divided into three parts: optics, dioptrics, and catoptrics. Now optics has been adequately treated by our predecessors and particularly by Aristotle, and dioptrics we have ourselves treated elsewhere as fully as seemed necessary. But catoptrics, too, is clearly a science worthy of study and at the same time produces spectacles which excite wonder in the observer.... That the rays proceeding from our eyes move with infinite velocity may be gathered from the following consideration. For when, after our eyes have been closed, we open them and look up at the sky, no interval of time is required for the visual rays to reach the sky. Indeed, we see the stars as soon as we look up, though the distance is, as we may say, infinite.... That our vision is directed along a straight line has, then, been sufficiently indicated. (Cohen and Drabkin, 1958, pp. 261–263)

Ptolemy (ca. 150): Now it has been shown ... (1) that this type of bending [refraction] of the visual ray does not take place in all liquids and rare media, but that a definite amount of bending takes place only in those media that have some likeness to the medium from which the visual ray originates, so that penetration may take place, (2) that a visual ray proceeds along a straight line and may be naturally bent only at a surface which forms a boundary between two media of different densities, (3) that the bending takes place not only in the passage from rarer and finer media to denser (as is the case in reflection) but also in the passage from a denser to a rarer, and (4) that this type of bending does not take place at equal angles but that the angles, as measured from the perpendicular, have a definite quantitative relationship. (Cohen and Drabkin, 1958, p. 272)

Galen (ca. 175): ... whatever is to be seen must be unobscured, with nothing interposed in the straight line that extends from the eye to the object. And if you have now understood this also, you will think those mathematicians not unreasonable who declare that objects seen are seen in straight lines. Then call these straight lines visual rays.... Now you have at some time, I suppose, seen the rays of the sun escaping through a narrow opening, advancing without being deflected or bending at any point, and pursuing a path that was perfectly straight and undeviating. Consider, please, the path of the visual rays to be like that too. (May, 1968, p. 493)

Heliodorus (ca. 200): Because if sight must reach the object to be seen in the shortest possible time, then it must travel along a straight line, since this is the shortest line joining two points. (Ronchi, 1970, p. 25)

Al-Kindi (ca. 860): Therefore I say that it is impossible that the eye should perceive its sensibles except [1] by their forms travelling to the eye, as many of the ancients have judged, and being impressed in it, or [2] by power proceeding from the eye to sensible things, by which it perceives them, or [3] by these two things occurring simultaneously, or [4] by their forms being stamped and impressed in the air and the air stamping and impressing them in the eye, which [forms] the eye comprehends by its power of perceiving that which air, when light mediates, impresses in it. (Lindberg, 1976, pp. 21–22)

Al-Farabi (ca. 950): Vision is a power and a tendency in matter and only potential vision before one can see; colors are only potentially visible and apparent until they are seen.... Thus by the light dispensed from the sun the vision becomes actually seeing. (Eastwood, 1979, p. 424)

Avicenna (ca. 1020): The organ of sight is the crystalline humour [lens] in the pupil. They are wrong who think that the act of seeing arises from something that goes out from the eye to the perceived object and collides with it. For if this something be material the difficulty would be that in the eye would have to be a body big enough to spread over half the vault of the heaven. (Singer, 1921, p. 393)

Avicenna (980–1037) after an illustration in Dumesuil and Bonnet-Roy (1947).

Ibn al-Haytham (ca. 1040): We find that the light of every self-luminous body radiates on every body opposite to it when there is not between them an opaque or nontransparent body that screens one from the other. For when the sun faces a body on the ground that is not screened from it, its light shines upon that body and is visible, and it simultaneously irradiates every place in all parts of the earth that face it at that time. It is similarly the case with the moon, and also with fire.... We also find that the radiation of all lights takes place only in straight lines and that no light radiates from a luminous object except in straight lines.... Moreover, all that mathematicians who hold the doctrine of the ray have used in their reasonings and demonstrations are imaginary lines which they call "lines of the ray." And we have shown that the eye cannot perceive any visible object except through these lines alone. Thus the view of those who take the radial lines to be imaginary lines is correct, and we have shown that vision is not effected without them. But the view of those who think that something issues from the eye other than the imaginary lines is impossible and we have shown its impossibility by the fact that it is not warranted by anything that exists, nor is there a reason for it or an argument that supports it. It is therefore evident from all that we have shown that the eye senses the light and colour that are in the surface of a visible object only through the form of that light and colour, which [form] extends from the object to the eye in the intermediate transparent body; and that the eye does not perceive any of the forms reaching it except through the straight lines which are imagined to extend between the visible object and the centre of the eye and which are perpendicular to all surfaces of the coats of the eye. (Sabra, 1989, pp. 13 and 81–82)

Averroes (ca. 1180): ... the faculty of sight is distinguished by the fact that, in addition to the medium [air], it has need of light. The proof of this is that it cannot see in the dark and when smoke or vapor arises in the air to prevent the light from entering the eye, the sight weakens. (1961, p. 8)

Grosseteste (ca. 1250): One should not assume that the emission of rays from the eye is only apparent and without reality. This is the opinion of those, who know parts, but do not consider the whole. One should understand that the visual species emitted from the eye is a substance which illuminates similarly as the sun. These emitted rays complete the act of vision when they join those from the objects in extrapersonal space. (Grüsser and Hagner, 1990, p. 64)

Albertus Magnus (ca. 1250): The light (*lumen*) which is ... generated by the luminous body in the transparent is related to the light (*lucem*), which is the form of the luminous body, as the *intentio* of the colour, generated in the transparent, is to the form of the colour, which is in the coloured body. (Dewan, 1980, p. 308)

Roger Bacon (ca. 1270): The principal requirement is simply that sight should perceive the object distinctly and sufficiently and with certitude, and this can occur through a pyramid in which there are as many lines as there are parts or

Robert Grosseteste (ca. 1168–1253) aften an effigy on his burial stone in Lincoln Cathedral.

Albertus Magnus (ca. 1198–1280) after an engraving in Freher (1688).

points in the visible body, along which individual species come from individual parts [of the object] until they reach the anterior glacial humor [the lens] in which the visual power resides. And those lines are terminated on individual parts of the glacial humor, so that the species of the parts of the object are arranged on the surface of the sentient organ exactly as are the parts of the visible object. (Lindberg, 1976, pp. 109–110)

Witelo or Vitellonis (ca. 1275): ... because light has the actuality of corporeal form it makes itself equal to the corporeal dimensions of the bodies into which it flows and extends itself to the limits of capacious bodies, and nonetheless since it always contemplates the source from which it flows according to the origin of its power, it assumes *per accidens* the dimension of distance, which is a straight line, and thus it acquires the name "ray." (Lindberg, 1976, p. 119)

Pecham (ca. 1280): ... any point of a luminous or colored object is visible in any part whatsoever of the adjacent medium. But the point is seen only by making an impression on the eye. Therefore the point makes an impression on every part of the medium.... If it should be supposed that rays issue from the eye and fall on the visible object as if to seize it, either they return to the eye or they do not. If they do not return, vision is not achieved through them, since soul does not issue from body. If they do return, how do they do so? Are they animated? Are all visible objects mirrors by [virtue of the property of] reflecting rays? Furthermore, if the rays return to the eye with the form of the visible object, they go out in vain, since light itself (or the form of the visible object through the power of light) diffuses itself throughout the whole medium. Therefore the visible object need not be sought out by rays as by messengers. Moreover, how would any power of the eye be extended all the way to the stars, even if the whole body were transformed into spirit. (Lindberg, 1970, pp. 63 and 129)

Dante Alighieri (ca. 1300): Colour and light are however visible because we can perceive these by sight alone. These visible things because they are visible come within our eyes—I do not mean the things themselves but rather their forms—through the diaphanous medium as if it were through transparent glass. It is the water which exists in the pupil of the eye which renders the form visible and this happens because this water is enclosed in the eye, rather like a mirror which is glass backed with lead so that the form cannot travel any further.... From the pupil the visual spirit which continues from there to the fore part of the brain where is situated the sensitive power, immediately depicts the image and thus we see. (Ronchi, 1970, p. 64)

Dante Alighieri (1265–1321) after an engraving in Knight (1833a).

Reisch (1486): In the visible bodies we must distinguish between light (*lux*) itself, illuminating light (*lumen*) and colour. *Lux* is the natural property of luminous bodies that imparts a motion similar to that of the body to which it belongs.... *Lumen*, namely the illuminating light, is the image of the light itself that is to say of *lux*, and its derivation is of a primary nature. . . . Colour is the extremity of the transparent body in the limited body, that is to say that colour is a quality that resides at the surface of a body that is both limited and opaque and which touches the transparent medium. (Ronchi, 1970, p. 66)

Porta (1593): The incident line is that along which the *lumen* of the source arrives or the simulacrum flows through a transparent substance. . . . The refracted line is that along which travels the ray or the image in the second medium of different transparency. (Ronchi, 1970, pp. 83–84)

Kepler (1604): *Proposition I*: Light has the property of flowing or being emitted by its source towards a distant place.
Proposition II: From any point the flow of light takes place according to an infinite number of straight lines.
Proposition III: Light itself is capable of advancing to the infinite.
Proposition IV: The lines of these emissions are straight and are called rays.
(Ronchi, 1970, p. 87)

Franciscus Maurolico (1494–1575) after an illustration in Schmitz (1981).

Maurolico (1611): 1. Every point of a luminous body radiates in a straight line. 2. The denser rays illuminate more intensely; rays of equal density illuminate equally. (Ronchi, 1970, p. 100)

Snel (ca. 1621): If the eye at O (in the air) receives a light ray coming from a point R in a medium (for example, water) and refracted at S on the surface A of the medium, then O observes the point R as if it were at L on the line RM [perpendicular to the] surface A. then $SL:SR$ is constant for all rays. (Struik, 1975, p. 501)

Willebrord Snel (ca. 1581–1626) after an engraving in Freher (1688).

Descartes (1637): I would have you consider light as nothing else, in bodies that we call luminous, than a certain movement or action, very rapid and very lively, which passes toward our eyes through the medium of air and other transparent bodies, in the same manner that the movement or resistance of the bodies that this blind man encounters is transmitted to his hand through the medium of his stick. . . . Neither will you find it strange that by means of it we can see all kinds of colors; and you may perhaps even be prepared to believe that these colors are nothing else, in bodies that we called colored, than the diverse ways in which these bodies receive light and reflect it against our eyes: you have only to consider that the differences which a blind man notes among trees, rocks, water, and similar things through the medium of his stick do not seem less to him than those among red, yellow, green, and all the other colors seem to us. . . . In consequence of which, you will have occasion to judge that there is no need to assume that something material passes from the objects to our eyes to make us see colors and light, nor even that there is anything in these objects which is similar to the the ideas or the sensations that we have of them. (1965, pp. 67–68)

Hobbes (1640): The said image or colour is but an *apparition* unto us of the *motion*, agitation, or alteration, which the *object* worketh in the *brain*, or spirits, or some internal substance of the head. (1840, p. 4)

Grimaldi (1665): Light can be considered analogous to a liquid which can also spread out in waves, namely, when it passes round an object. (Mach, 1926, p. 134)

Franciscus Maria Grimaldi (1613–1663) after an illustration in Mach (1926).

Huygens (1690): Now there is no doubt at all that light also comes from the luminous body to our eyes by some movement impressed on the matter which is between the two ... If, in addition, light takes time for its passage ... it will follow that this movement, impressed on the intervening matter, is successive; and consequently it spreads, as Sound does, by spherical surfaces and waves: for I call them waves from their resemblance to those which are seen to be formed in water when a stone is thrown into it, and which present a successive spreading as circles.... each little region of a luminous body, such as the Sun, a candle, or a burning coal, generates its own waves of which that region is the centre. Thus in the flame of a candle, having distinguished the points A, B, C, concentric circles described about each of these points represent the waves which come from them. And one must imagine the same about every point of the surface and of the part within the flame. (1912, pp. 4 and 17)

Christiaan Huygens (1629–1695) after an illustration in Lenard (1933).

Berkeley (1709): In vain shall any Man tell me, that I perceive certain *Lines* and *Angles* which introduce into my Mind the various *Ideas* of *Distance*; so long as I my self am conscious of no such thing.... The Truth of this Assertion will be, yet, farther evident to any one that considers those *Lines* and *Angles* have no real Existence in Nature, being only an *Hypothesis* fram'd by *Mathematicians*, and by them introduced into *Optics*, that they might treat of that Science in a *Geometrical* way. (pp. 7–8)

Malebranche (1712): The science of optics in fact teaches only how to deceive the eyes, and its technique consists only of finding ways of imposing on us at inappropriate moments those compound or natural judgments. (1980, p. 36)

Newton (1730): Are not the Rays of Light very small Bodies emitted from shining Substances? For such Bodies will pass through uniform Mediums in right Lines without bending into the Shadow, which is the Nature of Rays of Light. (p. 345)

Waves of light emanating from points on a candle flame.

Hartley (1749): One may conjecture, indeed, that the Rays of Light excite Vibrations in the small Particles of the Optic Nerve, by a direct and immediate Action. (p. 22)

Condillac (1754): If, as is the case, the principles of optics are insufficient to explain vision, they are all the more insufficient to teach us to see. Moreover, this science does not instruct us at all on the way in which we must move our eyes. It supposes only that the eyes are capable of different movements and that they must change shape according to the circumstances. (1982, p. 274)

Melvill (1756): Those who suppose that light is nothing else than vibrations or pulses propagated thro' a subtile elastic *medium* from the visible object to the eye, may perhaps remove the difficulty by ascribing a sufficient minuteness to the particles of that *medium*; since we see, by experience, that sound in the air, and waves in the water, are conveyed in different directions, without sensibly interfering. (pp. 13–14)

Young (1801): . . . for if the focus be adapted to collect the red rays to a point, the blue will be too much refracted, and expand into a surface; and the reverse will happen if the eye be adapted to the blue rays; so that, in either case, the line will be seen as a triangular space. This observation is confirmed, by placing a small concave speculum in different parts of a prismatic spectrum, and ascertaining the utmost distances at which the eye can collect the rays of different colours to a focus. By these means I find, that the red rays, from a point at 12 inches distance, are as much refracted as white or yellow light at 11. . . . I cannot observe much aberration in the violet rays. This may be, in part, owing to their faintness; but yet I think their aberration must be less than that of red rays. (pp. 50–51)

Young (1807): Supposing the light of any given colour to consist of undulations, of a given breadth, or of a given frequency, it follows that these undulations must be liable to those effects which we have already examined in the case of waves of water, and pulses of sound. It has been shown that two equal series of waves, proceeding from centres near each other, may be seen to destroy each other's effects at certain points, and other points at to redouble them; and the beating of two sounds has been explained from a similar interference. We are now to apply the same principles to the alternate union and extinction of colours. . . . Two equal series of waves, diverging from the centres A and B, and crossing each other in such a manner, that in the lines tending towards C, D, E, and F, they counteract each other's effects. (pp. 464 and 777)

Etienne Louis Malus (1775–1812) after an illustration in Mach (1926).

Malus (1810): Optical phenomena have the special advantage of being able to be measured with great precision and of being linked by a small number of mathematical laws. These laws are independent of the hypotheses which can be advanced concerning the nature of light, because, whether we agree with Newton that it consists of a very rarefied fluid emitted by all parts of luminous bodies, or whether we agree with Huygens that it is produced by vibrations of an ethereal fluid, the path of the rays is always the same. . . . I will adopt Newton's opinion, not as an unquestionable truth, but as a starting point and in order to interpret the operations of the analysis. It is a simple hypothesis

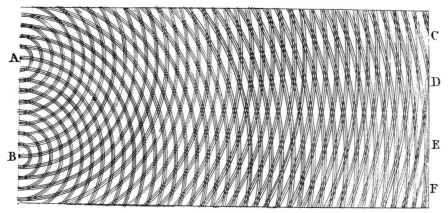

Interference between waves from two sources.

which in reality has no influence on the results of the calculations. (Ronchi, 1970, p. 232)

Fresnel (1815): It seems to me that the wave theory lends itself better to the explanation of the complex behaviour of the phenomena of light. (Ronchi, 1970, p. 243)

Augustin Jean Fresnel (1788–1827) after an engraving in Mach (1926).

2.2 Eye and Camera

The principles of shadow casting were known to scientists in China as early as the fifth century B.C., and the practical optics of the camera obscura, or dark chamber, were described by Chinese scientists in the ninth century A.D. (Needham, 1954; Hammond, 1981). Little in the way of a theory of image formation is considered to have derived from these experimental inquiries. Ibn al-Haytham described an inverted image in a dark chamber, and pinhole experiments were conducted by late medieval students of optics (Waterhouse, 1902; Lindberg, 1983a). However, the equation of such optical image forming devices with the eye appeared much later. Leonardo da Vinci conducted experiments with a camera obscura and drew an analogy between its operation and that of the eye. In this regard, as in many others, his ideas were neither published nor widely known until long after his death. Another artist, Daniele Barbaro (1569), appreciated the assistance that the camera obscura could offer to the painter, particularly when a convex lens was placed in an enlarged aperture. Despite these earlier accounts, it is Porta who has most frequently been accorded the distinction of equating the optics of a camera with those of the eye, not because he was the first to reach that conclusion, but because his description in *Natural Magick* (1589) was the most widely read. This book, which was an amalgam of mysticism, folklore, and science, was reprinted in many editions and translated into several languages (see Ronchi, 1970). Perhaps an additional factor was that both Priestley (1772) and Wilde (1838), in their histories of optics, acclaimed Porta as the discoverer of the camera obscura. Porta concluded that forming an image on a surface was an adequate account of "how vision is made," but he still considered that the image was formed on the rear surface of the lens. A few years earlier, Felix Platter (1583) had suggested that the retina was the sensitive organ of vision, and a few years later, Kepler gave the correct description of image formation on the retina. Kepler achieved this analysis without an adequate appreciation of the laws of refraction; Snell's sine law was not made known until later in the century, and Kepler had based his analysis on an approximation to the law.

A more precise equation of artificial and natural image formation was described and illustrated by Scheiner. In his book *Oculus* (1619), he not only gave the first correct illustration of its structure (see section 2.6) but also described how an artificial eye could be constructed. Eleven years later, Scheiner presented a pictorial analysis of the optical image formation in the camera and the eye—with both inverted and upright images due to the addition of convex and concave lenses. He noted that an upright retinal image resulted in inverted vision. Furthermore, Scheiner described how an image could be seen on the exposed surface of an excised animal's eye, an experiment "he had often performed." The most familiar illustration of image formation in an excised eye is that from Descartes's *Dioptrique* (1637), in which the cosmic observer ponders on the spatial arrangement of the inverted and reversed retinal image. Descartes was able to provide a more precise analysis of image formation because he could apply Snell's law to the refractions taking place. The analysis was essentially repeated in his *Treatise of Man*, but slightly different diagrams were printed in the Latin and in the French editions.

It is clear from Scheiner's analysis that the equation of eye and camera raised issues of focusing or accommodation within the optical system. These were given more detailed consideration by Rohault, who constructed a large artificial eye, and conducted a series of experiments with it. The limitations of an optical system with a lens of fixed curvature were plain to see: objects at only one distance can be adequately focused; otherwise the artificial retina needed to be moved closer to or farther from the lens. This point was also appreciated by Newton, who added that any color added to the ocular media as a consequence of disease will also influence the perceived color of objects. It is also evident,
in the quotation taken from Smith's *Opticks*, the extent to which Newton's views were both accepted and repeated with authority but without attribution.

However, it was the issue of inversion and reversal of the retinal image that taxed the philosophers and physiologists (see also section 7.2), but the very existence of a picture in the eye was soon under attack. Bishop Berkeley's objections were logical, arguing that any such picture is only imagined to be present when looking into the eyes of others. Le Cat's misgivings were physiological: the retina should not be considered as a passive screen. Few appreciated the shortcomings of equating vision with the retinal image more than Thomas Reid, founder of the Scottish "common-sense" school of philosophy; he expressed his reservations that a picture could be a part of perception with characteristic forcefulness. Reid's views were leavened by the physiological knowledge of Thomas Brown, another member of the common-sense school. The picture in the eye is a conceptual convenience that should not be seen as synonymous with vision. Nonetheless, the optics of image formation, as stated by Young, in a form essentially equivalent to that of Descartes, remained a cornerstone of visual science in the early nineteenth century.

Ibn al-Haytham (ca. 1040): . . . if there is a fire facing a hole that leads into a dark chamber, the light of that fire will appear in the chamber opposite the hole. (Sabra, 1989, p. 14)

Leonardo da Vinci (ca. 1500): Experience which shows that objects transmit their images or likeness intersected within the eye in the albugineous humour, shows what happens when the images of the illuminated objects penetrate through some small round hole into a dark habitation. You will then receive these images on a sheet of white paper inside this habitation somewhat near to this small hole, and you will see all the aforesaid objects on this paper with their true shapes and colours, but they will be less, and they will be upside down because of the said intersection. (Keele, 1955, p. 386)

Vesalius (1543): This humour is most transparent, like the very best crystal. Even when removed [from the eye], it still magnifies enormously everything on which it is put like a piece of glass, just as certain lenses do which curve outwards on both sides. (Koelbing, 1968, p. 221)

Barbaro (1569): If you wish to study the outlines, colors, and shadows of things as nature spaces them in distance, make a hole in a window shutter and set in it a thick lens from an old man's eyeglasses (not a thin lens made for a young man). Now close all the shutters and doors until no light enters the camera except through the lens, and opposite it hold a sheet of paper, which you move forward and backward until the scene appears in the sharpest detail. There on the paper you will see the whole view as it really is, with its distances, its colors and shadows and motion, the clouds, the water twinkling, the birds flying. If you partly cover the lens to leave only a small aperture, the image grows sharper. By holding the paper steady you can trace the whole perspective outline with a pen, shade it, and delicately color it from nature. (Mayor, 1946, p. 18)

Daniele Barbaro (1513–1570) after an engraving in Freher (1688).

Porta (1589): Before I part from the operations of this Glass [lens], I will tell you some use of it, that is very pleasant and admirable, whence great secrets of

Nature may appear to us. As, *To see all things in the dark, that are outwardly done in the Sun, with the colours of them.* You must shut all the chamber windows, and it will do well to shut up all holes besides, lest any light breaking in should spoil all.... Now will I declare what I ever concealed till now, and thought to conceal continually. If you put a small centicular Crystal glass to the hole, you shall presently see all things clearer.... Hence you may, *If you cannot draw a Picture of a man or any thing else, draw it by this means*; If you can but onely make the colours. This is an Art worth learning.... Hence it may appear to Philosophers, and those that study Optics, how vision is made; and the question of intromission is taken away, that was anciently so discussed; nor can there be any better way to demonstrate both, than this. The image is let in by the pupil, as by the hole in the window; and that part of the Sphere, that is set in the middle of the eye, stands instead of a Crystal Table. (1669, pp. 363–364 and 365)

Andreas Laurentius (1558–1609) aften an engraving in Du Laurens (1613).

Laurentius (1599): Finally, the eye is like vnto the looking glasse, and this receiueth all such shapes as are brought vnto it, without sending any thing of it owne vnto the object. They differ onely in this, that the looking glasse hath no power to recommend his formes and shapes vnto their judge, as the eye doth vnto the common sense by the nerue opticke. (1938, pp. 41–42)

Kepler (1604): I say that vision occurs when the image (*idolum*) of the whole hemisphere of the world which is in front of the eye, and a little more, is formed on the reddish white concave surface of the retina (*retina*). I leave it to natural philosophers to discuss the way in which this image or picture (*pictura*) is put together by the spiritual principles of vision residing in the retina and in the nerves, and whether it is made to appear before the soul or tribunal of the faculty of vision by a spirit within the cerebral cavities, or the faculty of vision, like a magistrate sent by the soul, goes out from the council chamber of the brain to meet the image in the optic nerves and retina, as it were descending to a lower court. (Crombie, 1964, pp. 147–148)

Maurolico (1611): The forms of luminous objects seen through a hole are upside down on a plane. (Ronchi, 1970, p. 100)

Scheiner (1630): Art and nature, tubes and eyes, in viewing the sun. (Plate 3)

Descartes (1637): But you can be even more certain of this if, taking the eye of a newly deceased man, or, for want of that, of an ox or some other large animal, you carefully cut through the membranes which enclose it, in such a manner that a large part of the humor *M* which is there remains exposed without any of it spilling out because of this. Then, having covered it over with some white body thin enough to let daylight pass through it, as for example with a piece of paper or with an eggshell, *RST*, place this eye in the hole of a specially made window such as *Z*, in such a manner so that it has its front,

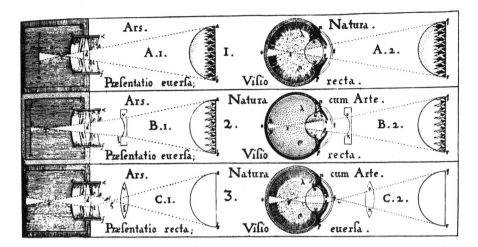

Focusing by cameras and eyes in order to form inverted or erect images. The figure represents three of the seven parts of the whole plate printed; the remaining four have more complex systems of lenses in front of the eyepieces or eyes (Scheiner, 1630).

BCD, turned toward some location where there are various objects, such as *V*, *X*, *Y*, illuminated by the sun; and the back of it, where the white body *RST* is located, toward the inside of the chamber *P* (where you will be), into which no light is allowed to enter except that which will be able to penetrate through the eye, all of whose parts, from *C* to *S*, you know to be transparent. For when this has been done, if you look at the white body *RST*, you will see there, not perhaps without admiration and pleasure, a picture which will represent in natural perspective all the objects which will be outside of it toward *VXY....* Neither can we doubt that the images which we cause to appear on a white cloth in a dark chamber are formed there in the same way and for the same reasons as on the back of the eye. (1965, pp. 91–93 and 97)

Kircher (1646): ... demonstrating images passing through the pupil and humours of the eye to be represented in the fund of the eye, which is located in darkness. (p. 162)

Descartes (1664): They [rays] assemble at precisely one part of the nerve that is there [the retina]; and, by the same means, other rays that enter the eye are prevented from touching the same part of this nerve. For example, when the eye is so arranged as to look at point *R*, the arrangement of the crystalline humor makes all the rays *RNS*, *RLS*, and so forth, reassemble exactly at point *S* and, by the same means, prevents any of those that come from points *T* and *X* and so forth, from arriving there. It also assembles all those from point *T* in the neighborhood of point *V*, those from point *X* in the neighborhood of point *Y*, and so on. Whereas if no refraction occurred in this eye, object *R* would send

Athanasius Kircher (1602–1680) after an engraving in *The Historic Gallery of Portraits and Paintings*, Vol. 6. London: Vernor, Hood, and Sharpe, 1810.

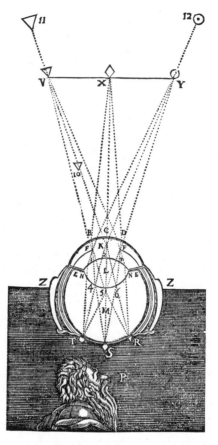

Scheiner's procedure for viewing the retinal image with an excised eye, as illustrated in Descartes (1637/1902).

only one of its rays to point *S* and the others would spread here and there throughout the space *VY*; and similarly the points *T* and *X*, and all those in between, would each send its rays toward this same point *S*. (Hall, 1972, pp. 54–55)

Rohault (1671): But because there are some Difficulties to make this Experiment [using an egg-shell] succeed well; I have thought that the same Thing might be done, by making a large artificial Eye, which I accordingly tryed: The opake Coats, or Tunicks, were all made of thick Paper, except the *Retina*, which was made of a very white thin Piece of Vellum; in the Room of the *Tunica Cornea*, I put a transparent Glass, and instead of the Chrystalline Humour, was a Piece of Chrystal of the Figure of a Lens, but more flat than this Humour; for since there was nothing in this Machine but Air, in the Places of the aqueous and vitreous Humours, a little less Convexity was sufficient to produce the Refractions required: And because it was very difficult to flatten

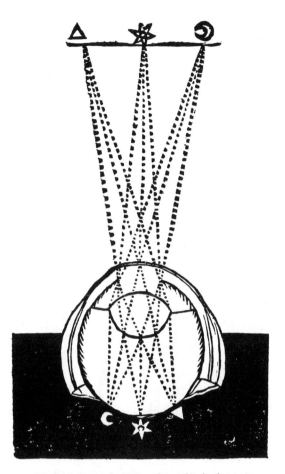

Kircher's diagram of image formation in the eye.

or lengthen this artificial Eye, in the manner the natural Eye is done by the Muscles, I placed the Vellum in such a manner, that it could be moved backward and forward, at pleasure. This artificial Eye being so placed in the Window of a Room, that the Glass which represents the *Tunica Cornea*, may be directly against some Objects that are very much illuminated; we shall not only see the Images of them impressed on the Vellum, but we may also observe all the most minute Particularities, which we before collected from Reason. Thus we may observe, *First*, That it is at one particular Distance only of the Vellum from the Chrystal Lens, that the Image will appear the most distinct that is possible. *Secondly*, That this Image is not so distinct in the extreme Parts, as in the Middle. *Thirdly*, That if the Vellum be too near the Lens; the Image will be less, and very much confused. *Fourthly*, That if it be too far, the Image will be larger, but all confused likewise. *Fifthly*, That the distinct Image of any Object, is so much less, as the Object is more remote. *Sixthly*, If a certain

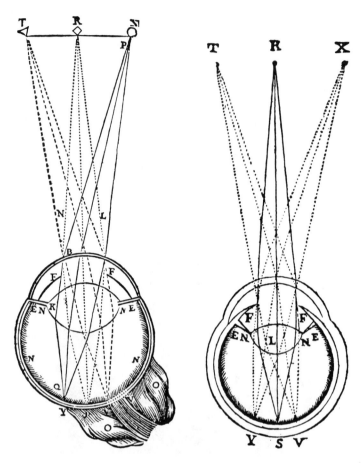

Diagrams of the dioptrics of the eye from Descartes (1662) on the left and Descartes (1664/1909) on the right.

Distance between the Lens and the Vellum, be requisite to make a distinct Image of an Object at a moderate Distance; the Vellum must be moved a little nearer, so that the Distance of the Lens from it may be less, if we would have a distinct Image of another Object, which is at a considerably further Distance. *Seventhly*, When the Vellum is at a proper Distance, to represent distinctly an Object which is at a great Distance, suppose an Hundred, or Two hundred Yards; there is no need of altering it, in order to represent, as distinct as is possible, any Objects that are at a still greater Distance. *Eighthly*, The nearer the Object is to this artificial Eye, the further must the Vellum be removed from the Lens. *Ninthly*, When the Object is too near this articial Eye, it is impossible to get any distinct Image, let the Vellum be removed to what Distance we will. (1723, pp. 243–244)

Molyneux (1692): We are likewise to observe, that the Representation of the Object *a b c* on the Fund of the Eye *f e d* is *Inverted*. For so likewise it is on the Paper in a dark Room; there being no other way for the Radious Cones to enter the Eye or the dark Chamber, but by their Axes *a o, b o, c o,* crossing in the Pole *o* of the Crystalline or Glass. (p. 105)

Newton (1704): In like manner when a Man views any Object PQR, the Light which comes from the several Points of the Object is so refracted by the transparent skins and humours of the Eye, (that is by the outward coat EFG called the *Tunica Cornea,* and by the crystalline humour AB which is beyond the Pupil *mk*) as to converge and meet again at so many Points in the bottom of the Eye, and there to paint a Picture of the Object upon that skin (called the *Tunica Retina*) with which the bottom of the Eye is covered. For Anatomists when they have taken off from the bottom of the Eye that outward and most thick Coat called the *Dura Mater,* can then see through the thinner Coats the Pictures of Objects lively painted thereon. And these Pictures propagated by Motion along the Fibres of the Optick Nerves into the Brain, are the cause of Vision. For accordingly as these Pictures are perfect or imperfect, the Object is seen perfectly or imperfectly. If the Eye be tinged with any colour (as in the Disease of the Jaundise) so as to tinge the Pictures in the bottom of the Eye with that Colour, then all Objects appear tinged with the same Colour. (p. 10)

Berkeley (1709): What greatly contributes to make us mistake in this Matter is, that when we think of the Pictures in the Fund of the Eye, we imagine our selves looking on the Fund of another's Eye, or another looking on the Fund of our own Eye, and beholding the Pictures Painted thereon. (pp. 135–136)

Smith (1738): This account of the eye and of the cause of vision is farther confirmed by these arguments; that anatomists when they have taken off from the bottom of the eye that outward and thickest coat called the dura mater, can see through the thinner coats the pictures of objects lively painted thereon. And these pictures propagated by motion along the fibres of the optick nerves into the brain, are the cause of vision. For according as these pictures are perfect or imperfect, the object is seen perfectly or imperfectly. If the eye be tinged with any colour (as in the disease of the jaundice) so as to tinge the pictures in the bottom of the eye with that colour, then all objects appear tinged with the same colour. (p. 27)

Le Cat (1744): . . . the Retina receives the Impression, moderates it, and fits it, if I may be allowed the Expression, to the Unison of the genuine Organ. But, in receiving this Impression, it is no ways sensible of it. The Image is represented on the Retina, as on an oiled Paper. It is not the oiled Paper that discerns the Image, it is the Eye, the Organ that is behind the paper. (1750, p. 168)

Diderot (1749): . . . the miniatures painted on the retina are so beautiful and exact that no painter, however skilful, could hope to reproduce them. There is

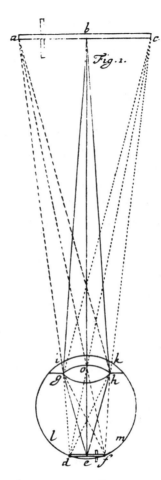

The dioptrics of the eye as illustrated by Molyneux.

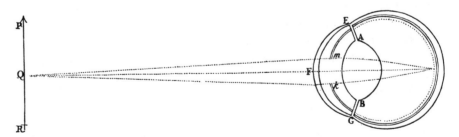

The dioptrics of the eye as illustrated by Newton.

nothing more precise than the resemblance between this representation and the object represented, and the size of the canvas on which they are painted is after all not so small as to cause confusion between the shapes on it, being a full half-inch square. (Morgan, 1977, p. 53)

Reid (1764): There is not the least probability that there is any picture or image of the object either in the optic nerve or brain.... Nor is there any probability, that the mind perceives the pictures upon the retina. These pictures are no more objects of our perception, than the brain is, or the optic nerve. No man ever saw the pictures in his own eye, nor indeed the pictures in the eye of another, until it was taken out of the head, and duly prepared.... We acknowledge, therefore, that the retina is not the last and most immediate instrument of the mind in vision. There are other material organs, whose operation is necessary to seeing, even after the pictures on the retina are formed. (pp. 284 and 310)

Erasmus Darwin (1794): If our recollection or imagination be not a repetition of animal movements, I ask, in my turn, What is it? You tell me it consists of images or pictures of things. Where is this extensive canvas hung up? or where are numerous receptacles in which those are deposited? or to what else in the animal system have they any similitude? That pleasing picture of objects, represented in miniature on the retina of the eye, seems to have given rise to this illusive oratory! It was forgot that this representation belongs rather to the laws of light, than to those of life; and may with equal elegance be seen in the camera obscura as in the eye; and that the picture vanishes for ever, when the object is withdrawn. (p. 29)

Bell (1803): We have seen that the picture of an object is formed in the bottom of the eye. It was formerly sufficient to say, that the mind contemplates this image. We should say now, that this image is conveyed into the sensorium by the optic nerve. This is an hypothesis merely; and we have no more consciousness of the object being in the brain or sensorium than in any other part of the body: we may rather say, that the impression made on the organ, nerves, and brain, is followed by sensation, and that the intelligence is the joint operation of the whole. (pp. 349–350)

Young (1807): The rays of light, which have entered the cornea, and passed through the pupil, being rendered still more convergent by the crystalline lens, are collected into foci on the retina, and form there an image, which, according to the common laws of refraction, is inverted, since the central rays of each pencil cross each other a little behind the pupil; and the image may easily be seen in a dead eye, by laying bare the posterior surface of the retina.... A picture painted on the retina in an inverted position, seen by dissecting off the sclerotica and choroid behind it. (pp. 448 and 787)

Brown (1820): There is truly, when we look at external things, a miniature image of them on the retina,—an image which, from our knowledge of the fabric of the eye, and of the laws that regulate the motion of light, might have been optically predicted, and which may be made distinctly visible, in a dissected eye, after separation of its posterior coats. The peculiar distinctness of the visual image, and the power of thus exhibiting it to others, give it an importance, in our conception of it, to which physiologically it is not entitled; and to those who do not consider the circumstances very minutely, it may seem a very reasonable supposition, that the figured surface of light, which we are capable of perceiving so distinctly in a dissected eye, must equally have formed a part of the visual perception of the individual whose eye thus exhibits it to us even after death. (pp. 155–156)

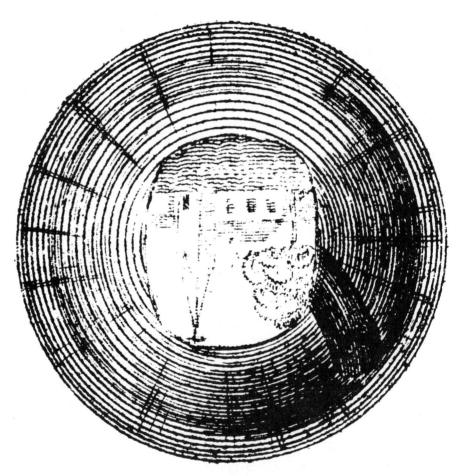

Young's representation of the inverted retinal image.

2.3   Accommodation

The analogy between eye and camera, together with an appreciation that the retina was the receptive organ, introduced a new set of problems into the study of vision. If the camera can only focus on objects at a particular distance, how is the eye able to focus on objects over a wide range of distances? This is the problem of *accommodation*, the term that Porterfield (1738) used for what had previously been referred to as the adjustment or adaptation of the eyes to different distances. Indeed, from the time of Kepler to the middle of the nineteenth century, accommodation was one of the most intensively studied and controversial topics in vision. Not surprisingly, the solutions were often derived from characteristics of cameras. A camera with a small aperture has a much greater depth of focus than one with a larger aperture; moving the camera lens toward or away from the screen onto which images are projected will vary the distance at which objects are sharply focused; conversely, moving the screen itself will have the same effect. Each of these physical speculations was advanced, together with others that were physiological. Kepler favored the view that the lens moved forward and backward in the eye. Scheiner supported this proposal largely on the basis of observations with a camera, but he did also mention that the lens could vary in shape. However, Scheiner's greatest contribution to this area was his experiment with closely spaced pinholes: when their separation was less than the diameter of the pupil, objects seen through them were multiplied at all but one distance of the card from the eye. The example he described was of viewing the spire on a tower, but the illustration used here is taken from later in his book.

La Hire considered that the eye did not need to accommodate to objects at different distances, because it could function well by ignoring blurred images. However, Descartes's physiological speculations were to prove more astute than the analogies with cameras. The lens itself was considered to change its curvature, becoming more convex for focusing on near objects, and less convex for more distant ones. He even suggested that accommodation provides a source of distance information for objects that are close to the eye (see section 7.11). To these speculative mechanisms could be added another, that the cornea increased its curvature to focus on near objects. This was advanced by Desaguliers, and later supported by Everard Home. Another possibility that was entertained concerned the elongation of the eye as a consequence of the action of the extraocular muscles. Associated with such elongation would have been an increase in cornal curvature. Porterfield was able to discount both of these speculations by recourse to the vision following removal of the crystalline lenses; such an aphakic individual was unable to accommodate at all without the aid of a convex lens. Porterfield concluded that since elongation of the eye was still possible for such a person, the crystalline lens must be involved in accommodation, although he remained unsure of the manner in which it functioned.

Writers on the eye and vision selected one or more of these hypotheses as their candidates for accommodation until the late eighteenth century, when experiment supplanted speculation. The engine for these experimental enquiries was Thomas Young. His logical and physiological conclusions were initially presented in a paper to the Royal Society of London in 1793, upon which was founded his election as a Fellow. There followed a remarkable series of experiments that were published in 1801, supporting changes in lens curvature. Such support was not derived from direct evidence, but rather from the rejection of all alternative hypotheses. Changes in corneal curvature were excluded in two ways: the images of candle flames reflected from the cornea did not change with variations in accommodation, and immersion of the eye in water did not abolish accommodation. Elongation of the eye was rendered untenable because accommodation was still possible when considerable external pressure was applied to the eye. Nonetheless, alternative interpretations were still advanced, and the general frustration of not having a solution to a clearly defined problem was voiced by Brewster. The cautious Johannes Müller, in 1838, still sat on the fence, awaiting definitive evidence for one or other of the hypotheses.

A related issue, not pursued in the quotations, concerned the structure of the crystalline lens itself, and the possibility of its supporting a change in curvature. Leeuwenhoeck had described its fibrous structure which, according to Soemmerring was an artifact of preparing the sections. An alternative structure was proposed by Young, who considered that the fibers acted like muscles. However, the most accurate examination of its formation was carried out by Brewster (1833). He observed the diffraction patterns produced by light from a candle passing through a small section of a lens, and calculated the spacing between the fibers. His values and his model of the fiber arrangement in the lens have been supported by modern microscopy (Duncan, 1984).

The association of accommodation of the eye to convergence of the eyes was made by many writers, and it was discussed principally in the context of depth or distance perception (see section 7.11). In the seventeenth century, both Aguilonius and Descartes discussed them as cues to distance, and they formed a cornerstone of Berkeley's theory of muscular involvement in distance perception (see Baird, 1903; Boring, 1942). However, their close physiological connection was emphasized by both Porterfield and Wells.

Kepler (1611): It is not possible that the retina maintaining the same position in the eye should receive a defined image both from near and from remote objects. . . . they accommodate for different distances by altering the form of the eye. (Donders, 1864, p. 445)

Scheiner (1619): Make a number of perforations with a small needle in a piece of pasteboard, not more distant from one another than the diameter of the pupil of the eye ... if it is held close to one eye, while the other is shut, as many images of a distant object will be seen as there are holes in the pasteboard ... at a certain distance, objects do not appear multiplied when they are viewed in this manner.... Some youngsters see everything sharply because they have a flexible crystalline lens and a very movable retina, so that it can from choice be drawn nearer or further away: like we experience with a screen and a selected convex lens; because the movement of the screen alone gives rise to the pictures of things appearing clearly on it: if in addition the crystalline lens becomes sometimes flatter, sometimes more curved, and the retina has the ability to approach and recede, then this would allow us to see the chosen objects at their best. (Rohr, 1919, pp. 56 and 125)

Christoph Scheiner (1571–1650) after an illustration in Polyak (1957).

Descartes (1664): The refraction that occurs in the crystalline humor [the lens] serves to strengthen the vision and at the same time render it clearer. For you must know that the shape of this humor is accommodated to refractions occurring elsewhere in the eye and to the distances of different objects, so that when vision is trained on a particular point of an object all the rays that come from this point, and that enter the eye through the pupil, are caused to reassemble at another point at the back of the eye.... The change of shape that occurs in the crystalline humor permits objects at different distances to paint their images distinctly on the back of the eye ... if, for example the humor *LN* [the lens] is of such a shape that it causes all the rays from point *R* to strike the nerve precisely at point *S*, the same humor without being changed will be able to make the rays from point *T* (which is closer) or those from point *X* (which is farther away) come there too. But it will make the ray *TL* go toward *H*, and *TN* toward *G*; and *XL* contrarily, toward *G*, and *XN* toward *H*, and so with the others. Whence in order to represent point *X* distinctly, it is necessary that the whole shape of this humor *LN* be changed and that it become slightly flatter, like that marked *I*; and to represent point *T* it is necessary that it become slightly more arched like that marked *F*. (Hall, 1972, pp. 54 and 56)

Rohault (1671): Now we may observe, that if the Object be removed further from the Eye, in such a manner that the Point B continues always in the Line BD, and the Shape or Disposition of the Eye be no ways altered; the Rays which come from the Point B to the Pupil, will not diverge so much.... This is what would happen, if the Figure of the Eye could not be altered; but to remedy all these Inconveniencies Nature has so formed the Eye, that it become flatter or longer to such a Degree, as to adjust it self to the different Distances that we would view the Object at. (1735, p. 240)

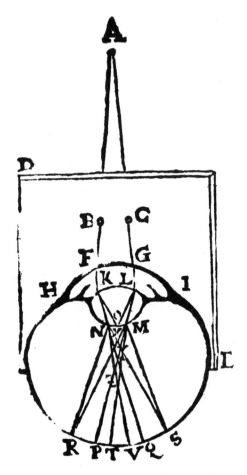

The Scheiner experiment from Scheiner (1619).

La Hire (1685): ... the Crystalline does not change its Figure or Situation, and ... the Eye receives no new Figure or Conformation, in viewing Objects at different Distances. (Porterfield, 1759a, p. 394)

Molyneaux (1692): 'Tis therefore contrived by the *Most Wise and Omnipotent Framer of the Eye*, That it should have a Power of adapting it self in some Measure to *Nigh* and *Distant* Objects. For they require different Conformations of the Eye.... But whether this variety of Conformation consist in the Crystallines approaching nigher to, or removing farther from the Retinas; Or in the Crystallines assuming a different Convexity, sometimes greater, sometimes less, according as is requisite, I leave to the scrutiny of others, and particularly of the curious Anatomist. (p. 104)

Malebranche (1712): ... for when a man sees an object nearby, his eyes are necessarily longer, or the crystalline lens is farther from the retina, than if the object were farther away. This is so because in order for the rays of this object

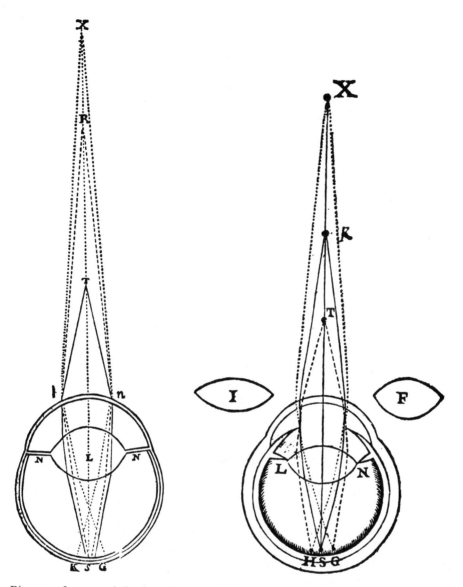

Diagrams of accommodation from Descartes (1662) on the left and Descartes (1664/1909) on the right.

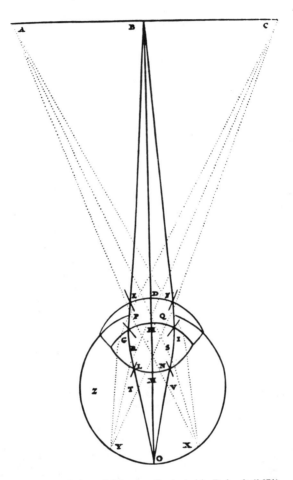

Accommodation of the eye as illustrated in Rohault (1671).

to converge on the optic nerve (which is necessary to see it distinctly, especially when the object is poorly lighted), the distance between this nerve and the crystalline lens must be greater. It is true that if the crystalline lens became more convex with the proximity of the object, this would have the same effect as if the eye were elongated. But it is hard to believe that the crystalline lens can easily change its convexity, and we have, on the other hand, a rather likely proof that the eye does elongate. Anatomy teaches us that there are muscles surrounding the middle of the eye whose effort is felt as they press and lengthen the eye when we wish to see something very near. But the way in which this happens need not be known at this point—it is enough that the change in the eye does occur, whether because the muscles surrounding the eye press it or because the small nerves that correspond to the *ciliary* ligaments (which hold the crystalline lens suspended between the other humors of the eye) relax in order to increase the convexity of the crystalline lens and tighten to decrease it, or, finally, because the pupil dilates and contracts, for there are many people whose eyes undergo no other change. (1980, pp. 42–43)

Desaguliers (1719): If then the *Eye* kept one and the same Figure always, there would be no distinct Vision, but when *Objects* are at one determinate Distance from the *Eye*, which would be inconvenient for *Animals*: And therefore to remedy this, the *Eye* has the power of changing its Figure, whereby the *Cornea* is sometimes part of the Surface of a larger Sphere, and sometimes of a lesser, and it is on this Account that the *Eye* is made to consist of various flexible Humours and Parts, the most moveable of all which is the watery Humour, lying immediately under the *Cornea*, next to which is the Christalline of the firmest Consistence; the Christalline is closely embraced by the *Ligamentum Ciliare*, by which it is suspended, and the Fibres of the Ligament by their Contraction or Dilatation bring the Christalline backwards or forwards. When the Christalline is brought forwards, it forwards the aqueous Humour, and makes the *Eye* more protuberant, or the Segment of a lesser Sphere: On the contrary, when the Christalline is brought back, the aqueous Humour returns also, and the *Eye* becoms more flat, or the Segment of a larger Sphere; so that by the Motion of the Christalline the *Cornea* is made more or less convex, the greatest Refraction being made on the *Cornea*. 'Tis by this Mobility or Changeableness of the *Eye* that we are made to see *Objects* at different Distances from us. (pp. 168–170)

Taylor (1727): . . . it [the aqueous humor] serves moreover for the *Crystalline* Humor to move forward or backward in, as Occasion requires us to see Objects nearer or farther from us. . . . For since the Rays proceeding from the luminous Points of nigh Objects diverge more than those from more remote Objects, the Power of Making them equally converge at the *Retina* is very probably lodg'd in the *Crystalline* humor; which, according to the different Distances of Objects, approaches nigher to, or removes farther from the *Retina*, and perhaps too assumes a different Convexity. (pp. 33 and 44)

Porterfield (1738): A Man having a Cataract in both Eyes, which intirely deprived him of Sight, committed himself to an Oculist, who finding them ripe, performed the Operation, and couched the Cataracts with all the Success could be desired; but after they were couched, he could not see Objects distinctly, even at an ordinary Distance, without the Help of a very convex *Lens*. . . . for it was observed, that the same *Lens* was not equally useful for seeing all Objects distinctly, but that he was obliged, for seeing them distinctly, to use Glasses of different Degrees of convexity, still the more convex the nearer the Object. . . . Seeing that nothing happens in the Eye, in couching the Cataract, but that the Crystalline is depress'd, it follows that the Change made in our Eyes, according to the Distance of Objects, must be attributed to this Humour. (pp. 182–183, 184, and 186).

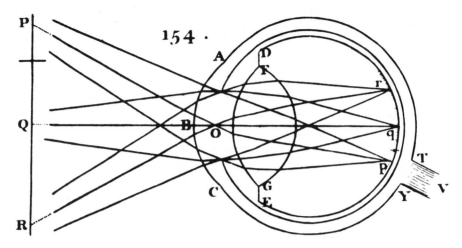

Accommodation of the eye according to Smith.

Smith (1738): Besides this there was greater need of the lens *FG* upon another account; namely to help the eye to conform it self for seeing objects distinctly at all distances, which was wanting in the fictitious eye. There are two ways of doing it by the help of this lens *FG*, in order to see things near at hand; either by moving it nearer to the outward cornea, or by increasing its convexity, or perhaps by doing both at once. If it is moved towards the cornea, this may be effected by the pressure of the muscles against the sides of the eye, and consequently against the vitreous humor; but if the crystalline lens alters its figure and becomes rounder for seeing near objects, the filaments *DF*, *EG*, whose greater tension helps to flatten it, may perhaps be slackened by the lateral pressure aforesaid. (pp. 26–27)

Hartley (1749): Thus let us suppose a Candle to be brought nearer and nearer to the Child's Eye. It is evident, that the Quantity of Light which falls upon the Eye will grow greater and greater. It will therefore agitate all the circular Fibres of the *Iris* more powerfully, and particularly the greater and lesser Rings; *i.e.* it will bend the *Cornea* into greater Convexity, bring the Origin of the Ciliar Ligament nearer to its Insertion in the *Capsula* of the Crystalline, *i.e.* suffer the *Capsula* to become more convex also, and narrow the Pupil, *i.e.* lessen the *Radius* of Dissipation. The Image of the Candle upon the *Retina* may therefore continue to be distinct, as it approaches, by this mechanical Influence of Light upon the Eye. (p. 219)

Haller (1767): But yet there is no power in the human eye which can either move the crystalline humour out of its place, or compress it. But we do not perceive this faculty in ourselves: for we move a book nearer to our eyes when it is too far off, so as to appear confused; which we would have no occasion to do, if by changing the internal figure of the eye we could correct the fault of the distance: and through a small hole, we perceive an object only single in the point of distinct vision, but double in every other. Perhaps the contraction of the pupil may do something, by which we can perceive more distinctly such objects as are near. (1786, p. 27)

Wells (1792): When the [optic] axes are parallel to each other, the eyes are in their lowest refracting state; but in their highest, when the axes are mutually intersected within two or three inches of the face; every intermediate inclination being also conjoined with an intermediate degree of refracting power. (p. 82)

Young (1793): I conceive, therefore, that when the will is exerted to view an object at a small distance, the influence of the mind is conveyed through the lenticular ganglion, formed from the branches of the third and fifth pairs of nerves, by the filaments perforating the sclerotica, to the orbitus ciliaris, which may be considered as an annular plexus of nerves and vessels; and thence by the ciliary processes to the muscles of the crystalline, which, by the contraction of its fibres, becomes more convex, and collects the diverging rays to a focus on the retina. (p. 174)

John Hunter (1728–1793) after an engraving in Knight (1834).

Hunter (1794): ... I saw no power that could adapt the eye to the various distances of which we find it capable in the human body, unless we suppose the crystalline humour to be varied in figure, which can only be effected by a muscular action within itself. (p. 25)

Hosack (1794): ... by presenting to the eye different objects at different distances, I soon perceived that its contraction and dilatation were irregular. ... I have endeavoured, first, to point out the limited action of the iris, and of consequence the insufficiency of this action for explaining vision. Secondly, to prove that the lens possesses no power of changing its form to the different distances of objects. Thirdly, that to see objects at different distances, corresponding changes of distance should be produced between the retina and the anterior part of the eye, as also the refracting powers of the media through which the rays of light are to pass. And, fourthly, that the combined action of the external muscles is not only capable of producing these effects, but that from their situation and structure they are also peculiarly adapted to produce them. (pp. 196 and 215)

David Hosack (1769–1835) after a lithograph in Williams (1845).

Home (1795): In this man we had all the circumstances combined, which seemed to be required to determine how far the crystalline lens was the principal agent in adjusting the eye. The man himself was in health, young, intelligent, and his left eye perfect; the other had been an uncommonly short time in a diseased state, and appeared to be free from every other defect but the loss of the crystalline lens.... As these experiments were made with a view to determine whether the eye, when deprived of its crystalline humour, had a power of adjusting itself to different distances; that being ascertained, they were not prosecuted further, on account of the tender state of the man's eye....

Everard Home (1756–1832) after a frontispiece engraving in Home (1814).

The result of these experiments convinced us that the internal power of the eye, by which it is adjusted to see at different distances, does not reside in the crystalline lens.... That in changing the focus of the eye from seeing with parallel rays to a near distance, there is a visible alteration produced in the figure of the cornea, rendering it more convex. (pp. 5, 8, 9, and 19)

Young (1801): I shall now finally recapitulate the principal objects and results of the investigation which I have taken the liberty of detailing so fully to the Royal Society. First, the determination of the refractive power of a variable medium, and its application to the constitution of the crystalline lens. Secondly, the construction of an instrument [an optometer] for ascertaining, upon inspection, the exact focal distance of every eye, and the remedy for its imperfections. Thirdly, to show the accurate adjustment of every part of the eye, for seeing with distinctness the greatest possible extent of objects at the same instant. Fourthly, to measure the collective dispersion of coloured rays in the eye. Fifthly, by immerging the eye in water, to demonstrate that its accommodation does not depend on any change in the curvature of the cornea. Sixthly, by confining the eye at the extremities of its axis, to prove that no material alteration of its length can take place. Seventhly, to examine what inferences can be drawn from the experiments hitherto made on persons deprived of the lens; to pursue the inquiry, on the principles suggested by Dr. Porterfield; and to confirm his opinion of the utter impossibility of such persons to change the refractive state of the organ. Eighthly, to deduce, from the aberration of the lateral rays, a decisive argument in favour of a change in the figure of the crystalline; to ascertain, from the quantity of this aberration, the form into which the lens appears to be thrown in my own eye, and the mode by which the change must be produced in that of every other person. (pp. 82–83)

Wells (1811): Having discovered that my own eyes were unfit for the experiments, which I wished to make with Belladonna, I instructed an ingenious young physician ... in the manner elsewhere described by me, of ascertaining the range of perfect vision by means of luminous points. This he found, in consequence, to begin, with respect to his left eye, at the distance of six inches, and not to terminate at the distance of eight feet, beyond which he could not see clearly the object, with which he had hitherto made his experiments, the image of the flame of a candle in the bulb of a small thermometer. The flame of a lamp, distant about sixty yards, gave a faint indication of its rays meeting before they fell on the retina; the rays from a star had very evidently their focus a little before the membrane. He now applied the juice of Belladonna to his left eye. Half an hour later, when his pupil was but little dilated, perfect vision commenced at the distance of seven inches; in fifteen minutes more, it began at the distance of three feet and a half. When his pupil had aquired its greatest enlargement, the rays from the image of the flame of a candle, in the bulb of a small thermometer at a distance of eight feet, could not be prevented from converging to a point behind the retina. The rays from lamps still more distant, and from stars, had their focuses at the same time on the retina. This state of vision continued, in its greatest extent, to the following day; and it was not till the ninth day after the application of Belladonna, that he completely recovered the power of adapting his eye to near objects. While his left eye was thus affected, the vision of his right eye remained unaltered. (pp. 382–383)

Purkinje (1823b): If we place the candlelight about six inches from someone's eye in order that we can see the flame on the cornea when we are sitting to the side of the visual axis of the eye, within the circle of the pupil nearer the periphery, we will see in the back of the pupil a blinking flame, still smaller in its diameter but reversed and of feeble illumination, which we can easily judge, by comparing it with the one on the artificial lens, that it is reflected from the posterior wall of the lens. The front surface of the lens, and partly its inner matter, under the conditions of full transparency we can make accessible for observation if, by looking into the pupil from the side and by placing the light on the opposite side of the eye, the straight lines from the eye to the observer and from the light of the candle shining into the pupil form an obtuse angle. Here one will see an elongated image of the flame, which, because it is straight, shows that it is reflected from the convex surface of the lens. . . . Both of these methods for the observation of the surfaces of the lens will not be without use, I think, in therapeutic investigation, especially where one wants to differentiate precisely whether only the capsule of the lens is involved, the lens itself, its posterior surface, or the vitreous humor. From the exact measurements of the flame reflections on the lens of a living human subject one can determine with considerable labor its shape and its relation to the acuity of vision.

Fig. 1. Candlelight reflection from anterior and posterior cornea and from the anterior and posterior portion of the lens.

Fig. 2. Candlelight reflection from the anterior surface of the cornea and the posterior surface of the lens where the image is reversed.

Fig. 3. Candlelight reflection from the anterior surface of the cornea and from the anterior surface of the lens where the reflection is erect.

Fig. 4. Semicircular umbrula which projects from the iris to the anterior surface of the lens.

Fig. 5. A light from the substantia albuginea to the center of the anterior chamber.

(John, 1959, pp. 59 and 61)

Brewster (1824): There is no part of the physiology of the eye which has excited more discussion than the power by which it accommodates itself to different distances. Although the most distinguished philosophers have contributed their optical skill, and the most acute anatomists their anatomical knowledge, yet,

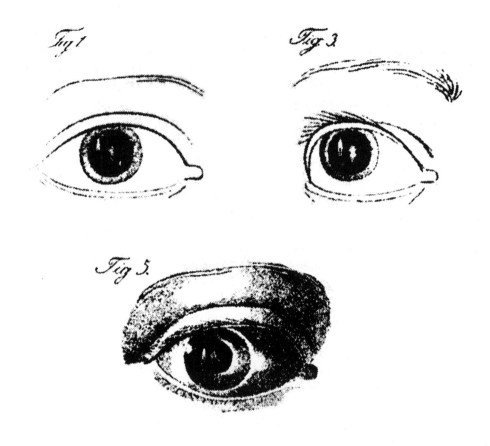

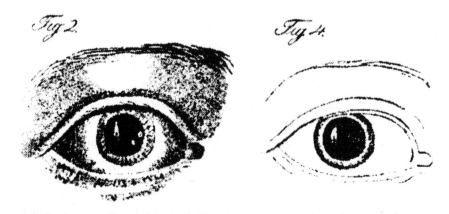

Purkinje images.

notwithstanding all this combination of science, the subject is as little understood at the present moment as it was in the days of Kepler.... it appears to me impossible to avoid the conclusion, that the power of adjusting the eye depends on the mechanism which contracts and dilates the pupil; and, since this adjustment is independent of the variation of its aperture, it must be effected by the parts which are in immediate contact with the base of the iris [the ciliary muscles]. (pp. 77 and 81)

Treviranus (1828): The distance at which a light point appears clearest to the eye is not one and the same, if the point is observed with both eyes. We see more clearly and more sharply with both eyes than with each alone, and in many people the two eyes have different strengths, which is not always a consequence of habit. This is often, though not always, founded in the original development of the eyes, in that the lens in one eye does not always have the same form as that in the other. (p. 50)

Volkmann (1836): ... one is necessarily led to the conclusion that the crystalline lens is a refracting medium, which possesses the capability of uniting all light rays in one *focus* without noticeable dispersion. (p. 117)

Müller (1838): The adaptation of the eye to distinct vision at different distances may be attributed to changes in several different parts, namely, to movements of the iris, to change of place of the lens, to elongation of the axis of the eye, or to ateration in the convexity of the lens or cornea.... It appears, therefore, most probable that the faculty of the eye which enables it to adjust itself to different distances depends on an organ which has certainly a tendency to act by consent with the iris, but yet is in a certain degree independent of it. Reasoning *per exclusionem*, it is certainly most probable that the ciliary body has this motor power. and this influence on the position of the lens, but we have no positive proof of its possessing contractility. (1843, pp. 1140 and 1150)

2.4 Errors of Refraction and Their Correction

Interest in the history of eye corrections for errors of refraction derives from the use of optical instruments as aids to scientific inquiry (Crombie, 1967; Singer, 1921). Telescopes and microscopes use the same basic materials, ground-glass lenses, to explore the upper and lower limits of nature inaccessible to the naked eye. Yet there was a long period before the connection between eyeglasses and other optical instruments was made. Ronchi (1968) has argued that the delay was occasioned by a basic medieval distrust of the senses, though this view has been questioned by Lindberg and Steneck (1972). Historical accounts of the development of eyeglasses themselves can be found in Hill (1915), Needham and Gwei-Djen (1967), Rosen (1956), and Schmitz (1982, 1983).

Individual differences in vision were remarked upon by Aristotle and Pliny, and Seneca gave an account of the magnification of letters seen through a lens or burning glass (a water-filled glass ball). Weakness of vision in the nearsighted and the aged was commented on by Aristotle, who speculated that they might have different foundations: the former inspected objects at close distances whereas the latter viewed them from afar. He also noted that nearsighted people reduce the aperture of the pupil in order to see more clearly. However, a distinctly emissive theory of vision appears to be adopted in interpreting this, unlike the receptive theory attributed to Aristotle previously. The sections on nearsightedness are to be found in *Problemata*, which is often cited as pseudo-Aristotle; it was derived from a genuine work of that title, since lost, and it was probably a compilation from several sources. There are many repetitions and contradictions in Book XXXI, which is concerned with problems connected with the eyes, and it is likely that the reference to vision proceeding forth in a more concentrated form through a small aperture was an addition by another hand. Suspect as the interpretation might be, the observations are acute.

Optical corrections for presbyopia are considered to have been adopted from the late thirteenth century, initially by Roger Bacon and thereafter avidly exploited by many; the anonymous reference to Spina, a Dominican monk, indicates that there was an element of the magical or the mercenary in their production. The inscription on a burial slab in Florence, marking the death of one Salvino degli Armati in 1317, noted that he was the inventor of eyeglasses, for which God was asked to forgive him his sins! Eyeglasses were soon depicted in paintings, the earliest of which is said to be by Tommaso da Moderna in 1352 (Rosen, 1956). Leonardo considered that convex lenses assisted vision by reducing convergence; that is, the lenses were thought of as prisms rather than magnifiers. Porta, on the other hand, interpreted their benefits to the aged in terms of reduced control over pupil diameter. Maurolico came much closer to appreciating the correct relationship; while still maintaining that the crystalline lens was the receptive organ, he described changes in its shape and associated these with myopia and presbyopia. Moreover, he advocated the use of concave lenses for the nearsighted, in contrast to convex lenses for aged eyes. As was noted above, Maurolico's book, though written in the middle of the sixteenth century, was not published until 1611, and so was unlikely to have been read by Kepler, when he was addressing similar problems.

Thus, although the assistance of convex lenses in presbyopia was readily appreciated in the thirteenth century, the integration of the lenticular optics with vision, and their relation to accommodation was to wait another three centuries. Two factors retarded such integration: ignorance both of the dioptrics and of the anatomy of the eye. When these were more clearly understood, early in the seventeenth century, corrections for both near- and farsightedness became routine, notwithstanding the doubts that remained concerning their causes. Kepler considered that these conditions were a consequence of experience; those whose work involved detailed observation of near objects became incapable of seeing distant objects, and vice versa. Descartes's analysis was much more mechanistic and pragmatic. He attributed near- and farsightedness to the shape of the eyeball itself, and sought to determine the appropriate optical correction by, essentially, employing different lenses to define the near and far points of distinct vision. Despite his assertion that the lenses selected should be the easiest to make, those shown in the accompanying diagram would not comply with that requirement. It is noteworthy that in his *Dioptrics*, from which the quotation and figures are taken, detailed directions for grinding lenses are given.

Thereafter, the corrections for myopia and presbyopia were amplified and illustrated by many writers. Molyneux added that the assistance of convex lenses in presbyopia was not simply a consequence of magnification: in old age even large print could not be read without the aid of spectacles. The optical corrections for these conditions were readily apparent, but the causes remained obscure. In order to reconcile the obvious correction with the possible cause a variety of hypotheses was entertained. Newton proposed that the cornea shrunk and the lens grew flatter in old age. Thus, convex lenses corrected the defect, resulting in distinct vision for nearer objects. Myopes, who required concave lenses in their youth, were able to see distant objects more distinctly in old age; this demonstrated an appreciation of the developmental interaction of myopia and presbyopia. Newton's views were repeated by his accolytes Desaguliers and Smith. Indeed, Newton's views were repeated virtually verbatim by Smith, who also provided illustrations of optical corrections for both far- and nearsightedness. Le Cat pursued the Cartesian analysis with greater diligence, providing in a single illustration the focal planes for normal, myopic, and presbyopic eyes. The pragmatism of the New World was evident in Franklin's introduction of bifocal glasses: since reading was normally accomplished with downward gaze and scanning the scenery with an upward gaze, the two halves of the spectacles could be so constructed to accommodate both functions.

Throughout the period of using optical corrections there was an implicit acknowledgement of the development of farsightedness: convex lenses were generally used by the aged. However, more concern with the time course of this development is evident in the writings of Haller, Wells, and Ware. Haller noted that children are naturally myopic, whereas Wells charted the onset of presbyopia from the age of forty. Mackenzie was aware that the aged eye has a reduced depth of focus, because distant as well as near objects required lenses for their distinct vision. Ware's observations of the onset of myopia were particularly astute. Not only did he recognize that the onset is often in adolescence but he also appreciated that farsightedness can occur in the young. For such individuals, prescribing convex lenses resulted in distinct vision of near and far objects, and he favored the crystalline lens as the source of the refractive error.

Aristotle (ca. 330 B.C.): Why do the short-sighted bring their eyelids close together when they look at anything? . . . They do so in order that the vision may proceed forth in a more concentrated form, since it passes through a narrower opening, and that it may not be immediately dispersed by passing out through a wide aperture. . . . Why is it that though both a short-sighted and an old man are affected by weakness of the eyes, the former places an object, if he wishes to see it, near the eye, while the latter holds it at a distance? Is it because they are afflicted with different forms of the weakness? For the old man cannot see the object; he therefore removes the object at which he is looking to the point at which the vision of his two eyes meets, expecting them to be able to see it best in this position; and this point is at a distance. The short-sighted man, on the other hand, can see the object but cannot proceed to distinguish which parts of the thing at which he is looking are concave and which convex, but he is deceived on these points. Now concavity and convexity are distinguished by means of the light which they reflect; so at a distance the short-sighted man cannot discern how the light falls on the object seen; but near at hand the incidence of light can be more easily perceived. (Ross, 1927, pp. 959a and 959b–960a)

Seneca (ca. 63): Letters, however tiny and obscure, are seen larger and clearer through a glass ball filled with water. (1971, pp. 57–59)

Pliny (ca. 77): Moreover some people have long sight but others can only see things brought close to them. The sight of many depends on the brilliance of the sun, and they cannot see clearly on a cloudy day or after sunset; others have dimmer sight in the day time but are exceptionally keen-sighted at night. (1940, p. 521)

Pliny the Elder (ca. 23–79) after an engraving in *The Historic Gallery of Portraits and Paintings*, Vol. 6. London: Vernor, Hood, and Sharpe, 1810.

Roger Bacon (ca. 1220–1292) after an engraving in *Cabinet Portrait Gallery of British Worthies*, Vol. 1. London: Charles Knight, 1845.

Roger Bacon (ca. 1270): If any one examine letters or other minute objects through the medium of crystal or glass or other transparent substance, if it be shaped like the lesser segment of a sphere, with the convex side towards the eye, and the eye being in the air, he will see the letters far better, and they will seem larger to him. . . . *For this reason such an instrument is useful to old persons and to those with weak eyes, for they can see any letter, however small, if magnified enough.* . . . For we can so form glasses and so arrange them with regard to our sight and to objects that the rays are refracted and deflected to any place we wish, so that we see the object near at hand or far away beneath whatever angle we desire. And so we can read the smallest letters or count grains of sand or dust from an incredible distance owing to the magnitude of the angle beneath which we see them, and again the largest objects close at hand might seem to be scarcely visible owing to the smallness of the angle beneath which we view them; for *it is on the size of the angle on which this kind of vision depends, and it is independent of distance save per accidens.* (Singer, 1921, pp. 395–396)

Pecham (ca. 1280): Wherefore some old people see better at a great than at a small distance.... There are others who ... see [best] when very close [to the object]. (Lindberg, 1970, p. 131)

Anon (ca. 1289): I am so weighted with years that without glasses called *occhiali* I could neither read nor write. These have been lately invented to the convenience of poor old people whose sight is enfeebled. (Singer, 1921, p. 399)

Anon (1313): ... Alessandro Spina, a monk of most excellent character and most acute mind, whatever had been made, when he [Spina] saw it with his own eyes, he too knew how to make it; and when it happened that somebody else was the first to invent eyeglasses and was unwilling to communicate the invention to others, all by himself he made them and goodnaturedly shared them with everybody. (Rosen, 1956, p. 14)

Leonardo da Vinci (ca. 1500): A demonstration how lenses help vision. Let *a b* be the lenses and *c d* the eyes. When these become aged the object which they are accustomed to see easily at *e* with the axes diverted from the straight line of the optic nerves [can no longer be seen so near]. For owing to the advance of age this power of turning becomes weakened so that it is not possible to converge without great pain to the eyes, and it is then necessary to move the object further off, that is from *e* to *f*, where it can be seen better but in less detail. Now, interposing the lens, the object can easily be seen at the proximity customary for youth, i.e. at *e*. This is so because the image of the object passes to the eye through a compound medium both rare and dense, *rare* as to the air between object and lens, and *dense* as to the thickness of the glass of the lens itself. Wherefore the direction turns in its course through the glass and diverts the line so that the object is seen at *e* as though it were at *f*, with the added advantage of not diverting the axis of the eye from its optic nerve, and as it is nearer it can be seen and recognized better at *e* than at *f*, especially if minute. (Singer, 1921, p. 401)

Porta (1593): There are two reasons why older people by using convex lenses can see better and more clearly. First because with age the pupil becomes slack and not only the pupil but all the organs and the control of the organs of the body, which becomes incontinent. Because of the slackening of the pupil the rays wander more freely and carry to the crystalline lens the object less well defined. By means of the converging lenses the rays of the simulacrum are once again re-united and the pyramid is more closely composed ... so that converging lenses by constricting the simulacrum compensate the defect. The second reason is because in old people the vitreous humour becomes altered and less pure ... and when light enters the eye through a crystal it becomes clearer and brighter. (Ronchi, 1970, p. 72)

Kepler (1604): Those who see remote objects distinctly, and near objects confusedly, require glasses that are in relief [convex]. However those who see

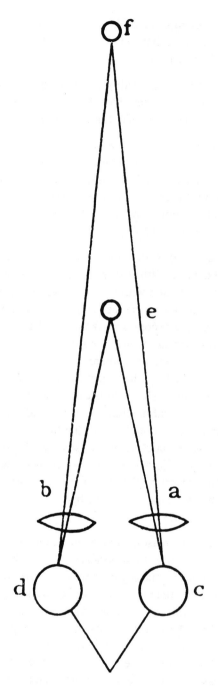

The benefits of convex lenses according to Leonardo.

remote objects confusedly, and near objects distinctly, are helped by depressed [concave] glasses. (Plehn, 1921, p. 103)

Maurolico (1611): . . . from its [the crystalline lens's] shape depends the quality of sight either short or long. (Ronchi, 1970, p. 104)

Descartes (1637): So that it seems that the eyes are formed, in the beginning, a bit longer and narrower than they ought to be, and that afterwards, as we grow old, they become flatter and wider. Thus, in order for us to be able to remedy these deficiencies through art, it will first be necessary that we seek what shapes the surfaces of a piece of glass or some other transparent body must have, in order to curve the rays falling on them in such a manner that all those rays coming from a certain point of the object are disposed, in passing through these surfaces, as if they had come to another point, which was nearer or farther away, that is, a point that is nearer to aid those who are nearsighted, and one that it farther away for the aged as well as, generally, for all those who wish to see objects nearer than the shape of their eyes will permit. Because, for example, if the eye B or C is disposed so that all the rays coming from point H or I are gathered together in the middle of its back, and if it is not able to make those from point V or X assemble there, it is evident that, if we put in front of it lens O or P, which cause all the rays from point V or X to enter inside in the same way as if they came from H or I, by this means we make up its deficiency. And because there can be lenses of many different shapes which in this case have exactly the same effect, it will be necessary, in order to choose those most suited to our purpose, that we still bear in mind two primary conditions: first, that these shapes be the simplest and the easiest possible to describe and to fashion; and second, that through these means the rays which come from other parts of the object, such as E, E, enter into the eyes in approximately the same manner as if they came from as many other points, such as F, F. (1965, pp. 116–117)

Molyneux (1692): And this is the Fault of their Eyes, who are called *Myopes*, *Purblind*, or *Short-sighted*. For in them the Crystalline is too Convex (as in *Fig. 2.* both the Convex Glass and Crystalline joyn'd together make too great a Convexity) uniting the Rays before they arrive at the *Retina*. and therefore they are helped by Concave glasses. . . . On the contrary, the Eyes of Old Men have their Crystalline too Flat (*Fig. 3.*) and cannot correct the Divergence of the Rays *b i, b k*, to make them meet on the *Retina r t*, but beyond the Eye at *e*. Wherefore for their Help 'tis requisite they add the Adventitious Convexity of a Glass; that both it and the Crystalline together, may be sufficient to unite the Rays just at the *Retina*: And from hence it appears, that Spectacles help Old Men, not by magnifying the Object, but by making its Appearance Distinct; for Old Men cannot read the largest Print without Spectacles, and yet with Spectacles, they read the smallest, though these with Spectacles do not appear so large, as those without Spectacles. (pp. 108–109)

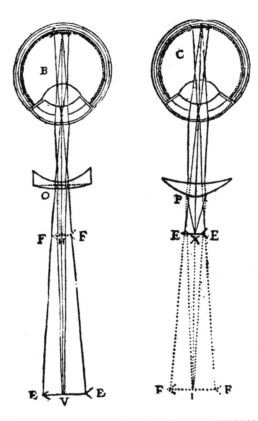

Diagrams of optical corrections from Descartes (1637/1902).

Boerhaave (1703): How foolish it is to treat a deviation in the eye, resulting from a wrong refraction of the rays of light, by means of eyedrops or by a medicinal draught! How successfully is this complaint alleviated by means of glasses. (1983, p. 111)

Newton (1704): If the humours of the Eye by old Age decay, so as by shrinking to make the *Cornea* and Coat of the *Crystalline humour* grow flatter than before, the Light will not be refracted enough, and for want of sufficient Refraction will not converge to the bottom of the Eye but to some place beyond it, and by consequence paint in the bottom of the Eye a confused Picture, and according to the indistinctness of this Picture the Object will appear confused. This is the reason of the decay of Sight in old Men, and shews why their Sight is mended by Spectacles. For those Convex-glasses supply the defect of plumpness in the Eye, and by encreasing the Refraction make the Rays converge sooner so as to convene distinctly at the bottom of the Eye if the Glass have a due degree of convexity. And the contrary happens in short-sighted Men whose Eyes are too plump. For the Refraction being now too great, the Rays converge and convene in the Eyes before they come at the bottom; and therefore the

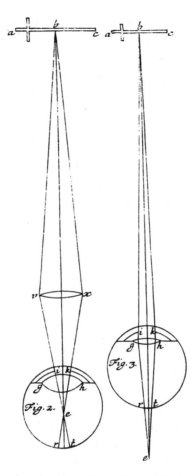

Myopia and presbyopia according to Molyneux.

Picture made in the Bottom and the Vision caused thereby will not be distinct, unless the Object be brought so near the Eye as that the place where the converging Rays convene may be removed to the bottom, or that the plumpness of the Eye be taken off and the Refractions diminished by a Concave-glass of a due degree of Concavity, or lastly that by Age the Eye grow flatter till it come to a due Figure: For short-sighted Men see remote Objects best in Old Age, and therefore they are accounted to have the most lasting Eyes. (pp. 10–11)

Desaguliers (1719): Some can't see *Objects* but when they are very near them, or close to their Eye, which being very convex, or the Segment of a small Sphere, will unite the Rays of *Objects* at a *Distance* before they come to the *Retina*. They who have this Fault in their Sight are called *Myopes*. On the contrary, there are those whose Eyes are very flat, or Segments of large Spheres, who can't see unless the *Objects* be at a good *Distance* from them, and the Rays which come from one Point to fall into the Eye are *quam proxime* Parallel—Because old Men have generally their Eyes very flat, so that they can't see but at a Distance; therefore those who are troubled with this Fault, are called *Presbytæ*: Both the Faults of Vision may be helpt by *Lens*'s; for those who are Short-sighted, and can't see any Object but what is very near them, by looking thro' a Concave Lens, will see distinctly Objects, which at the same Distance without the Lens, they could not see but very confusedly. (pp. 176–177)

Smith (1738): If the humours of the eye decay by old age so as by shrinking to make the cornea and coat of the crystalline humour grow flatter than before, the light will not be refracted enough, and for want of sufficient refraction will not converge to the bottom of the eye, but to some place beyond it; and by consequence will paint in the bottom of the eye a confused picture; and according to the indistinctness of the picture the object will appear confused. This is the reason of the decay of sight in old men, and shews why their sight is mended by spectacles. For the convex glasses supply the defect of plumpness in the eye and by increasing the refractions make the rays converge sooner, so as to convene distinctly at the bottom of the eye, if the glass has a due degree of convexity. And the contrary happens in short-sighted men, whose eyes are too plump. For the refraction being now too great, the rays converge and convene in these eyes before they come at the bottom; and therefore the picture made in the bottom and the vision caused thereby will not be distinct, unless the object be brought so near the eye as that the place where the converging rays convene may be removed to the bottom; or that the plumpness of the eye be taken off and the refraction diminished by a concave glass, of a due degree of concavity; or lastly that by age the eye grows flatter till it comes to a due figure. For short-sighted men see remote objects best in old age, and therefore they are accounted to have the most lasting eyes. (pp. 27–28)

Le Cat (1744): The Myopes, or those that can only see Objects very near have Choroides too far off from the Crystalline Humour, or from the Crossing of the Rays; either because their transparent Cornea projects too much, the Crystalline Humour is too convex, and too strong a Refraction makes the

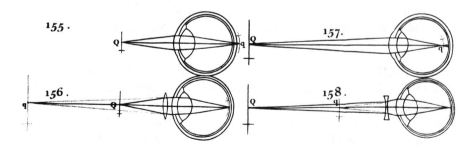

Convex and concave lenses for optical correction (Smith, 1738).

Rays cross too soon: or else because, with an ordinary Refraction, their Globe of the Eye is too big, and too much distended, or the Space of the vitreous Humour too large.... Those who discern nothing but at a great Distance, have the Choroides, H, I, K, too near the Crossing, d, d, of the Rays; either because they have the transparent Cornea or the Crystalline Humour too little convex, or else the vitreous Space is too small. (1750, pp. 258–259)

Haller (1767): Age itself advancing, gives some relief to the short-sighted; for children are in a manner naturally myoptical: but, as the eye grows older, it becomes flatter, in proportion as the solids grows stronger; and, contracting to a shorter axis, the converging powers of the lens and cornea are diminished.... Such persons are, in some measure, relieved by looking through a black tube held before the eye; by the use of which the retina grows tenderer, and the rays come to the eye in a parallel direction. The remedy here is a convex lens of glass. (1786, pp. 28–29)

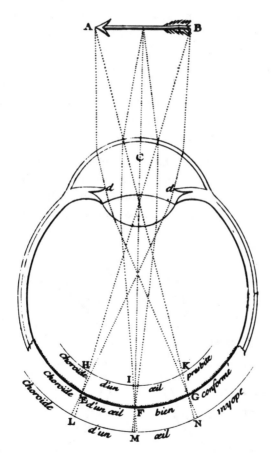

The dioptrical characteristics of the normal, myopic, and presbyopic eye as illustrated in Le Cat (1744).

Benjamin Franklin
(1706–1790) after an
engraving in Knight
(1834).

Franklin (1785): I imagine it will be found pretty generally true, that the same Convexity of Glass, through which a Man sees clearest and best at the Distance proper for Reading, is not the best for greater Distances. I therefore had formerly two Pair of Spectacles, which I shifted occasionally, as in travelling I sometimes read, and often wanted to regard the Prospects. Finding this Change troublesome, and not always sufficiently ready, I had the Glasses cut, and half of each kind associated in the same Circle, thus. By this means, as I wear my Spectacles constantly, I have only to move my Eyes up or down, as I want to see distinctly far or near, the proper Glasses being always ready. (1970, pp. 337–338)

Wells (1792): If it were asked, then, what is the real foundation of the common reproach against spectacles for long-sighted people? I should answer, a very different one from that, which is, for the most part, assigned.—For the change, in the conformation of the eyes, which renders them useful, seems to be one of those which nature has destined to take place at a particular age, and to which there is no gradual approach through the preceding course of life. A person, for instance, at forty, sees an object distinctly, at the same distance that he did at twenty. When he draws near fifty, the change I have spoken of commonly comes on, and obliges him in a short time to wear spectacles. As it proceeds, he is under the necessity of using others with a higher power. But, instead of supposing that his sight is thus gradually becoming worse, from a natural process, he attributes the increase of the defect in it to his too early and frequent use of glasses. (pp. 126–127)

Bell (1803): How the changeable state of the rays produces the indistinctness of the near-sighted eye, may be understood from this (11.) diagram. When the rays A strike the convex surface of the cornea, part of them will be reflected from the surface of the cornea, in the direction of the lines B B, when they

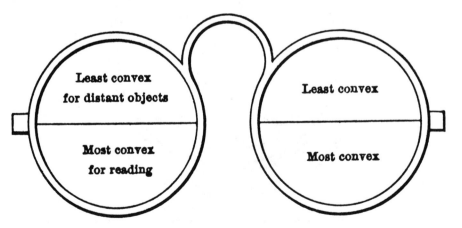

Franklin's bifocal spectacles.

will consequently strike upon the convex surface of the lens in luminous rings, these rings will be still farther multiplied and diminished in diameter, in being in part transmitted, in part reflected, from the surface of the lens *c*, and vitrious humor D D. These effects of the alternate disposition of the rays for transmission and reflection would not be perceptible, did the converging powers of the cornea and lens bring the focus of the rays exactly to the surface of the retina; but as the focus is formed at E, some way before the retina, the rays have decussated and spread out again before they form the image upon the bottom of the eye. (pp. 242–243)

Ware (1813): Near sightedness usually comes on between the ages of ten and eighteen.... There are also instances of young persons, who have so

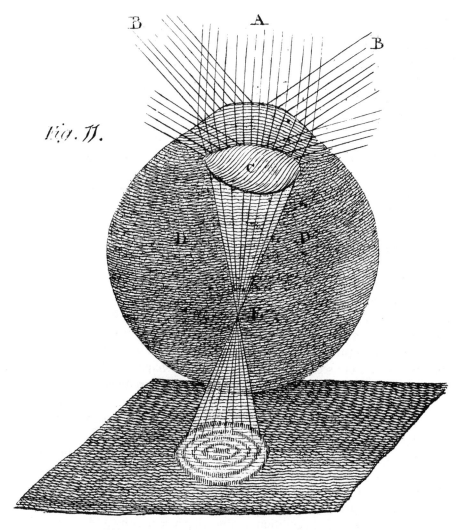

Dispersion of light through the eye as represented by Bell.

disproportionate convexity of the cornea or crystalline, or of both, to the distance of these parts from the retina, that a glass of considerable convexity is required to enable them to see distinctly, not only near objects, but also those that are distant; and it is remarkable, that the same glass will enable many such persons to see both near and distant objects; thus proving that the defect in their sight is occasioned solely by too small a convexity in one of the parts abovementioned, and that it does not influence the power by which their eyes are adapted to see at distances variously remote. In this respect such persons differ from those who had the crystalline humour removed by an operation; since the latter always require a glass to enable them to discern distant objects, different from that which they use to see those that are near. This circumstance, in my apprehension, affords a convincing proof that the crystalline humour is indispensably necessary to enable the eye to see at different distances. (pp. 31 and 43–44)

Mackenzie (1830): Although the eye, after middle life, loses the power of distinguishing near objects with correctness, it generally retains the sight of those that are distant. Instances, however, are not wanting of persons of advanced age requiring the aid of convex glasses to enable them to see distant, as well as near objects. (p. 729)

Volkmann (1836): It appears from my experience a rare occurrence that children up until the tenth and twelfth year suffer from short sightedness.... Almost all people whose eyes are exclusively directed to near objects, scholars, mechanics, engravers, etc., suffer from short sightedness. (pp. 163–164)

2.5   Astigmatism

Optical aberrations of the eye received less attention than errors of refraction, and were described much later. Chromatic aberration in optical systems, as detailed by Newton, was a problem for instrument making, but it was not thought of as a problem for vision until Young measured it in his own eyes. Young was also instrumental in measuring astigmatism in his own eye. A small spot of light was not seen as such, and it varied in shape according to the distance that it was from the eye. The sequence of such variations was illustrated in Young's article. Despite the clarity of the description and the illustrations, little heed was taken of astigmatism, and it was essentially rediscovered by Airy. He realized that he did not use his left eye in reading, and so sought to read with it alone. On so doing, he found it very difficult to read because of the indistinctness of the letters. When comparing the perceived shape of a bright spot with that of his other eye he remarked on their differences. The changes were similar to those given by Young, although for Airy the astigmatic axis was inclined to the vertical. In addition to observing a spot of light, Airy viewed perpendicular black lines on a white ground, inclined to the axis of astigmatism, and described the difference in distinctness with viewing distance. This type of stimulus was to form the basis of subsequent tests of astigmatism. He also determined the cylindrical lens correction required for astigmatism. Airy's paper was originally presented to a meeting of the Cambridge Philosophical Society in 1825 and was published in their *Transactions* two years later. It was republished in the *Edinburgh Journal of Science* (edited by Brewster) in 1827, and the editor added some observations on the paper: the corneal or crystalline basis for astigmatism could be determined from the symmetrical or asymmetrical image of a candle flame reflected from the cornea.

Young (1801): When I look at a minute lucid point, such as the image of a candle in a small concave speculum, it appears as a radiated star, as a cross, or as an unequal line, and never as a perfect point, unless I apply a concave lens inclined at a proper angle, to correct the unequal refraction of my eye. If I bring the point very near, it spreads into a surface nearly circular, and almost equably illuminated, except some faint lines, nearly in a radiating direction. For this purpose, the best image is a candle, or a small speculum, viewed through a minute lens at some little distance, or seen by reflection in a larger lens. If any pressure has been applied to the eye, such as that of the finger keeping it shut, the sight is often confused for a short time after the removal of the finger, and the image is in this case spotty and curdled. The radiating lines are probably occasioned by some slight inequalities in the surface of the lens, which is very superficially furrowed in the direction of the fibres ... When the point is further removed, the image becomes very evidently oval, the vertical diameter being longest, and the lines a little more distinct than before, the light being strongest in the neighbourhood of the centre.... Removing the point a little further, the image becomes a short vertical line; the rays that diverged horizontally being perfectly collected, while the vertical rays are still separate. In the next stage, which is the most perfect focus, the line spreads in the middle, and approaches nearly a square, with projecting angles, but is marked with some darker lines towards the diagonals. The square then flattens into a rhombus, and the rhombus into a horizontal line unequally bright. At every greater distance, the line lengthens, and acquires also breadth, by radiations shooting out from it, but does not become a uniform surface, the central part remaining always considerably brightest.... This may also be verified, by observing the image delineated by a common glass lens, when inclined to the incident rays. (see Fig. 28–40.) (pp. 43–44)

Airy (1827): ... I observed that the image formed by a bright point (as a distant lamp or a star) in my left eye was not circular, as it is in the eye which has no other defect than that of being near sighted, but elliptical, the major axis making an angle of 35° with the vertical, and its higher extremity being inclined to the right. Upon putting on concave spectacles, by the assistance of which I saw objects distinctly with my right eye, I found that to my left eye a distant

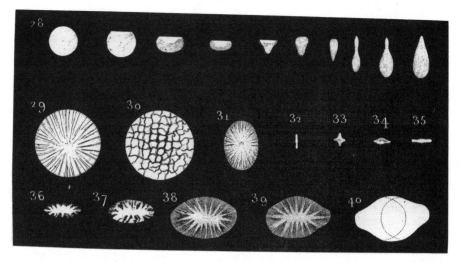

Astigmatic images of a point source of light (Young, 1801).

George Biddell Airy
(1801–1892) after a
frontispiece engraving in
Nature vol. 18, 1878.

lucid point had the appearance of a well-defined line, corresponding in direction and nearly in length to the major axis of the ellipse above-mentioned. I found also that if I drew upon a paper two black lines crossing each other at right angles, and placed the paper in a proper position, and at a certain distance from the eye, one line was seen perfectly distinct, while the other was barely visible. Upon bringing the paper nearer to the eye, the line which was distinct now disappeared, and the other was seen very well-defined.... My object was now to form a lens which should refract more powerfully in one certain plane, than those in the plane at right angles to it; and the first idea was to employ one whose surfaces should be cylindrical and concave, the axis of the cylinders crossing each other at right angles, and their radii being different. (pp. 322–323)

Brewster (1827): Mr. Airy does not seem to have ascertained in what part of the eye this curious defect exists—whether in the cornea or the crystalline lens. By examining the image of a taper reflected from the outer surface of the cornea, he will readily discover whether its form is spherical or cylindrical. If it is spherical, there can be little doubt that the crystalline is at fault, and it will remain to be determined whether the differences of refraction in different planes arise from the lens having one or both of its surfaces cylindrical, or what is more probable, from a want of symmetry in the variation of its density—an effect which is very common at that period of life when the eye begins to feel the approach of age. (pp. 325–326)

2.6   Anatomy of the Eye

Some knowledge of ocular anatomy must have been available to the Babylonian, Mesopotamian, and Egyptian specialists who carried out eye operations. However, the records that have survived usually relate to the fees they charged and the penalties they suffered for faulty operations (Shastid, 1917; Duke-Elder, 1961). Their skills and understanding would have been passed on to Greek scholars, who both developed and recorded them. While illustrations have not survived, Magnus (1901) has reconstructed the diagrams of the eye that reflect the written accounts of ocular anatomy in the Greek period; these are shown here in black and white, although Magnus's plates are colored. It should be borne in mind that the reconstructions were made by one versed in both anatomy and perspective, and that they would have appeared very strange to the authors to whom they are attributed. Sudhoff (1907) has counseled caution in interpreting these reconstructions, and it is instructive to compare Galen's eye (after Magnus) with the fragment of a manuscript drawing that is reproduced in May's (1968) translation of Galen. The latter is a much cruder representation that does not bear a great deal of similarity to the reconstruction by Magnus.

The initial Greek speculations about the anatomy of the eye, like those advanced by Empedocles in the sixth century B.C., were fueled by philosophy: the four elements of earth, air, fire, and water led to the proposition that there must be four coats to the eye. At about the same time, Alcmaeon is credited with discovering the optic nerve, which was thought of as a hollow tube, enabling humors to pass from the brain to the eye. A more concrete view of the eye was supplied by the materialist, Democritus; the eye consisted of two coats enclosing a humor that could pass along the hollow optic nerve, after the manner proposed by Alcmaeon. The dominance of philosophy over observation was reversed for one school of Greek medicine that emerged in the fifth century B.C., of which Hippocrates was a member. Naturalistic observation superseded superstition, but the examination of anatomical organs was still eschewed. The moral strictures of the time did not countenance dissection of dead bodies, although this was soon to change with the Platonic dissociation of the body from the soul. It is known that Aristotle did dissect the eyes of animals, and he is believed to have written two books (now lost) on the eye. Drawing on the evidence from dissection marked the dawning of more exact knowledge of the structure of the eye. The diagram reconstructed by Magnus shows three coats enclosing the humor, supplied by three ducts; two led to the cerebellum and the nasal one was connected to that from the other eye, and passed to the brain. The lens was probably not included because its appearance was assumed to be post mortem—an artifact of dissecting a dead eye.

The lens did appear in the eye of Celsus, although it was located in its center. The anterior chamber of the eye was separated from the posterior by a membrane, and to the lens was assigned the faculty of vision. This notion was to survive for many centuries. The lens was more accurately located in the representation attributed to Rufus of Ephesus, and the vitreous humor lay between it and the retina. The vitreous was completely enclosed and the optic nerve was not continuous with it, unlike Galen's diagram. In the latter the anterior and posterior curvatures of the lens were distinguished, and two of the extraocular muscles were shown. Vision was considered to reside in the lens, but Galen provided evidence for this assertion, namely clouding due to cataract resulted in blindness. Thus we find a clear attempt to assign the function of vision to a structure in the eye.

Galen based his anatomy on dissections of animals, particularly monkeys, but most of his ocular anatomy was derived from dissecting the eyes of freshly slaughtered oxen (Siegel, 1970). The restrictions that were subsequently placed on dissections of humans and animals resulted in a reliance on Greek works on anatomy, and they were recounted until the time of Vesalius over one thousand years later. The journey from Galen to Vesalius was not straight. The disinterest in science and medicine after the sacking of Rome in the fifth century left Europe in the "Dark Ages," but Greek wisdom was retained by Islamic scholars, who translated many books into Arabic and eventually transmitted them to late medieval students. Galen's medical works were translated by Hunain ibn Is-hâq (ca. 807–877). The earliest surviving diagrams of the eye are to be found in Islamic manuscripts (see Meyerhof, 1928; Polyak, 1942, 1957), of which that by Hunain ibn Is-hâq is probably the oldest. It is essentially a functional diagram, since it adopts different viewpoints for different parts of the eye. This could be the reason why the pupil and the lens are shown in circular form and the lens is situated in the middle of the eye. Hunain ibn Is-hâq's illustration was copied several times in the centuries that followed. Thus, Arabic accounts of the eye drew on Galen for inspiration, but their illustrations reflect a greater concern with geometry than anatomy. This is also the case for the diagrams corresponding to Ibn al-Haytham's text. The Arabic manuscript does represent two eyes, and incorporates the meeting of the optic nerves at the optic chiasm. Ibn al-Haytham added greatly to the understanding of binocular vision, which was probably the reason for representing two eyes. The illustration of the eye that was printed in Risner's (1572) translation of Alhazen and Witelo is essentially similar to that of Vesalius, and shows a single eye.

As was noted above for optics, scholars in the late Middle Ages derived much of their knowledge from manuscript translations of Ibn al-Haytham into Latin, and the diagrams of the eyes by both Bacon and Pecham showed a similar preoccupation with geometry (see Polyak, 1942). Printed figures of the eye were published from the beginning of the sixteenth century, and Reisch's diagram is perhaps the oldest version. However, this is unlikely to have been based on observation of actual eyes, but derived from earlier manuscript drawings; it does bear a close resemblance to a fifteenth-century manuscript drawing based on concentric circles (see Sudhoff, 1907; Choulant, 1945). Reisch, who was a Carthusian prior, wrote his encyclopedia as a guide for the monks in his order. The section including the diagram of the eye is but a small part of the work, and does not suggest any active pursuit of ocular anatomy. A similar diagram was printed in a specifically anatomical book by Ryff (1541), with an improvement in the representation of the crystalline lens: it took on a lenticular rather than a spherical shape. Very shortly thereafter, the genius of Vesalius was brought to bear on the topic, and the modern era of anatomy was founded.

Following the relaxation of sanctions prohibiting dissection of human bodies in the fourteenth century, knowledge concerning anatomy in general slowly began to be based on more secure ground, although the descriptions were not always accurate and observation often remained a slave to Galenic dogma. The dissecting skills of the anatomist were critical, and the major advances came with practitioners like Leonardo and Vesalius. Leonardo's detailed drawings of dissections did not make any immediate impact because they remained both in manuscript form and in private hands. Unlike his anatomical drawings of the musculature, those of the eye reflected a conflation of dissection and dogma: his rather crude drawings reflected a reliance on Galen, even though he did prepare the excised eye (by boiling it in the white of an egg) for dissection. The lens was represented as spherical and central in the eye, and the optic nerves passed to the cerebral ventricles (see figure, p. 103). The renaissance of anatomy is associated with Vesalius, who published his book *De humani corporis fabrica* (On the Structure of the Human Body) in 1543. The blocks from which the woodcuts were printed survived into the twentieth century, and they were reprinted in Saunders and O'Malley (1950); it is this source that has been used for the portrait and figures shown here. Vesalius presented an account of anatomy that was almost free from the legacy of Galen. While Vesalius could examine the structure of the eye with his own rather than Galen's eyes, he did not pay too much attention to it. His drawings did not match the detail or accuracy of those of the skeletal musculature and internal organs: a symmetrical lens was still located in the center of the eye and the optic nerve was situated on the optic axis. Platter moved the lens toward the pupil and recorded the differences between the curvatures of its front and back surfaces. Fabricius ab Aquapendente placed the lens appropriately, and defined the optical centers of several surfaces of the eye. The optic nerve left the eye centrally in these diagrams, but there is a hint of its lateral shift in the figure by Aguilonius. A few years later (in 1619), Scheiner gave the first accurate diagram of the eye; the lens and its curvatures are appropriately represented and the optic nerve leaves the eye nasally. This figure has frequently been reprinted, but it is rarely acknowledged that it is not a human eye. Scheiner stated that he did not have the opportunity of dissecting a human eye, and so his evidence was based on the eyes of domestic animals; he applied reasoning in extrapolating the structure to the human eye. Thereafter appear a series of more detailed diagrams of the eye, integrating its dioptrical properties with its gross anatomy. This is evident in the diagrams of Descartes and Le Cat. Hosack's illustration of extraocular muscles is included to show that this aspect of ocular anatomy had also progressed apace, even though it has not been pursued here.

Emphasis on the dioptrical properties of his own eye was possible for Young as a consequence of his estimates using his optometer and reflected images. Soemmerring was concerned principally with anatomy and with the preparation of the eye prior to representing it. In 1801 he produced a set of eight detailed plates of the eye; most of the dissections depicted on the plates were reproduced three times, with the figures in them shown first in outline, then shaded, and finally colored. In addition to the sections of the eye, representations of the eyelids, ducts, vascular supply, musculature, and bones of the orbit are also printed. Nonetheless, dissatisfaction was voiced by some, like Mackenzie, concerning the accuracy of anatomical drawings of sections of the eye. He excluded Soemmerring from his strictures, but it was this disquiet that resulted in the beautiful engravings that are shown here. Mackenzie commissioned a colleague, Wharton Jones, to make the drawings, the left one of which was the frontispiece of his book.

Celsus (ca. 25 B.C.–A.D. 29) after an engraving in Shastid (1917).

Democritus (ca. 400 B.C.): The structure of the eye after the description of Democritus. (Magnus, 1901, Plate 1, Figure 1)

Aristotle (ca. 330 B.C.): From the eye there go three ducts to the brain: the largest and the medium-sized to the cerebellum, the least to the brain itself; and the least is the one situated nearest to the nostrils. The two largest ones, then, run side by side and do not meet; the medium-sized ones meet—and this is particularly visible in fishes,—for they lie nearer than the large ones to the brain; the smallest pair are the most widely separate from one another, and do not meet. (Smith and Ross, 1910. p. 495a)

Celsus (ca. 29): The eye, then, has two external coats; the exterior coat of which by the Greeks is called ceratoides; and this, where it is white, is pretty thick, but before the pupil is thinner. The interior coat is joined to this, in the middle where the pupil is, and is concave, with a small aperture; round the pupil it is thin, but at a distance from it, something thicker; and by the Greeks is called chorioides. And these two coats surround the internal part of the eye, they again join behind it, and becoming finer, and uniting together, pass through the opening, which is between the bones, to the membrane of the brain, and are fixed to it. Under these, in the part where the pupil is, there is a void space; then again below is an exceeding fine coat, which Herophilus called arachnoides, the middle part of which subsides, and in that cavity is contained somewhat, which, from its resemblance to glass, the Greeks call hyaloides. This is neither liquid nor dry; but seems to be a concreted humor; from the color of which, that of the pupil is either black or gray, though the external coat be white. This is enclosed by a small membrane, which proceeds from the internal part of the eye. Under these is a drop of humor, resembling the white of an egg, from which proceeds the faculty of vision. By the Greeks it is called chrystalloides. (Shastid, 1917, p. 8581)

Rufus of Ephesus (fl. 100) after an illustration in Walsh (1935).

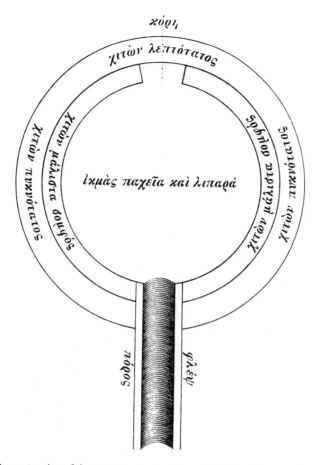

κόρη
χιτὼν λεπτότατος
χιτὼν ἀραχνοειδὴς ὁμαλότατος
χιτὼν μάλιστα σαρφός
ἰκμὰς παχεῖα καὶ λιπαρά
χιτὼν μάλιστα σαρφός
χιτὼν παχυνότατος ὁμαλότατος
πόρος
φλέψ

Reconstruction of the eye according to Democritus (after Magnus, 1901).

Rufus of Ephesus (ca. 100): The fourth tunic encloses the crystalline humour; at first this had no special name, but later it was named *lentil-like* on account of its form, and *crystalline* on account of the character of its humour. (Singer, 1921, p. 389)

Galen (ca. 175): ... to each eye there extends from the encephalon an outgrowth [*n. opticus*], which is compressed where it passes through the bones in order to make it resistant to injury, but which upon reaching the eyes themselves is resolved again, flattens out, embraces the vitreous humor like a tunic, and is inserted into the crystalline humor [the lens].... The crystalline humor itself is the principal instrument of vision, a fact clearly proved by what physicians call cataracts, which lie between the crystalline humor and the cornea and interfere with vision until they are couched. (May, 1968, pp. 463–464)

Claudius Galen (ca. 130–200) after an engraving in Pettigrew (1840b).

Hunain ibn Is-hâq (ca. 850): So we find that the eye is composed of many different parts, but that the vision is not in all the parts but only in that humour

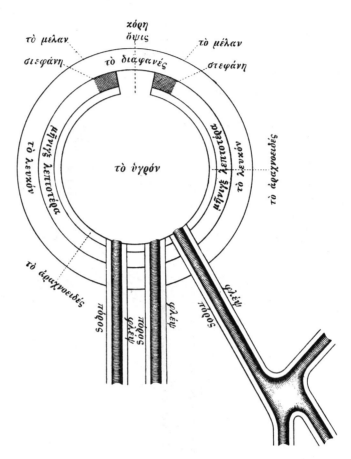

Reconstruction of the eye according to Aristotle (after Magnus, 1901).

which resembles ice and which is called in Greek ... the *ice-like*. ... it is white, transparent, luminous and round; its roundness, however, is not perfect (globe-shaped), but there is a flattening in it. (Meyerhof, 1928, p. 3)

Ibn al-Haytham (ca. 1040): (*1*) From the fifth chapter of the first book of the *Book of Optics* by Ibn al-Haitham ... (*2* and *33*) ... pupil ... (*3* and *34*) ... lower lid ... (*4*) ... cornea ... (*5* and *31*) ... aqueous [humor] ... (*6* and *30*) ... iris ... (*7* and *29*) ... conjunctiva ... (*8*) ... upper lid ... (*9* and *25*) ... anterior capsule of the lens ... (*10*) ... crystalline lens ... (*11* and *24*) ... optic nerve ... (*12* and *26*) ... vitreous [humor] ... (*13* and *23*) ... eyeball ... (*14*) ... chiasma ... (*15* and *20*) ... optic foramen ... (*16*) ... one of the two nerves which originate from the brain ... (*17*) ... the anterior portion of the brain. (Polyak, 1942, Figure 8)

Each of the eyeballs consists of a number of coats. The first is a white grease which fills the concavity of the bone; it is the larger part of the eye and it is

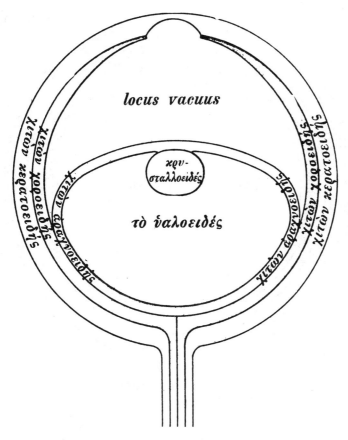

Reconstruction of the eye according to Celsus (after Magnus, 1901).

called "conjunctiva." Within this grease there is a round and hollow sphere which is black in most cases or blue or grey in some eyes. The body of this sphere is thin but of close texture and not frail; its exterior surface is fastened to the conjunctiva; the inside is hollow and lined with something like velvet. The conjunctiva encloses this sphere except at its front, for instead of covering the front of this sphere the conjunctiva circles round it. This coat is called "uvea" because it is like a grape. In the middle of the front of the uvea a circular hole leads into its cavity; it lies opposite the end of the cavity of the nerve on which the eye is mounted. This hole and the entire front of the uvea (which is encircled by the conjunctiva) are covered on the outside with a firm, white coat called "cornea" because of its likeness to white horn [in colour and] also in transparency. There exists in the forepart of the uvea's concavity a small, delicate, white and humid sphere, of cohesive humidity. Though transparent, it is not perfectly so, but somewhat opaque. Being similar in transparency to ice, it is called "crystalline." It is mounted on the extremity of the nerve's cavity. There is in the front of this sphere a slight flattening similar

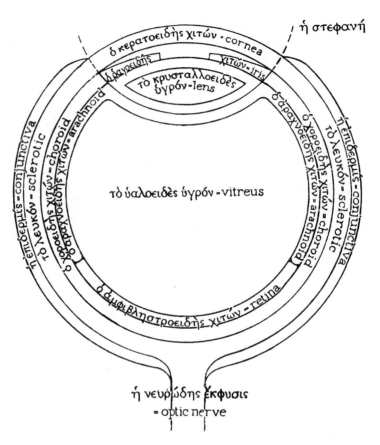

Reconstruction of the eye according to Rufus of Ephesus (after Magnus, 1901, with English names added by Singer, 1921).

to that of a lentil's exterior. Thus the surface of its front is a portion of a spherical surface which is larger than the spherical surface surrounding the rest of it. The flattened part lies opposite the aperture in the front of the uvea and is symmetrically situated with respect to it. This humor is divided into two parts of different transparencies, one towards the front, the other towards the back. The posterior part resembles crushed glass in transparency and is therefore called "vitreous humour." The shape of these two parts together is the round shape we mentioned. Enclosing them together is an extremely thin and frail tissue called "aranea," because it resembles a spider's web. (Sabra, 1989, p. 56)

Averroes (ca. 1180): As for the organ of the faculty of sight, it is the eye. This organ is distinguished by the fact that the predominant element in its composition is water, which is the lustrous, transparent substance. (1961, p. 6)

Pecham (ca. 1280): This is evident, for the part in which the power of sight resides is very delicate and sensitive because it is transparent and watery and

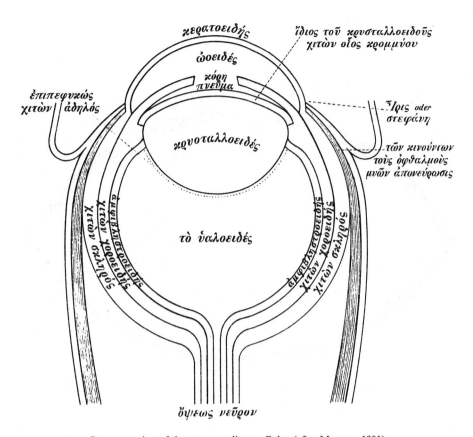

Reconstruction of the eye according to Galen (after Magnus, 1901).

of the most delicate composition. If this were not so, it would be unsuited to the subtlety of the visual spirits coming from the brain. (Lindberg, 1970, p. 113)

Leonardo da Vinci (ca. 1500): Species that penetrate the albugineous [vitreous] humor by way of the pupil meet in the sphere of the crystalline humor, and either the *virtu visiva* is in it or at the extremity of the optic nerve, which receives the species and refers them to the common sensorium.... If the *virtu visiva* is in the center of the crystalline lens, this either receives the species on its surface, they being sent to it from the surface of the cornea where the objects see themselves, or they are reflected to it from the surface of the *uvea* (retina) which surrounds and covers the albugineous humor.... In all animals the pupil dilates or contracts according to the amount of light.... The pupil of the human eye in its dilation and contraction, dilates or contracts by half its size, and in nocturnal animals it contracts and dilates more than a hundredth part of its size. And this may be seen in the eye of the owl, a nocturnal bird, by bringing toward its eye a lighted torch, or better if you make it look at the

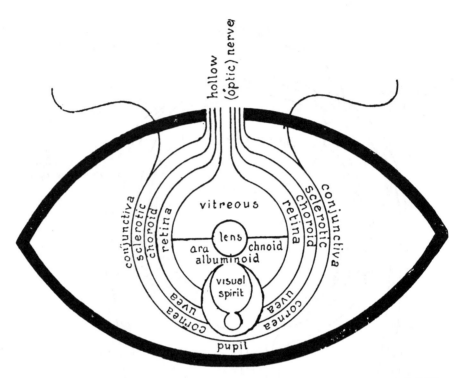

Copy of the eye according to Hunain ibn Is-hâq and labeled in English (after Meyerhof, 1928).

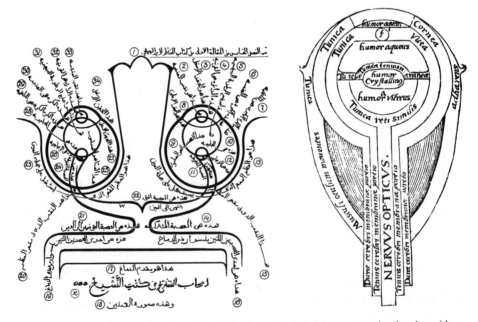

Two diagrams of the eye according to Ibn al-Haytham. On the left is a manuscript drawing with Arabic numerals added by Polyak (1942). On the right is the printed illustration from Risner's translation of Alhazen (1572); it is essentially derived from the engraving of the eye by Vesalius.

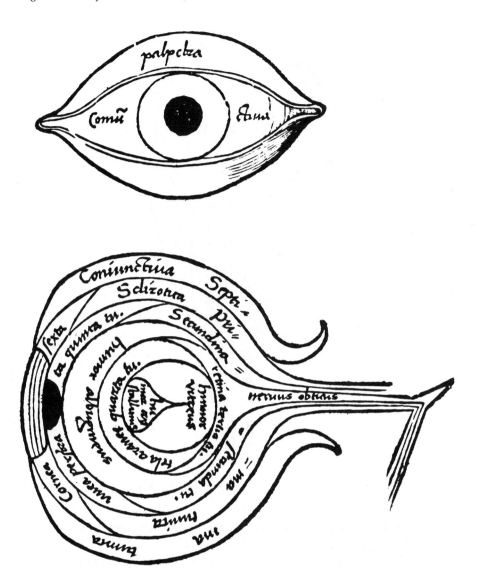

Diagram of the eye as represented by Reisch.

sun, for then you will see the pupil, which at first occupied all the eye, contract
to the size of a grain of millet, and in this contraction it equals the pupil of
man and clear and lustrous objects appear of the same color as, in such time,
they appear to man, and so much the more that the brain of such an animal is
less than the brain of man. Whence it happens that such a pupil increasing in
the night-time a hundred times more than that of man, sees a hundred times
more light than man, so that the power of vision is not then overcome by the
darkness of night. (McMurrich, 1930, pp. 220 and 224)

Reisch (1503): The visual organ and the natural eye, which has seven tunics
and four humours. (Book X, tract II)

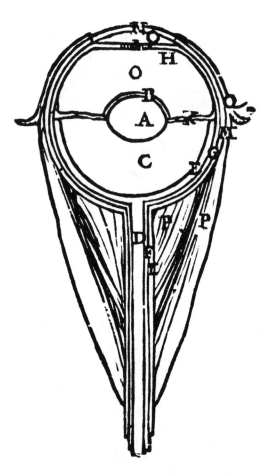

The anatomy of the eye according to Vesalius.

Vesalius (1543): A, the crystalline lens; B, a portion of the capsule; C, the vitreous body; D, the optic nerve; E, the retina; F, pia-arachnoid coat of optic nerve; G, the choroid; H, the iris; I, the pupil; K, the ciliary processes; L, the dural coat of the optic nerve; M, the sclera; N, the cornea; O, the aqueous humor; P, ocular muscles; Q, the conjunctiva. (Saunders and O'Malley, 1950, p. 200)

Platter (1583): Membranes of the human eye.

a	Crystalline humour.
b	Vitreous humour.
c	Aqueous humour.
d	Related coat.
e	Opaque part of the sclerotic.
f	Choroid.
g	Retina.
h	Hyaloid.

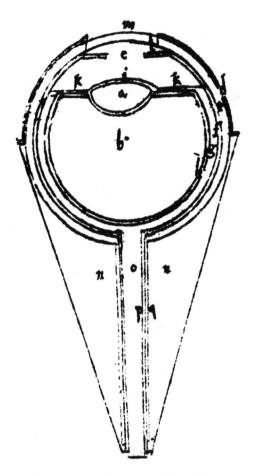

The anatomy of the eye according to Platter.

i   Crystalline capsule.
k   Ciliary processes.
l   Boundary of the choroid on the sclerotic.
m   Cornea.
nn   Ocular muscles.
o   Optic nerve.
p   Thin nerve membranes.
q   Thick nerve membranes.

(Plate 49)

Fabricius ab Aquapendente (1600): Section through the human eye.

1.   Centre of the eye.
2.   Centre of the front surface of the lens.
3.   Centre of the cornea.

(p. 105)

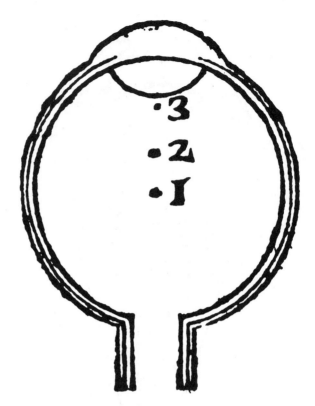

The anatomy of the eye according to Fabricius ab Aquapendente.

Hieronymus Fabricius
ab Aquapendente
(1537–1619) after an
engraving in Freher
(1688).

Aguilonius (1613): A glacial humor, B araneum, C optic nerve, D retina, E vitreous humor, F delicate part of the membrane, G choroid, H hard part of the membrane, I scleroid and cornea, K aqueous humor, L pupil, P ocular muscle, Q fatty tissue surrounding the muscles. (p. 3)

Scheiner (1619): SECTION OF THE EYE THROUGH THE AXIS AND THE OPTIC NERVE. Let line segment AB, which is the transverse diameter of the eye, be divided into two equal portions from the medium point C from which the semicircle ADB is described; at the same time let the straight line DE, taken perpendicularly through the centre C of the diameter AB, be the optical axis of which the semiaxis CE will be divided into three equal parts CF, FG, and GE; then an imaginary line HGI, which is parallel to the diameter AB, will be taken and an arc HEI, which refers to the external surface of the corneal membrane, will be described from the centre F at the distance FE; if this surface will be continued from the arcs HA and IB and described from the centres K and L more or less to the distances KA and LB, a plane HADBI of the consolidative membrane will stand out and will be prolonged towards the corneal membrane as far as H and I. Where these membranes join in H and I, the ciliary processes continue directly toward the liquid of the crystalline MN, the central area of which is hidden by the retinal membrane and by its radii

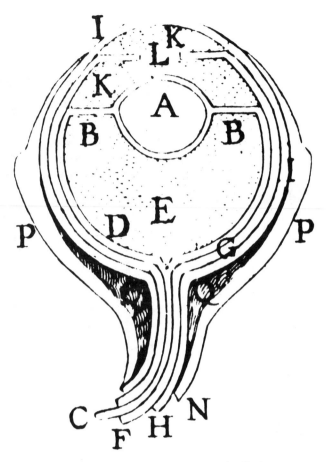

The anatomy of the eye according to Aguilonius.

which are shaped like a comb. The sclerotic membrane is a very thick one and
it is made up of a white ring; this membrane is cut behind by the HADBI
margin and it is circumscribed in front by the cornea by means of the HEI arc.
Opposite there is the ciliary body and the uvea which is represented by two
black lines slightly curved along the two parts; after them there is the choroide
which expands as a internal white ring from M to N along O: then there is
the retina represented by a white internal ring MON; the retina is supported
by the very fragile hyaloide membrane which is placed in front of the eyes and
is represented by the curved line MON. The ocular nerve OPQR, from which
all the membranes develop, does not lie on the optical axis ED, but it originates
from the left side in the right eye and from the right side in the left eye. The
observation of most animals' eyes tells us all these things; indeed these processes
happen in the eyes of cows, sheep, goats, and pigs, on which I have done
many experiments in the presence of other people; logical reasoning leads me
to suppose a similar process for the human eyes as well, because in every man's

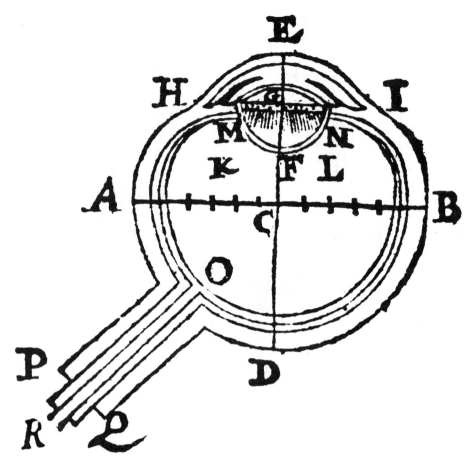

The first correct anatomical drawing of the eye (Scheiner, 1619).

eye there is a hole, through which the optical nerve comes out, placed in the same position as in animals; indeed the cavities of each eye are placed in the skull along the sides of the bone which shapes the nasal projection, although in the case of man we have to rely on reasoning more than on observation, because I have never had the opportunity to test a human eye. (pp. 17–18)

Descartes (1637): If it were possible to cut the eye in half, without any of the liquids with which it is filled escaping, and without any of its parts changing their places, so that the place of the section passed right through the middle of the pupil, it would appear such as it is represented in this diagram. *AB.C.D* is a rather hard and thick membrane which constitutes something like a round vessel in which all the eye's interior parts are contained. *DEF* is another membrane, a finer one, which is stretched in the manner of a tapestry inside of the former. *ZH* is the optic nerve, which is composed of a great number of small fibers whose extremities are extended throughout the space *GHI*, where, mingling with an infinity of small veins and arteries, they compose a sort of

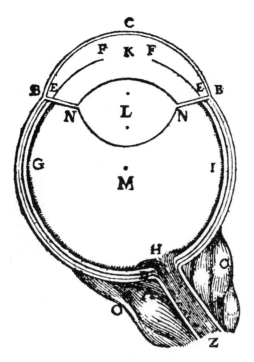

Diagram of the eye from Descartes (1637/1902).

extremely tender and delicate flesh, which is like a third membrane that covers
the entire inside of the second. *K*, *L*, *M*, are three kinds of transparent glairs
or humors which fill the entire space contained within these membranes, and
each of them has the shape which you see represented here. And experiment
shows that the one in the middle, *L*, which we call the crystalline humor, causes
almost the same refraction as glass or crystal, and that the other two, *K* and
M, cause slightly less, about the same as ordinary water, so that rays of light
pass more readily through that of the middle than through the two others, and
yet more easily through these two than through the air. In the first membrane,
the part *B.C.B* is transparent and slightly more bulging than the rest, *BAB*.
In the second, the interior surface of the part *EF*, which faces toward the base
of the eye, is completely black and obscure; and in the middle it has a small
round hole *FF*, which we call the pupil, and which appears quite black in the
middle of the eye, when we look at it from the outside. The hole is not always
the same size, and the part *EF* of the membrane in which it is found, swimming
freely in the humor *K*, which is quite liquid, seems to be like a small muscle
which can contract and enlarge as we look at objects which are nearer or
farther, or more or less lighted, or else when we wish to see them more or less
distinctly. (1965, pp. 84–85)

Descartes (1664): *ABC* is a rather hard and thick membrane [the sclera] that
consitutes a round vase, as it were, in which all parts of the eye are contained.

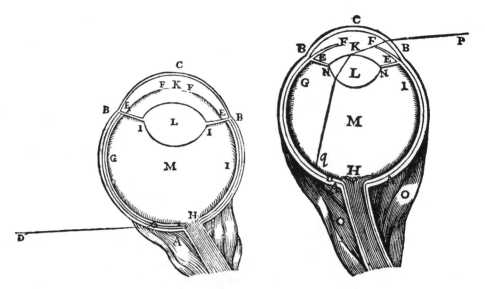

Diagrams of the eye from Descartes (1662) on the left and Descartes (1664/1909) on the right.

DEF is another, thinner membrane [the choroid], spread like a tapestry inside the preceding. *GHI* is the nerve [the retina] whose little threads (*HG, HI*)—spreading in all directions from *H* as far as *G* and *I*—entirely cover the back of the eye. *K*, *L*, and *M* [the aqueous humor, lens, and vitreous humor] are three sorts of extremely clear and transparent albumins or humors which fill all the space in the interior of these membranes and which have, respectively, the shapes pictured here. (Hall, 1972, p. 50)

Le Cat (1744): The dura Mater D D, ... on thus expanding itself, forms the first, or exterior, Membrane D b c of the Globe of the Eye, called the Cornea. The anterior Portion b c b of this Cornea is transparent, and corresponds with the Pupil. All the rest is opaque.... Towards the fore Part of the Eye the Choroides is doubled: and this exterior Complication forms what is named the Iris H H, in the middle of which is the Perforation of the Pupil. This Iris is furnished with muscular Fibres in the Form of Rays and Circles, by means of which the Pupil dilates and contracts itself. It dilates itself in the Shade and paralytic State of the optic Nerves, by the Repose or sinking of its Fibres, and is contracted when affected by Light, particularly a strong one.... The whole Space of the Eye, that is before the Corona Ciliaris I I, and the Crystalline Humour K, is filled with limpid Water, called the aqueous Humour, in the Center of which swims the Iris H, H, or Pupil.... The medullary and inner Part A ... of the optic Nerve is expanded as well as the preceding Coats, and forms a flabby Texture marked by small Points in the Plate. This Texture constitues the innermost Membrane of the Globe of the Eye, named the Retina, and terminates at the Corona Ciliaris I I. This soft Substance of the Nerve, at the

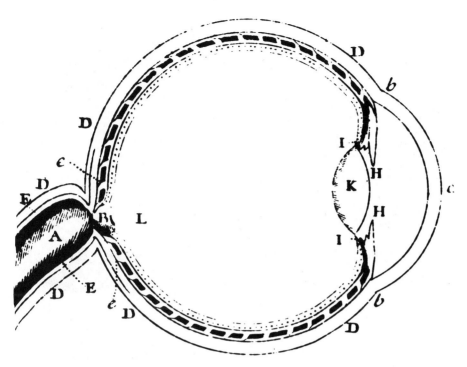

Diagram of the eye from Le Cat (1744).

Beginning of its Expansion, forms the small medullary Button B. (1750, pp. 141, 143, 144–145)

Hosack (1794): Upon carefully removing the eyelids, with their muscles, we are presented with the muscles of the eye itself, which are six in number; four called recti, or straight, and two oblique; so named for their direction, (see fig. 1) AAAA, the tendons of the recti muscles, where they are inserted into the sclerotic coat, at the anterior part of the eye. B, the superior oblique, or trochlearis, as sometimes called, from its passing through the loop or pulley connected to the lower angle of the orbiter notch in the os frontis; it passes under the superior rectus muscle, and backwards to the posterior part of the eye, where it is inserted by a broad flat tendon into the sclerotic coat. C, the inferior oblique, arising tendinous from the edge of the orbiter process of the superior maxillary bone, passes strong and fleshy over the inferior rectus, and backwards under the abductor to the posterior part of the eye, where it is inserted by a broad flat tendon into the sclerotic coat. DDD, the fat in which the eye is lodged. In fig. 2. we have removed the bones forming forming the external side of the orbit, with a portion of the fat, by which we have a distinct view of the abductor. ABC, three of the recti muscles, arising from the back part of the orbit, passing strong, broad, and fleshy over the ball of the eye, and inserted by flat, broad tendons into the sclerotic coat, at its anterior part. D, the tendon

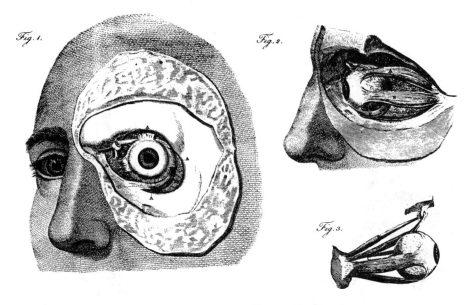

The extraocular muscles as illustrated by Hosack.

of the superior oblique muscle. E, the inferior oblique, fig. 3. A, the abductor of the eye. B, the fleshy belly of the superior oblique, arising strong, tendinous, and fleshy from the back part of the orbit. C, the optic nerve, D and E, the recti muscles. (pp. 207–208)

Young (1801): I have endeavoured to express ... the form of every part of my eye, as nearly as I have been able to ascertain it.... Vertical section of my right eye, seen from without ... Horizontal section seen from above. (pp. 49 and 86)

Soemmerring (1801): Lower part of a horizontally sectioned left eye from an adult man in his prime. This eye was extracted in a fresh condition and set out without any artificial preparation and drawn in that position. (p. 64)

Bell (1803): The eye with the cornea cut away, and the sclerotic coat dissected back.

a.   THE OPTIC NERVE.
b.   The SCLEROTIC COAT dissected back, so as to show the vessels and nerves of the choroid coat.
cc.   THE CILIARY NERVES seen piercing the sclerotic coat, and passing forward to be distributed to the iris.
d.   A small nerve passing from the same source to the same destination, but appearing to give off no branches.
ee.   Two of the VENÆ VORTICOSÆ.

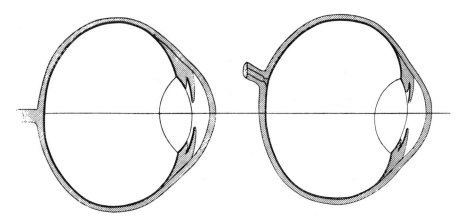

Vertical (left) and horizontal (right) sections of Young's eye.

f.   A point of the sclerotic coat through which the trunk of one of the veins had passed.

g.   A lesser venous trunk.

h.   The orbiculus ciliaris of Zinn; the ciliary ligament of others.

i.   The IRIS.

k.   The streight fibres of the iris.

l.   A circle of fibres or vessels which divide the iris into the larger circle k, and the lesser circle m.

m.   This points to the lesser circle of the iris.

n.   The fibres of the lesser circle.

o.   The pupil.

(pp. 481–482)

Young (1807): The eye is an irregular spheroid, not very widely differing from a sphere; it is principally composed of transparent substances, of various refractive densities, calculated to collect the rays of light, which diverge from each point of an object, to a focus on its posterior surface, which is capable of transmitting to the mind the impression of the colour and intensity of the light, together with a distinction of the situation of the focal point, as determined by the angular place of the object.... A section of the human eye. A is the cornea; B the aqueous humour, in which the uvea hangs; C the crystalline lens; the ciliary processes being between it and the uvea; D the vitreous humour; E F G the choroid coat, lined by the retina; H I K the sclerotica, and L the optic nerve. (pp. 447 and 787)

Mackenzie (1835): Horizontal section of the right human eye.

1.   Sclerotica.

2.   Sheath of the optic nerve.

William Mackenzie (1791–1868) after an illustration in Comrie (1932).

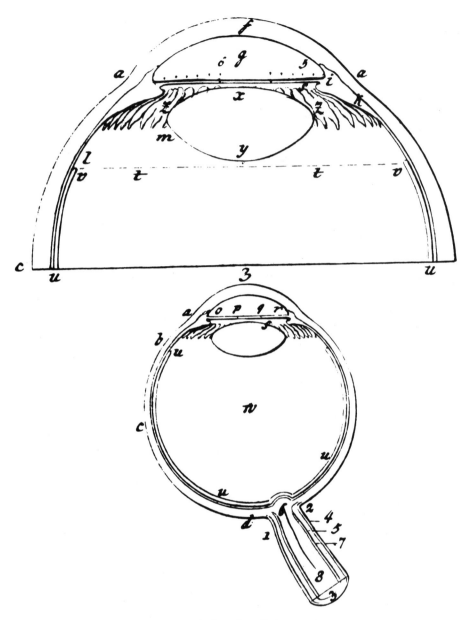

Soemmerring's section of a human eye.

Fig . 1 .

Bell's diagram of the eye.

3. Circular venous sinus of the iris.
4. Cornea.
5. Arachnoidea oculi.
6. Membrane of the anterior chamber of the aqueous humour.
7. Choroid.
8. Annulus albidus.
9. Ciliary ligament.
10. 10′. Ciliary body, consisting of (10) a pars-fimbriata or ciliary processes, and (10′) a pars non-fimbriata.
11. Ora serrata of the ciliary body.
12. Iris.
13. Pupil.
14. Membrane of the Pigment.

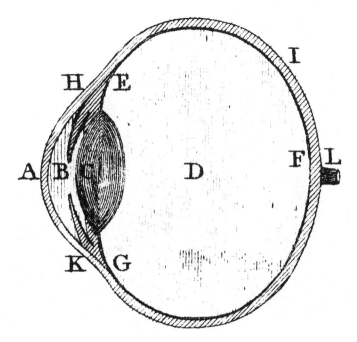

Young's (1807) diagram of the eye.

15. Delicate membrane lining the posterior chamber of the aqueous humour.
16. Membrane of Jacob.
17. The optic nerve surrounded by its general neurilema.
17′. The fibrils of the optic nerve.
18. Central artery of the retina.
19. Papilla conica of the optic nerve.
20. Retina. The situation of its vascular layer is indicated by a dotted line.
21. Central transparent point of the retina.
22. Vitreous body.
23. Hyaloid Membrane.
24. Canalis hyaloideus.
25. Zonula Ciliaris. In the plate none of its fimbriated part is seen, being concealed by the ciliary processes.
26. Canal of Petit.
27. Crystalline lens.
28. Anterior wall of the capsule of the lens.
29. Posterior wall of the capsule of the lens.
30. Posterior chamber of the aqueous humour.
31. Anterior chamber of the aqueous humour.

(Frontispiece and pp. xxxiv–xxxv)

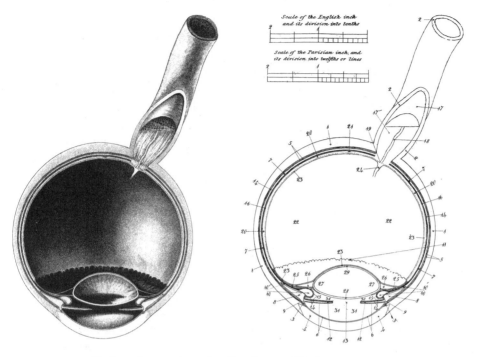

Mackenzie's engraving and outline diagram of the human eye.

2.7   Retina

The retina was considered by Galen to be an outgrowth of the brain; it had a netlike structure, and it provided nourishment for the vitreous, which nourished the lens—the "principal instrument of vision." In this way the pneuma, or visual spirits, could communicate between the brain and the lens: the pneuma traveled along the optic nerves to interact with images of external objects carried in the air to the lens. The visual spirits were carried to the cerebral ventricles where they interacted with the animal spirits. The retina was thus relegated to a nutritional role in this theory of vision. The difficulty with reconciling such a theory with the transmission of light through the transparent lens is evident in the statement by Averroes, although the text contains ambiguities, as does that of Paré. This ambiguity is removed in Platter's explicit location of the retina as the receptive organ, a view which was amplified by Kepler. How vision occurs was still a mystery, as Kepler acknowledged, and an appreciation of image formation on the retina was not the solution. Not surprisingly, the old ideas about species were retained by some, like Willis, to account for vision, even if the species were carried by the optic nerve to the brain. A truly mechanistic interpretation was given by Newton: light produced vibrations in the retina, and these were conducted to the brain along the optic nerve. It is evident from the text that Newton actually conducted experiments with cut sections of optic nerve, and concluded that vision, like hearing, is mediated by vibrations. For Hartley, Newton's vibrations became vibratiuncles in the brain, and he expanded on this concept to form a mechanistic theory of mind. Porterfield's purview was more circumscribed: he remained within the confines of vision and appreciated that the retina was a necessary but not sufficient component of perception. In this regard, he was able to draw on the experience of his own phantom limb, since he was often aware of feelings in the amputated part of his leg. These he attributed to the continued activities of the severed nerves in his stump, which would have transmitted signals to the brain.

A somewhat similar mechanical analogy was entertained by Leeuwenhoek, who examined the structure of the retina with his simple microscope. It consisted of many small "globuls," which could have been rods, cones, or optic nerve fibers. The retinal structure was analyzed in greater detail early in the nineteenth century, with the advent of powerful compound, achromatic microscopes, and Treviranus was able to provide drawings of retinal "papillae" from many species. Many of them showed the densely packed fibers in the retina, but there were several cross-sectional representations of retinal cells, of which the one shown here is the most detailed. The fibers he represented were retinal receptors and their associated structures, but the diagrams show these in the wrong orientation with respect to the eye: the receptors are directed toward the vitreous humor rather than toward the choroid. The yellow spot was described and illustrated by Soemmerring in 1801.

Galen (ca. 175): The sensory nerves [*nn. optici*] extending to the eyes are somewhat denser than the encephalon, but they do not seem to be very much harder. These are the only ones of all the nerves that will seem to you to have been made by compressing rather than by drying out the substance of the encephalon. Moreover, these nerves alone obviously contain perceptible channels. Hence many anatomists speak of them as channels, saying that two channels from the encephalon are inserted into the roots of the eyes, one into each eye, and that the netlike tunic [the retina] is formed by them as they break up and spread out. They say too, of course, that there are nerves leading to the muscles of the eyes.... The plexus called retiform by anatomists is the most wonderful of the bodies located in this region ... it is not a simple network but [looks] as if you had taken several fisherman's nets and superimposed them. (May, 1968. pp. 399–400 and 430)

Hunain ibn Is-hâq (ca. 850): The tunic which surrounds this vitreous humour is composed of two things: a hollow nerve through which the spirit passes by means of which vision is achieved, and veins and arteries.... When this net-like body is removed from the eye and its parts are collected, he who looks with persistent attention at them thinks that they are a part of the brain and cannot believe, when he sees them assembled, that they were in the eye. (Meyerhof, 1928, pp. 7 and 22)

Averroes (ca. 1180): The innermost coats of the eye [i.e., the retina] must necessarily receive the light from the humors of the eye, just as the humors receive the light from the air. However, inasmuch as the perceptive faculty resides in the region of this coat of the eye, in the part which is connected with the cranium and not in the part facing the air, these coats, that is to say, the curtains of the eye, therefore protect the faculty of the sense by virtue of the fact that they are situated in the middle between the faculty and the air. (1961, p. 9)

Paré (1579): Now followes the fourth coat called *Amphiblistroides* or *Retiformis*, the Net-like coat, because proceeding from the optick nerve dilated into a coat, it is woven like a net with veins and arteries which it receives from the grapy coat, both for the life and nourishment both of it self, and also of the glassie [vitreous] humor which it encompasses on the back part. The principall commodity of this coat is, to perceive when the Crystalline humor shall be changed by objects, and to lead the visive spirit instructed or furnished with the faculty of seeing, by the mediation of the glassie humor, even to the Crystalline being the principall instrument of seeing. (1649, p. 144)

Platter (1583): The principal organ of vision, namely the optic nerve, expands through the whole hemisphere of the retina as soon as it enters the eye. This receives and discriminates the form and colour of external objects which together with the light enter the eye through the opening of the pupil and are projected on it by the lens. (Koelbing, 1967, p. 72)

Kepler (1604): The sensation of vision follows the action of illumination, in manner and proportion. The retina is illuminated distinctly point by point from individual points of objects, and most strongly so at its individual points. Therefore in the retina, and nowhere else, can distinct and clear vision come about.... I say that vision occurs when the image of the whole hemisphere of the world which is in front of the eye, and a little more, is formed on the

Felix Platter (1536–1614) after an engraving in Freher (1688).

reddish white concave surface of the retina. I leave it to natural philosophers to discuss the way in which this image or picture is put together by the spiritual principles of vision residing in the retina and in the nerves, and whether it is made to appear before the soul or tribunal of the faculty of vision by a spirit within the cerebral cavities, or the faculty of vision, like a magistrate sent by the soul, goes out from the council chamber of the brain to meet this image in the optic nerves and retina, as it were descending to a lower court. (Crombie, 1964, pp. 142 and 147–148)

Willis (1664): In the mean time we shall take notice, that as in the smelling, so also in seeing, the sense is performed, not so much by the help of the nerve, as of the fibres, which are interwoven with the organ: to wit, the little fibres in the Membranes of the Eyes, and especially those inserted into the Sclerotick Coat, and disposed after the manner of a net, do receive the impression of the visible Species, and by representing the image of the thing, so as it is offered without, causes sight. But it is the office of the nerve it self to transmit inwardly, as it were the passage of the Optick Pipe, that image or sensible Species, and to carry it to the common Sensory. (1681, p. 140)

Leeuwenhoek (1675): I represent to my self a tall Beer-glass full of Water: This Glass I imagine to be one of the filaments of the *Optic Nerve*, and the Water in the Glass to be the globuls of which the filaments of that Nerve are made up, and then, the Water in the Glass being toucht on its surface with the finger, that to this contact did resemble the action of the visible object upon the Eye, whereby the outermost globuls of the fibres in the Optic Nerve next to the Eye are toucht. This contact of the Water made by the finger cannot be said to touch and move only the surface of the Water, but we must also grant, that all the water in the Glass is moved thereby, and that even the bottom of the Glass comes to suffer, and to be more pressed by it, than it was before the finger touched the Water, and that also all the parts of the Water are moved thereby. This motion then of the Water, said to be made by the contact of the finger, I imagine to be like the motion of a visible object made upon the soft Globuls, that lie at the end of the Optic Nerve next the Eye, which outermost globuls do communicate the like motion to the other globuls so as to convey it to the Brain. (p. 379)

Newton (ca. 1682): Light seldom strikes upon the parts of gross bodies, (as may be seen in its passing through them;) its reflection and refraction is made by the diversity of æthers; and therefore its effect upon the retina can only be to make this vibrate: which motion must be carried in the optic nerves to the sensorium, or produce other motions that are carried thither. Not the latter, for water is too gross for such subtile impressions; and as for the animal spirits, tho' I tied a piece of the optic nerve at one end, and warmed it in the middle, to see if any airy substance by that means would disclose itself in bubbles at the other end, I could not spy the least bubble; a little moisture only, and the marrow itself squeezed out. . . . And that vision is thus made, is very conformable to the sense of hearing, which is made by like vibrations. (Harris, 1775, p. 100)

Hartley (1749): Since the *Retina* is an Expansion of the Optic Nerve, we may conclude, from the Analogy of the other Senses, that it is the immediate Organ of Sight. Nor is the Want of Sensibility in the Button of the Optic Nerve, a sufficient Objection to this; as the minute Structure and Disposition of the Parts of the Button are not known. (p. 191)

Porterfield (1759b): The *Retina* then serves for receiving the Images of external Objects painted thereon, and to propagate the Impressions made thereby thro' the small Fibres of the optic Nerve into the common Sensory, or that Part of the Brain in which all the Fibres of our Nerves terminate, and in which our Mind resides, by which Means the Ideas of Objects are excited in the Mind. (p. 214)

Soemmerring (1801): *a.* Retina . . . *b.* Central hole in the middle of the retina. *c.* Yellow spot surrounding the central hole, becoming darker towards the centre, and appearing gradually paler towards the periphery. (p. 69)

Wardrop (1818): Besides the vascular and medullary lamina of the retina, there is a peculiar part of its structure called the *Macula Lutea*, discovered by Sömmerring, which has hitherto been only observed in the human eye, and

Samuel Thomas von Soemmerring (1755–1830) after an illustration in Garrison (1914).

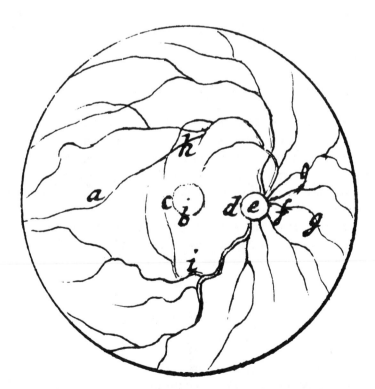

Soemmerring's diagram of the retina showing the yellow spot.

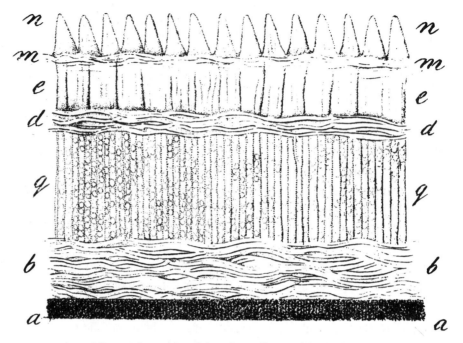

Microscopic structure of the retina as illustrated by Treviranus.

the functions of which have not been even conjectured. The macula lutea, which is observed near the centre of the retina, appears in the form of a spot of yellow colour, contiguous to which is also a fold and a small hole. (pp. 135–136)

Treviranus (1837): Fig. 3. A vertical section cut from outside to in of a very thin slice of the retina of a crow (Corvus Cornix), preserved in alcohol—aa. internally, from the dark pigmented sheet of the choroid—bb. above this level lies the layer of the organic element of the cellular structure; it covers the outermost surface of the retina and the extensive fibres, and ends in a network of (probably blood) vessels—qq. vertical extensions of these fibres, whose contents are transformed into rows of globules by the action of the alcohol—dd. second cellular layer, comprised of a network made up from the central artery of the optic nerve, and where the fibres qq become separated from the retina—ee. coarser papillae, arising from those in qq—mm. third layer of cells through which the fibres ee pass, and from which the papillae nn emerge. (pp. 105–106)

2.8   Blind Spot

Physicists, like Kepler, had demonstrated how an image is formed on the retina, and anatomists, like Platter, had argued for the retina as the receptive organ, but physiologists raised doubts about the conclusion. These doubts were based on a compelling phenomenon, first described by Mariotte in 1668, namely the blind spot. Mariotte found that the image of a small object falling on the base of the optic nerve is invisible when one eye alone is used. The phenomenon enabled Mariotte and others to locate the optic nerve with precision, either in terms of visual angles (Harris) or retinal dimensions (Young). Erasmus Darwin examined a strabismic child who had alternating suppression (see section 6.7), and he was surprised to find that the blind spot in each eye was half the value found for most other observers.

There remained considerable controversy concerning the interpretation of the phenomenon. Mariotte contended that the retina extended over the whole inner surface of the eye, including the optic nerve, and so the retina could not be the receptive organ. Almost everyone else disagreed, on the grounds that other obstructions in the eye to light do not disrupt vision (Pequet and Claude Perrault), that the blind spot is not seen with pressure images (Elliott), or that comparative studies indicate that the retina undergoes fewer species differences than the choroid (Haller). The dispute was instructive because it stimulated studies of this and related phenomena, particularly those concerned with subjective vision (see chapter 4).

Mariotte (1668): Having often observed in Anatomical Dissections of Man as well as Brutes that the *Optick Nerve* does never answer just to the Middle of the bottom of the Eye, i.e. to the place, where is made the picture of the Objects, we directly look on; and that in man it is somewhat higher, and on the side towards the Nose; to make therefore the Rayes of an object to fall upon the Optick Nerve of my Eye, and to find the consequence thereof, I made this Experiment; I fastn'd on an obscure Wall, about the hight of my Eye, a small round paper, to serve me for a fixt point of Vision; and I fastned such an other on the side thereof towards my right hand, at the distance of about 2 foot; but somewhat lower than the first, to the end that it might strike the *Optick Nerve* of my Right Eye, whilst I kept my Left shut. I then placed my self over against the First paper, and drew back by little and little, keeping my Right Eye fixt and very steddy upon the same; and being about 10 foot distant, the second paper totally disappear'd.... This Experiment hath given me cause to doubt, Whether Vision was indeed perform'd in the *Retina* (as is the Common opinion) or rather by that other Membrane, which at the bottom of the Eye is seen through the *Retina*, and is called the *Choroides*. For if Vision were made in the *Retina*, it seems that then it should be made wherever the *Retina* is. (pp. 668–669)

Edmé Mariotte (1620–1684) after an illustration in *Principales Découvertes de l'Église*. Paris: Librairie des Catéchismes.

Pecquet (1668): Every one wonders, that no person before you hath been aware of this Privation of Sight, which every one now finds, after you have given notice of it. But as to the *Sequele*, you draw from this Discovery, I see it not cogent to abandon the Plea of the *Retina* for being the principal Organ of Vision. For (not to insist here on other considerations) it will be sufficient,

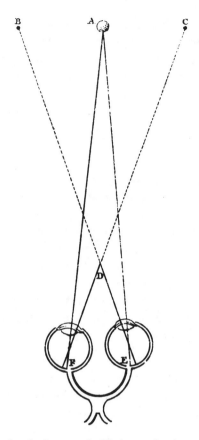

Pecquet's illustration of projection onto the blind spot of each eye (from Mariotte, 1717)

Antonius Pecquet
(1622–1674) after an
illustration in Dumesuil
and Bonnet-Roy (1947).

now to take notice, that at the place of the *Optick Nerve* there is something, that may very well cause this loss of the Object. *There* are the *Vessels* of the *Retina*, the trunks whereof are big enough to give a hindrance to Vision.... Fasten against a wall a round white paper [A], of the bigness of an inch or two, and on the side of this paper put two marks [B and C] one on the right, the other on the left side, each about two foot distant; then place your self directly before the Paper, at the distance of about nine foot, and put the End of your finger over against your both Eyes [at D], so that it may hide from the right Eye the left mark, and from the left Eye, the right mark. If you remain firm in that posture, and look steddily with both Eyes on the end of your finger, the paper, which is not at all cover'd thereby, will altogether disappear. (pp. 669–670 and 671)

Mariotte (1670): It must first of all be agreed, that in this Experiment, almost all men do loose sight of an intire circle of white paper, whose diameter is about the 9th or 10th part of its distance from the Eye.... because the Paper then wholly disappears, it follows, that all the basis of the Optick nerve is insensible of light; whence I conclude, that the *Choroeides* is the Principal Organ of sight; and that the *Retina* is not, seeing it is placed in that part, and is there apparently disposed in like manner to the rest of the bottom of the Eye. (pp. 1039–1040)

Perrault (1683): 1. If the *Choroide* were the seat of *Vision*, its function would be hindred by the branches of *Blood Vessels* lying in the *Retina*. 2. The *Choroide* should not be rugged and unequal; nor hard and thick; nor have a slimy or dirtiness upon it, to hinder the Impression of light, nor want a Communication with the *Optick Nerve*. 3. If the want of *Vision* in the foregoing Experiment may be salved by any of the Two probable reasons here offered; then there is no need of discharging the *Retina*. (pp. 265–266)

Claude Perrault (1613–1688) after an engraving in Charles Perrault (1696).

Mariotte (1683): ... there are defects in *Vision*, caused by the *Blood Vessels* in the *Retina* ... but these defects are not sensible when we look with both Eyes, for there are no *Vessels* that lye so near the *Axis Opticus* as to hinder a direct view. (p. 266)

La Hire (1709): Thus the whole difference between him [Mariotte] and me will be in the name of the principal organ, for he makes vision consist in a reflection of the luminous rays upon the *choroides*, and I in a shaking of the parts of the *choroides*, to be transmitted to the optic nerve or to the *retina*. (1742, p. 199)

Smith (1738): So far Mr. *Mariotte*; whose experiment I have tryed in a chamber from which all sensible light was excluded, except what came into it through a key-hole; and this also disappeared totally when it fell upon the base of the optick nerve; which shews it to be totally insensitive to light. Yet in looking at objects of an uniform colour with one eye, we are not sensible of any such defect or dark round spot ... This defect of sensation having been constantly supplied by the other eye, is now supplied by the imagination only. (*Remarks*, p. 7)

Haller (1767): Is it altogether false that the object is painted on the retina? Or is this picture made on the choroides? Is this new opinion confirmed by an experiment, by which it is found that the place where the optic nerves enters is blind? and which is thus explained, that there is in that place no choroides but the bare retina, and that thence there is no vision. But this is repugnant to a very well known observation, namely, that the retina is a most sensible nervous medulla; and that the choroides consists almost entirely of a few small nerves, and of vessels most certainly blind. This is likewise contradicted by the very great variety of choroides in animals; the equally great constancy of the retina; and the black spots, which, even in man, obscure the exterior surface of the retina. But by this experiment we know the reason why the optic nerve is not inserted into the axis of the eye, but into its side. For thus, except only in one single case, where there is an impediment in the concourse of lines drawn

Albrecht von Haller (1708–1777) after a mezzotint in Brucker (1741–1755).

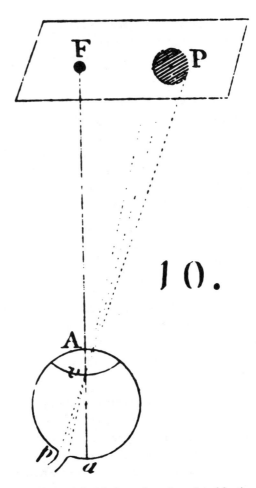

The angular separation of the blind spot from the point of fixation (Harris, 1775).

through the centre of the optic nerves, the one eye sees and assists that whose blind side is turned towards the object. (1786, p. 25)

Priestley (1772): Let three pieces of paper, A, B, and C, be fastened upon the side of a room, about 2 feet asunder, and let a person place himself opposite to the middle paper, and, beginning pretty near to it, retire gradually backwards, all the while keeping one of his eyes shut, and turned obliquely towards that outside paper which is toward the covered eye, and he will find a situation (which is generally about five times the distance at which the papers are placed from one another) where the middle paper will intirely disappear, while the two outermost continue plainly visible. (p. 193)

Harris (1775): Taking a medium of the several observations; the [angle] *avp* contained between the optic axis, and the line passing through the focal center of the eye, to the center of the insensible spot, is in Mr. *Short*'s eye about $15\frac{1}{3}$ *deg.* and in mine about $13\frac{1}{2}$ *deg.* (p. 115)

Elliott (1780): If an object be looked at whose image occupies the whole surface of the retina, one would imagine from thence that a hole or dark spot should be perceived in the part of the object answerable to the insensible spot in the retina; but no such spot or hole is seen. I have observed that images which fall near that spot are not perceived as properly defined. In the concave lumination excited by pressing the centre of the eye also, no such spot is discernable. There is no such spot or vacuity therefore in the retina of the sensory (if I may be allowed the expression); it seems to be filled up by the fibres of the optic nerve dispersed around the spot in the eye: hence the ill defined images there. (pp. 13–14)

Young (1801): To find the place of entrance of the optic nerve, I fix two candles at ten inches distance, retire sixteen feet, and direct my eye to a point four feet to the right or left of the middle of the space between them: they are then lost in a confused spot of light; but any inclination of the eye brings one or other of them into the field of view.... From the experiment here related, the distance of the centre of the optic nerve from the visual axis is found to be 16 hundredths of an inch; and the diameter of the most insensible part of the retina, one-thirtieth of an inch. (p. 47)

Treviranus (1828): When one calculates the various visual angles between the axis of the optic nerve and that of the eye, the average is 13° 32″. (p. 70)

2.9 Visual Pathways

The ignorance of the anatomy of the eye in antiquity was multiplied with respect to the pathways from the eyes to the brain. Indeed, the brain itself was often considered of relatively minor importance. Hippocrates did locate the pleasures, sensations, and thoughts in the brain, but Aristotle did not follow him in this speculation: the heart was the center of sentience. Galen believed that the origin of the visual pathways was located in the anterior ventricle of the brain, where the animal spirits could interact with the visual spirits, borne by the optic nerves. The optic nerves themselves came together at the optic chiasm, but each of the nerves remained on its own side. This error was to be repeated by Vesalius, and it was integrated into Descartes's analysis of vision. The anterior ventricle to which Galen referred was likely to have been the thalamus. Three ventricles were enumerated in Galenic anatomy, and Albertus Magnus incoporated them into late medieval philosophy as representing the sites of perception, reasoning, and memory. The prevalence of this notion is evident in Leonardo's drawing of the visual pathways: the optic nerves lead directly into the first of the three ventricles. However, some of his later drawings do depict the optic chiasm prior to the nerves entering the ventricles (see McMurrich, 1930; Keele, 1955). He attributed the conjugate movements of the eyes to the union of the nerves at the chiasm (see section 5.9).

The diagrams in Descartes's *Dioptrique* and in his *Traité de l'Homme* retain the ipsilateral projection of the optic nerves to the brain, but those from each eye are combined in the pineal body in the latter. The illustration from the *Traité* of 1664 has been reproduced many times, particularly in the context of historical analyses of binocular vision (van Hoorn, 1972; Held, 1976; Wade, 1987; Howard and Rogers, 1995). However, it is instructive to compare it with the monocular representation made for *De homine*: both engravings were derived from essentially the same text, which does not mention stimulation of two eyes. A number of similar illustrations from *De homine* (not reproduced here) all depict one eye only, whereas their corresponding figures from the *Traité* display two. Hence it might be an instance of an illustration having played a greater role than the text from which it was derived in historical interpretations; it is particularly significant in this case because neither of the series of diagrams was produced by the author of the text. Descartes did stress the correspondence between points on the object, those on the retina, and their projection to the brain, but it is unlikely that he was addressing the issue of corresponding points in the two retinas. His analysis of binocular vision was by analogy to a blind man holding two sticks (see section 6.2), and it was not physiological. The union that was depicted in the pineal body reflected an attempt to match singleness of vision with a single anatomical structure. Descartes's analysis of vision was based on his conception of light: when it strikes the eye it applies force to points on the retina which are transmitted along the optic nerve to the brain. Rohault, on the other hand, was specifically concerned with binocular projections, delineating sympathetic (or corresponding) points on each eye. Although he retained the independence of the two optic nerves, the fibers from corresponding points were united in an undefined part of the brain.

Briggs not only produced a delightful illustration of the visual pathways, retaining the independence of the optic nerves, but he also stimulated Newton's interest in them. Briggs sent his paper to Newton and their correspondence indicates the latter's reserve concerning it (see Brewster, 1855; Turnbull, 1960). In order to rise above the level of opinion, Newton conducted the experiment referred to above, made the first representation of partial decussation at the optic chiasm, and proposed a theory of binocular single vision based upon it. The subtlety of Newton's analysis was not, however, widely disseminated. He did make passing reference to it in Query XV of his *Opticks*, together with a telling reference to species differences, but the manuscript itself was not published. It passed into the possession of William Jones in the eighteenth century, and was later purchased by the Earl of Macclesfield (see Westfall, 1980). Prior to its purchase Harris saw the manuscript, and published a copy of it in his *Treatise of Optics* (1775). The engraving shown here was redrawn for Harris's book, but a copy of Newton's drawing can be found in Grüsser and Landis (1991). Newton was almost correct in his analysis: partial decussation was appropriate, but the nerves themselves united at the chiasm. That is, optic nerve fibers from corresponding points on each eye form single fibers in the optic tract. This detail was rectified by Taylor in an accurate representation of the partial crossing over and independence of the nerve fibers. Newton's conclusion that the fibers united in the optic chiasm had important implications for his theory of binocular vision (Wade, 1987).

John Taylor was a most colorful character (see James, 1933); he had been a student of William Cheselden in London, where he learned his surgical skills, such as they were. His opinion of himself was high, and his manner was flamboyant: the tours he made around the cities of Europe were more like circuses than surgeries. Samuel Johnson said of him that he was an instance of how far impudence will carry ignorance! It is also possible that he was responsible for blinding more of the nobility of Europe than any other single person. He published books in several languages, usually ennobling himself with such titles as Ritter von Taylor or Chevalier Taylor. Nonetheless, he must have divined some optical and ophthalmological knowledge through his travels because he did provide the first correct description of the partial decussation at the optic chiasm: fibers in the optic nerve diverged after the optic chiasm, with those from the left halves of each retina projecting to the left part of the brain, and vice versa.

Much of the remaining concern reflected here relates to the evidence supporting partial decussation at the chiasm. This tended to be derived from behavioral rather than anatomical studies. Elliott provided evidence from binocular color combination, and Wollaston gave a description of a hemianopia he periodically experienced, but he did pursue the anatomical consequences of it.

Hippocrates (ca. 400 B.C.): Men ought to know that from the brain, and from the brain only, arise our pleasures, joys, laughter and jests, as well as our sorrows, pains, griefs and tears. Through it, in particular, we think, see, hear, and distinguish the ugly from the beautiful, the bad from the good, the pleasant from the unpleasant, in some cases using custom as a test, in others perceiving them from their utility. (1923a, p. 175)

Pliny (ca. 77): The most learned authorities state that the eyes are connected with the brain by a vein; for my own part I am inclined to believe that they are also thus connected with the stomach: it is unquestionable that a man never has an eye knocked out without vomiting. (1940, p. 527)

Hippocrates (ca. 460–370 B.C.) after an engraving in Pettigrew (1840a).

Galen (ca. 175): Accordingly, the encephalon extends a part [*n. opticus*] of itself to tne crystalline humor in order to know how it is being affected, and this outgrowth is properly the only one to have a perceptible channel, because it alone contains a very large amount of the psychic pneuma.... the meeting of the optic nerves in the anterior part of the encephalon, from which they begin to pass forward in one plane to generate the eyes as a whole in their proper position and to bring it about that neither pupil of the eye is higher [than the other]. This is the reason why it was better that the nerves providing the eyes with the sensation of sight should start from a single source. (May, 1968, pp. 402 and 499)

Hunain ibn Is-hâq (ca. 850): Know that the brain is the source of all sensation and all motion and that from it the faculty of sensation and the faculty of motion proceed through the nerves into all the sensory and motor organs. (Now) the eye is both a sensory and a motor organ, and therefore it is controlled by two nerves from the brain. One is hard and effects the movements of the eyes ... The other nerve is soft and hollow; there is no hollow nerve in the body except this. The reason is that the eye needs the animal spirit in order to effect the vision by means of it.... When these two hollow nerves first join together in one place, then separate, at the same time their canals unite and join one to another until they become one only: it is here that the spirit is set free and sent into the second eye; it is here that the spirit from the brain is received, if one eye has been shut. (Meyerhof, 1928, pp. 7 and 30)

Ibn al-Haytham (ca. 1040): Two similar hollow tubes split off from the front of the brain, beginning at two points on either side of it. It is said that each of these nerves consists of two layers and that, originating from the two membranes of the brain, they extend to the surface of the brain's front at its middle; then they unite, forming one hollow tube. The nerve then divides again into two equal hollow nerves which subsequently continue to the vaulting of the concave bones surrounding the eyeballs. (Sabra, 1989, p. 55)

Leonardo da Vinci (ca. 1500): Saw a head in two between the eyebrows in order to find out by anatomy the cause of the equal movement of the eyes, and this practically confirms that the cause is the intersection of the optic nerves. (MacCurdy, 1938a, p. 186)

Vesalius (1543): The first of the two figures ... represents the base of the entire brain and cerebellum freed from the surrounding membranes so that the origin of the cerebral nerves may be suitably exposed to view. (Saunders and O'Malley, 1950, p. 144)

Descartes (1637): Further, not only do the images of objects form thus on the back of the eye, but they also pass beyond to the brain, as you will readily understand if you consider that, for example, the rays that come into the eye from the object V touch at point R the extremity of one of the small fibers of the optic nerve which has its origin in the point *7* of the interior surface of the brain *789*; and those of the object X touch at point S the extremity of another of these fibers, of which the beginning is point *8*; and those of Y touch another of them at point T, which corresponds to the spot of the brain marked *9*, and so with the others. And that, light being nothing but a movement or an action which tends to cause some movement, those of its rays which come from V

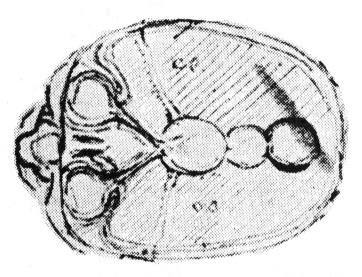

A vertical section of the eye (above) leading into the first of the three cerebral ventricles, and a horizontal section (below) showing the two optic nerves entering the first ventricle, according to Leonardo.

toward *R* have the power of moving the entire fiber *R7*, and as a result the point of the brain marked *7*; and those which come from *X* toward *S*, of moving the entire nerve *S8*, and even of moving it in a way other than that by which *R7* is moved, because the objects *X* and *V* are of two different colors. And similarly, those which come from *Y* move point *9*, from which it is manifest that the picture *789*, which is quite similar to the objects *V, X, Y*, is formed once more on the interior surface of the brain, facing toward its concavities. And from there I could again transport it to a certain small [pineal] gland which is found about the center of these concavities, and which is strictly speaking the seat of the common sense. (1965, p. 100)

Descartes (1664): ... the filaments 1–2, 3–4, 5–6, and the like that compose the optic nerve and extend to the back of the eye (1, 3, 5) to the internal surface of the brain (2, 4, 6). Now assume that these threads are so arranged that if the rays that come, for example, from point A of the object happen to exert

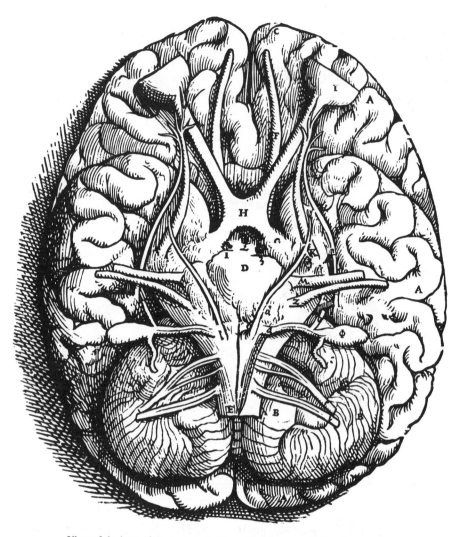

View of the base of the brain with the optic chiasm at *H* (Vesalius, 1543).

pressure on the back of the eye at point 1, they in this way pull the whole of thread 1–2 and enlarge the opening of the tubule marked 2. And similarly, the rays that come from the point B enlarge the opening of tubule 4, and so with the others. Whence, just as the different ways in which these rays exert pressure on points 1, 3, 5 trace a figure at the back of the eye corresponding to that of object AB.C., so evidently, the different ways in which tubules 2, 4, 6 and the like are opened by filaments 1–2, 3–4, and 5–6 must trace [a corresponding figure] on the internal surface of the brain. (Hall, 1972, p. 84)

Rohault (1671): I conceive that when we look upon an Object, we turn our Eyes to it in such a manner, that the two *Optick* Axes meet at the Point which we fix our Attention principally upon. Thus the Rays TE, VK, coming from

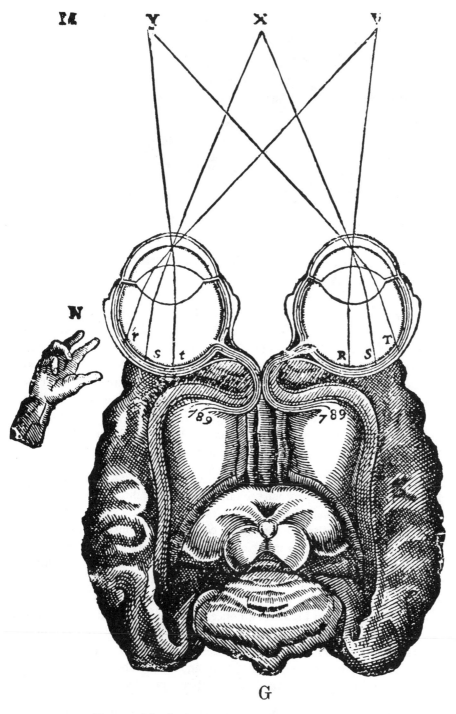

Diagram of the visual pathways from Descartes (1637/1902).

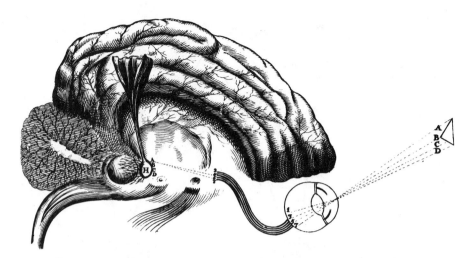

Diagrams of the visual pathways from Descartes (1662) on the left and Descartes (1664/1909) on the right.

that Point, and falling upon the *Sympathetick* Capillaments E and K, the two Impressions which they make there, are reunited in one Point only, viz in the Point Q. So likewise the Part of the Object which is on the right Hand, shakes the *Sympathetick* Capillaments D and I, the Impressions of which are carried to P. And again, the Part of the Object which is on the left Hand acts upon the *Sympathetick* Capillaments F and L, and their Impressions unite in the Point R, and so of the rest. So that though there be two Images impressed upon the Eyes, yet there is but one impressed upon the Part of the Brain X, which we here suppose to be the immediate *Organ* of Vision. (1723, p. 247)

Antonius van Leeuwenhoek (1632–1723) after an engraving in Stirling (1902).

Leeuwenhoek (1674): ... some Anatomists affirm'd the *Optic* Nerve to be hol'ow, and that themselves had seen that hollowness, through which they would have the Animal spirits, that convey the visible species, represented in the eye, pass into the Brain; I thereupon concluded with my self, that, if there were such a cavity visible in that Nerve, that it might also be seen by me, especially since, if it be so it must be pretty bigg, and the body pretty stiff, or else the circumjacent parts would press it together. And in order to this discovery, I sollicitously view'd three Optic Nerves of Cows; but I could find no hollowness in them; I only took notice, that they were made up of many filamentous particles, of a very soft substance, as if they only consisted of corpuscles of the Brain joined together, the threds were so very soft and loose: They were composed of conjoined globuls, and wound about again with particles consisting of other transparent globuls. (pp. 179–180)

Briggs (1682): aa, bb, cc, dd, &c. *Are the* Fibræ concordes *of the Optick Nerves as they arise alike from the* thalami Optici, llll, *which run also to* like parts *of the Eye and are described there by the same characters; where to note that* aa

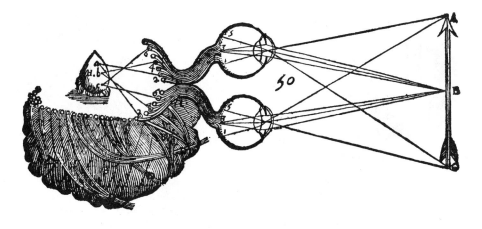

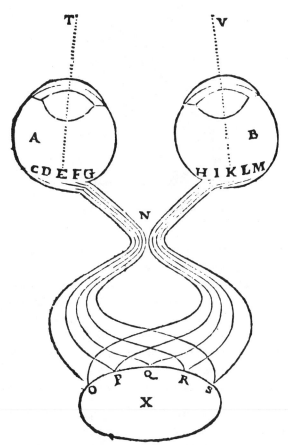

Pathways from the eyes to the brain from Rohault (1671).

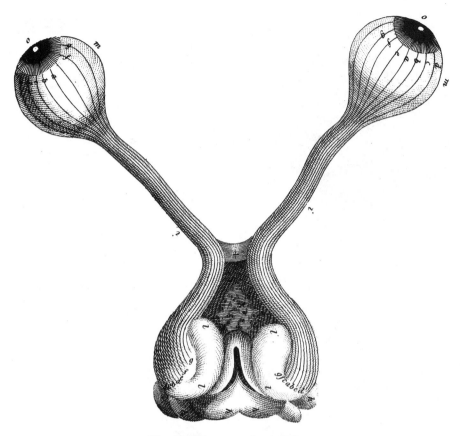

Visual pathways according to Briggs.

are the two uppermost Fibres, and are supposed to have the greatest flexure ... bb
cc dd, *The external-lateral fibres*; ee ff gg, *the internal-lateral*; llll, *The two*
protuberances (*called* thalami Nervorum Opticorum) *from whence the Optick
Nerves have their rise*; hhhhh, *The parts of the Brain that lie under them*; ii,
*The Optick nerves made here the larger to express as many of their fibres as
could be distinctly represented*; +, *The place of their union, from which they are
supposed to be separated a little, that the* distinct series *of their fibres may
appear the better* ... mm, *The Eyes divested of the* sclerodes *and* uvea nn, *The*
processus ciliares *lying just under the* Iris; oo, *the pupils*. (p. 178)

Newton (ca. 1682): Now I conceive that every point in the retina of one eye,
hath its corresponding point in the other; from which two very slender pipes
filled with the most limpid liquor, do without either interruption, or any
unevenness or irregularities in their process, go along the optic nerves to the
juncture EFGH, where they meet either betwixt G, F or F, H, and there unite
into one pipe as big as both of them; and so continue in one, passing either
betwixt I, L or M, K, into the brain, where they are terminated perhaps at the
next meeting of the nerves betwixt the cerebrum and cerebellum, in the same

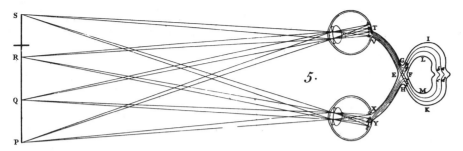

Newton's conception of the partial decussation at the optic chiasm was not published by him, but was to await its representation in Harris (1775), who read Newton's manuscript. A more detailed representation of the pathways can be seen in the figure on p. 282.

order that their extremities were situated in the retina's. And so there are a vast multitude of these slender pipes which flow from the brain, the one half through the right side nerve I L, till they come at juncture G F, where they are divided into two branches, the one passing by G and T to the right side of the right eye A B, the other half shooting through the space E F, and so passing by X to the right side of the left eye $\alpha \beta$. And in like manner the other half shooting through the left side nerve M K, divide themselves at F H, and their branches passing by E V to the right eye, and by H Y to the left, compose that half of the retina in both eyes, which is towards the left side C D and $\gamma \delta$. (Harris, 1775, pp. 109–110)

Newton (1704): Are not the Species of Objects seen with both Eyes united where the optick Nerves meet before they come into the Brain, the Fibres on the right side of both Nerves uniting there, and after union going thence into the Brain in the Nerve which is on the right side of the Head, and the Fibres on the left side of both Nerves uniting in the same place, and after union going into the Brain in the Nerve which is on the left side of the Head, and these two Nerves meeting in the Brain in such a manner that their Fibres make but one entire Species or Picture, half of which on the right side of the Sensorium comes from the right side of both Eyes through the right side of both optick Nerves to the place where the Nerves meet, and from thence on the right side of the Head into the Brain, and the other half on the left side of the Sensorium comes in like manner from the left side of both Eyes. For the optick Nerves of such Animals as look the same way with both Eyes (as of Men, Dogs, Sheep, Oxen, &c.) meet before they come into the Brain, but the optick Nerves of such Animals as do not look the same way with both Eyes (as of Fishes and of the Chameleon) do not meet, if I am rightly informed. (pp. 136–137)

Porterfield (1737): Suppose, as in *Fig.* 1, the Nerves composed of five Fibres, whose Extremities in the right Eye are A, B, C, D, E, and in the other Eye, *a, b, c, d, e*. The corresponding Fibres A*a*, B*b*, C*c*, D*d*, and E*e*, are supposed to meet

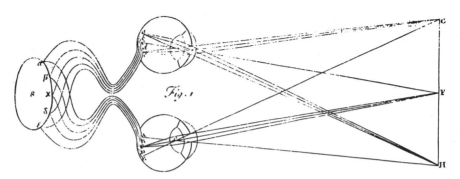

Binocular combination according to Porterfield.

in the *Sensorium* S, in the Points α, β, χ, δ, ε. Hence if both Eyes are directed to F, its Image will fall on the *Retina* at the Optick *Axes*, and there strike the sympathizing Fibres C and *c*; which motion being propagated to the single Point of the *Sensorium* χ, must there make but one Species or Picture. In like manner the Eyes retaining the same Direction, the Image of the Point G will fall upon the right Side of both Eyes; and by striking the correspondent Fibres E and *e*, will, in the *Sensorium*, make but one Impression at ε, where these Fibres terminate; and the Image of the Point H, by striking the corresponding Fibres A and *a*, will, in the *Sensorium*, make but one Impression at α: And thus, though both Eyes receive the same Impressions from Objects, yet they are not seen double, because of these two Impressions or Images, one is only formed in the *Sensorium*. (pp. 201–202)

John Taylor (1708–1772) after an engraving in Caulfield (1820).

Taylor (1738): a r T and b V T, represent the same fibres of the left eyes, and α T V α and β S V β, the fibres of the right eye. The part r s t which reside in the optic nerve of the left eye, after they meet in the head [at the optic chiasm], are not composed in the same way that they were when they left the eyes; but they are accompanied by the fibres which leave the left side of the fundus of the right eye; i.e., that a r T leaves b V T at the point of their meeting T in the head, and the fibres join themselves to α V T which leave the left side of the right eye, and are together the body Tutfr. (p. 174)

Le Cat (1744): The principle Nerves of the Eye, termed Optic Nerves A B, … make their Exit from the Cranium, one on each Side.... Then the two Nerves K K tend towards the fore Part of the Head, approaching again one towards the other; and unite as it were in a single Nerve A, without Crossing or Confusion. (1750, p. 140)

Elliott (1780): The most decisive experiment that I have met with against the junction of the respective fibres of both eyes is that blue and yellow thrown separately on the answerable parts of both retinæ do not cause a green. The corresponding fibres are not perhaps, united, because then if one nerve was disordered or destroyed, the other would also. They may only run by the sides

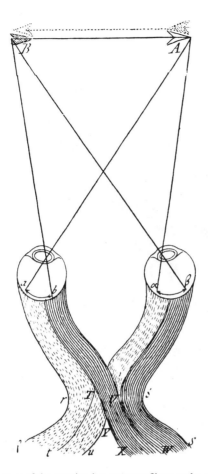

Taylor's diagram of the crossing but separate fibers at the optic chiasm.

of each other, or even in contrary directions, and terminate in the brain so as to form two different surfaces, concentric, and at a small distance from each. (pp. 5–6)

Wardrop (1818): But, though the optic nerves do not seem to cross each other, there is little doubt that there is an intermixture of some of their medullary fibres; for it is much more probable that the remarkable sympathy between the two eyes arises from this intermixture of the medullary fibres at the union of the nerves in the stella tunica, than from any intermixture of other nerves in the brain. (p. 134)

Wollaston (1824a): It is now more than twenty years since I was first attacked with the peculiar state of vision to which I allude, in consequence of violent exercise I had taken for two or three hours before. I suddenly found that I could see but half the face of a man whom I met; and it was the same with respect to every object I looked at. In attempting to read the name JOHNSON,

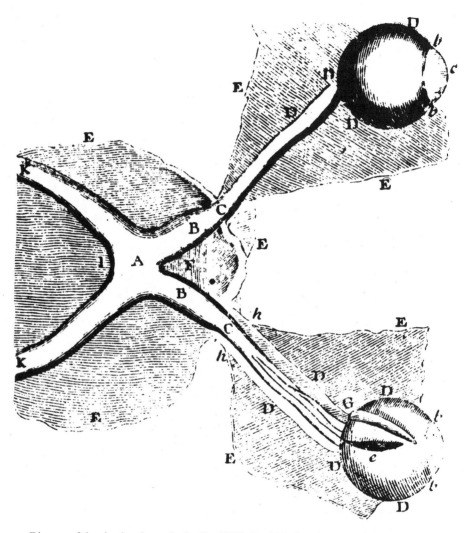

Diagram of the visual pathways by Le Cat (1744), in which there is no crossing at the chiasm.

William Hyde Wollaston (1766–1828) after an engraving in Knight (1833b).

over a door, I saw only SON; the commencement of the name being wholly obliterated to my view. In this instance the loss of sight was toward my left, and was the same whether I looked with the right eye or the left. The blindness was not so complete as to amount to absolute blackness, but was a shaded darkness without definite outline. The complaint was of short duration, and in about a quarter of an hour might be said to have wholly gone, having receded with a gradual motion from the centre of vision obliquely upwards towards the left.... Since the corresponding points of the two eyes sympathize in disease, their sympathy is evidently from structure, not from mere habit of feeling together, as might be inferred, if reference were had to the reception of ordinary impressions alone. Any two corresponding points must be supplied with a pair of filaments from the same nerve, and the seat of a disease in which similar parts of both eyes are affected, must be considered as situated at a distance from the eyes at some place in the course of the nerves where these filaments are still united, and probably in one or the other thalamus nervorum opticorum. It is plain that the cord which comes finally to either eye under the name of optic nerve, must be regarded as consisting of two portions, one half from the right thalamus, and the other from the left thalamus nervorum opticorum. According to this supposition, decussation will take place only between the adjacent halves of the two nerves. That portion of the nerve which proceeds from the right thalamus to the right side of the right eye, passes to its destination without interference; and in similar manner the left thalamus will supply the left side of the left eye with one part of its fibres while the remaining halves of both nerves in passing over to eyes of the opposite sides must intersect each other, either with or without intermixture of their fibres. (pp. 224–226)

3   Color

Isaac Newton (1642–1727) after an engraving in Birch (1743).

Newton (1704): *For let EG represent the Windowshut, F the hole made therein through which a beam of the Sun's Light was transmitted into the darkned Chamber, and AB.C. a Triangular Imaginary Plane whereby the Prism is feigned to be cut transversly through the middle of the Light. Or if you please, let AB.C. represent the Prism it self, looking directly towards the Spectator's Eye with its nearer end: And let XY be the Sun, MN the Paper upon which the Solar Image or Spectrum is cast, and PT the Image it self whose sides towards V and W are Rectilinear and Parallel and ends towards P and T semi-circular. YKHP and XLJT are the two Rays, the first of which comes from the lower part of the Sun to the higher part of the Image, and is refracted in the Prism at K and H, and the latter comes from the higher part of the Sun to the lower part of the Image, and is refracted at L and J, and the Refraction at L equal to the Refraction at H, so that the Refractions of the incident Rays at K and L taken together are equal to the Refractions of the emergent Rays at H and J taken together. . . . This Image or Spectrum PT was coloured, being red at its least refracted end T, and violet at its most refracted end P, and yellow green and blew in the intermediate spaces. Which agrees with the first Proposition, that Lights which differ in Colour do also differ in Refrangibility. . . . The homogeneal*

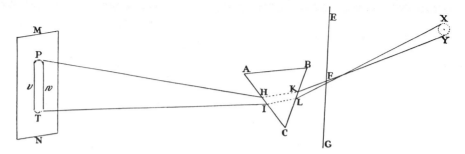

The prismatic spectrum of Newton.

light and rays which appear red, or rather make Objects appear so, I call rubrific or red-making; those which make Objects appear yellow, green, blue and violet, I call yellow-making, green-making, blue-making, violet-making, and so of the rest. And if at any time I speak of light and rays as coloured or endued with Colours, I would be understood to speak not philosophically and properly, but grosly, and according to such conceptions as vulgar People in seeing all these Experiments would be apt to frame. For the Rays to speak properly are not coloured. In them there is nothing else than a certain power and disposition to stir up a sensation of this or that Colour. For as sound in a Bell or musical String, or other sounding Body, is nothing but a trembling Motion, and in the Air nothing but that Motion propagated from the Object, and in the Sensorium 'tis a sense of that Motion under the form of sound; so Colours in the Object are nothing but a disposition to reflect this or that sort of rays more copiously than the rest; in the rays they are nothing but their dispositions to propagate this or that Motion into the Sensorium, and in the Sensorium they are sensations of those Motions under the form of Colours. (pp. 21–22 and 90–91)

A wide range of facts was available to Newton before he conducted his prismatic experiments, including the prismatic colors themselves. The facts had accumulated over centuries and spawned competing theories. Most theories were based on the Aristotelian view that color was a property of bodies, and that it was carried by light through the transparent medium to the eye. Newton was to replace them with a unified theory of light and colors: white light was not unitary but compounded, in precise proportions, of many components that differed in their refraction. Newton was able to analyze the prismatic colors, armed with Snell's law of refraction and wielding an armory of subtle experimental procedures. For example, small sections of the prismatic spectrum could be isolated and mixed with other parts, so that the rules of their mixture could be determined. However, he was not armed with a good-quality prism, since the one used initially in his experiments had been purchased for a shilling! Newton wished to avoid the speculations entered into by his arch rival, Robert Hooke, and to base his theory on experiment rather than hypothesis. Indeed, Newton opened his *Opticks* (1704) thus: "My Design in this Book is not to explain the Properties of Light by Hypotheses, but to propose and prove them by Reason and Experiments" (p. 1).

The first section in this chapter is entitled Color Vision, but the contents are broader than that. It is intended to provide a general platform from which the other sections can spring. It leads naturally to color mixing, which was preoccupied with primaries, then to color blindness and color zones. Thereafter, the topics reflect the growing awareness of the subjective dimensions of color perception. In one sense the progression mirrors the physical, physiological, and psychological analyses of light.

3.1 Color Vision

Newton was careful to distinguish between the refractions of light and their perception; that is, between the physical and psychological dimensions of color, between stimulus and sensation. Such distinctions could only be made when the nature of light itself was understood. Prior to that time, the physical and psychological dimensions were constantly confounded. To this confusion was added an ignorance of the structure of the eye itself. Within Greek science, color was considered in terms of an analogy with the four elements. The numeration of four basic colors recurs from Democritus in the fifth century B.C. to Leonardo in the sixteenth century, although Leonardo sandwiched them between white and black. As was noted in the commentary on light, both Plato and Aristotle considered that color was of paramount importance, but they stressed the importance of black and white, often contending that all colors could be made up from these two (see Beare, 1906). Perhaps it was the observation that no colors appeared as light as white or as dark as black that led to this speculation (Wasserman, 1978). Consequently, despite the equation of colors with the four elements, black and white tended to dominate the analyses of color. This belief in the purity of white was retained by some (like Goethe) long after Newton's experiments on the prismatic spectrum. Plato did raise the question of species differences in color vision, and the variety of iridescent colors in nature was a constant source of wonder. Lucretius referred to this, and noted how the colors are dependent on the direction of light striking the dove's plumage. However, it was the Aristotelian preoccupation with black and white that was repeated by Averroes, in the twelfth century. In the late medieval period one of the principal concerns was the colors seen in the rainbow, and the analysis of these resulted in some vital developments in the optics of refraction and internal reflection (see Boyer, 1959; Lindberg, 1983a).

The situation changed with investigations of prismatic colors. Although the occurrence of colors due to light passing through or reflected from glass or crystals had been known of since antiquity, their experimental examination had not been undertaken. A preliminary study of the prismatic spectrum was conducted by Thomas Harriot (ca. 1560–1621) at the beginning of the seventeenth century; he also formulated a relationship between angles of incidence

and refraction for the different colors, but did not divulge this to his correspondent, Kepler (Lohne, 1972). The spectrum was subjected to more detailed
analysis later in the century. Descartes treated the analysis of colors visible in
the rainbow in his discourse on *meteorology*. His approach was somewhat
different from that adopted later by Newton: sunlight fell normally on one
face of the prism and was refracted at the second face, upon which the aperture
was placed; he noted that the distinctness of the spectrum was dependent on
the size of the aperture. His mechanistic interpretation of visible colors was in
harmony with his concept of light generally: colors corresponded to different
rates of rotation of bodies in the medium. While his interpretation of color was
neither adopted generally nor developed by Descartes, it should not detract
from his analysis of the rainbow. The prismatic colors were clearly described by
Digby at about the same time, but his interpretation of them was Aristotelian:
color could be defined in terms of white and black, and the prismatic colors
made red where there was more white and blue where there was more black.
Gassendi was more circumspect, attributing prismatic colors to processes in
the perceiver, and color names to the vagaries of language. Accordingly, by the
middle of the seventeenth century the methods for the analysis of color were
in place, but the analysis itself was wanting. The disquiet associated with the
phenomena of color and their interpretation was nowhere as evident as in
Boyle's writing on the topic.

Newton conducted experiments with a prism from 1666, reporting the results
in a paper to a meeting of the Royal Society on 6 February 1672; it was published in the *Philosophical Transactions* later that year. The paper stimulated
widespread interest as well as opposition, particularly from Hooke and
Huygens, who argued that there were only two basic colors. Many of the
objections to Newton's theory were published in the *Philosophical Transactions*
in the following year, accompanied by Newton's rebuttals. The controversy
stirred by the theory so perturbed Newton that he did not publish his full
account of it until his *Opticks* appeared in 1704. By this time, he was president
of the Royal Society, and Hooke had died the year before. Newton's theory
underwent changes between its initial publication in 1672 and its mature
expression in 1704. Shapiro (1980) has remarked that historians of science
tend to base their analyses on the former, whereas scientists focus on the
latter; extracts from both are presented here. Newton was aware that the
range of refractions from a prism was continuous and yet the colors seen
were restricted in number. He reported seven colors—red, orange, yellow,
green, blue, indigo, and violet—and he arranged them in a particular circular
sequence, after the manner of the musical scale. Colored lights could be combined in such a way that their compound could be defined with respect to the
color circle. Combinations of primary colors that were opposite one another on
the circle produced a whitish compound that was positioned near the center.
Indeed, his contention that white could not be produced from two suitably

chosen primaries was probably occasioned by both Hooke's and Huygens's proposals that all colors could be derived from two colors alone.

Much of the subsequent debate focused on the nature and number of primary colors. For Newton, who introduced the term in 1672, they were the discrete colors that could be seen in the prismatic spectrum, despite his appreciation of continuity across the spectrum. With the formulation of the color circle they also became the colors from which compounds could be derived. Mariotte reduced the primaries to three, to which were added white and black. His primaries were red, yellow, and blue, as were those of Malebranche, Lomonosov, and Young (1802a): Young (1802b, 1807) later modified his selection to red, green, and violet. The vast range of colors that could be produced by appropriate combinations of a small number of primaries led to speculations regarding the physiological basis of color vision. Young (1802a) stated that color vision could be mediated by retinal mechanisms that responded selectively to each of the primaries. Similar views had been voiced earlier by Malebranche, Le Cat, Palmer, and Elliott, usually incorporating Newton's concept of vibratory motion of both light and the retinal response to it. Yet another aspect of color vision remarked upon by Newton related to brightness differences within the spectrum: yellow and orange appeared most intense, and he speculated that this reflected differences in sensitivity to these as compared to other colors. Descartes had remarked that green was the most pleasing color to view.

By the early nineteenth century Newton's corpuscular theory of light had been supplanted by wave theory, but his color theory reigned supreme. There was, however, one blemish that arose from a most unlikely source: Brewster, an ardent admirer of and biographer of Newton, proposed that white light consisted of three colors only—his theory of the triple spectrum. By passing prismatically separated sunlight through colored filters, Brewster reported that the colors could be changed in hue rather than attenuated, as proposed by Newton's theory. Because the theory of the triple spectrum challenged that of Newton, it was subjected to detailed experimental scrutiny. Helmholtz (1852) repeated the experiments, but with additional precautions to control for stray light, and produced results in accordance with Newtonian theory (see Sherman, 1981). It is tempting to speculate that Brewster's representation of the three overlapping sensitivity curves was to influence Helmholtz in deriving his speculative curves of the three color mechanisms involved in color vision (Wade, 1994a).

Empedocles (ca. 450 B.C.): The pores of the eye are arranged alternatingly of fire and of water. By passage through the fiery pores we perceive the white objects, whereas through the watery we perceive the black objects. Each sense perception has to fit into its end organ. Colors are carried by emanation to visual perception. (Siegel, 1959, p. 149)

Empedocles (ca. 493–433 B.C.) after an engraving in Runes (1959).

Democritus (ca. 400 B.C.): The simple colours ... are four. What is smooth is white; since what neither is rough nor casts shadows nor is hard to penetrate,—

all such substances are brilliant. . . . Black is composed of figures the very opposite [to those of white],—figures rough, irregular, and differing from one another. . . . Red is composed of figures such as enter into heat, save that those of red are larger. . . . Green is composed of both the solid and the void. . . . The other colours are derived from these by mixture. (Stratton, 1917, pp. 133–135)

Plato (ca. 350 B.C.): . . . black and white, and every other colour, will appear to us to be produced by the application of the eyes to a corresponding movement, and each thing that we say is colour, will neither be that which is applied, nor that to which it is applied, but some intermediate production peculiar to each. Would you positively maintain, that what each colour appears to you, such it also appears to a dog, and every other animal? (1896, p. 383)

Aristotle (ca. 330 B.C.): At present what is obvious is that what is seen in light is always colour. That is why without the help of light colour remains invisible. . . . If what has colour is placed in immediate contact with the eye, it cannot be seen. Colour sets in movement not the sense organ but what is transparent, e.g. the air, and that, extending continuously from the object of the organ, sets the latter in movement. (Ross, 1931, p. 419a)

Theophrastus (ca. 370–286 B.C.) after an engraving in *The Historic Gallery of Portraits and Paintings*, Vol. 4. London: Vernor, Hood, and Sharpe, 1809.

Theophrastus (ca. 300 B.C.): He [Democritus] should have given some distinctive [figure] to green, as he has to the other colours. And if he holds [green] to be the opposite of red, as black is of white, it ought to have an opposite shape; but if in his view it is not the opposite, this itself would surprise us that he does not regard his first principles as opposites, for that is the universally accepted doctrine. Most of all, though, he should have determined with accuracy which colours are simple, and why some colours are compound and others not. (Stratton, 1917, p. 141)

Lucretius (ca. 56 B.C.): For what colour can there be in blind darkness? Why, a colour is changed by the light itself, according as the brightness responds to a direct or oblique impact of light; in this way the dove's plumage shows itself in the sun, lying about the nape and encircling the neck; for at times it is red as the blazing carbuncle, again view it in a certain way and it comes to appear a fusion of emerald green with blue. (1975, p. 159)

Ptolemy (ca. 150): That which is seen mediately is colour because without light no colour is seen. . . . It is obvious that colour is really inherent in objects and is part of their nature but it is seen only if light cooperates with vision. (Lejeune, 1956, pp. 13 and 18)

Ibn al-Haytham (ca. 1040): For the sense of sight perceives the forms of visible objects from the forms that come to it from the colours and lights of those objects. An its perception of lights *qua* lights and of colours *qua* colours *is* by pure sensation. (Sabra, 1989, p. 130)

Averroes (ca. 1180): The color white arises from the mixture of pure fire with the element which is most translucent and that is air. The color black arises from impure fire which is mixed with the element that is least translucent and that is earth. The colors that are intermediate to white and black are derived in accordance with the smaller or greater degree of difference of these two properties, that is to say, the difference of the luminous body and the difference of the translucent body. The colours white and black are therefore regarded as the primary elements of colors. (1961, p. 10)

Albertus Magnus (ca. 1250): ... in the light [the visual process] accepts colours, in the dark only the spendour. (Grüsser and Landis, 1991, p. 31)

Alberti (1435): ... there are only four true colours—as there are four elements—from which more and more other kinds of colours may be thus created. Red is the colour of fire, blue of the air, green of the water, and of the earth grey and ash. (1966, pp. 49–50)

Leonardo da Vinci (ca. 1500): ... white is the first among the simple colours, and yellow the second, green the third of them, blue is the fourth, and red is the fifth, and black is the sixth. (Kemp, 1990, p. 268)

Gassendi (1624): For men have agreed to call the color of snow whiteness and to call whatever is infused with this same snow color white. But who knows whether I see it red, or another man sees it green? I do not mean that I see it the same color as I see a rose, nor that the other man sees it the same color as he sees grass; but I do say this, that it is possible that the points of view have been altered so that the color I see in a rose you see in snow, and the one you see in a rose, I see in snow. Nor does it make any difference that we both call the color of snow whiteness and the color of a rose redness. (Brush, 1972, p. 92)

Descartes (1637): Then remembering that a prism or triangle of crystal causes similar colors [as in a rainbow] to be seen, I considered one of them which was such as *MNP* is here, with its two surfaces, *MN* and *NP*, completely flat, and inclined to one another at an angle of around 30° or 40°, so that if the rays of the sun *ABC* cross *MN* at right angles, or nearly so, so that they do not undergo any noticeable refraction there, they must suffer a fairly great refraction in coming out through *NP*. And when I covered one of these two surfaces with a dark body, in which there was a rather narrow opening such as *DE*, I observed that the rays, passing through this opening and from there going to contact the cloth or white paper *FGH*, paint all the colors of the rainbow there, and that they always paint the color red at *F*, and the blue or violet at *H*.... After this, I tried to understand why these colors are different at *H* and at *F*, even though the refraction, shadow, and light concur there in the same way. And conceiving the nature of light to be ... the action or movement of a certain very fine material whose particles must be pictured as small balls rolling in the pores of earthly bodies, I understood that these balls

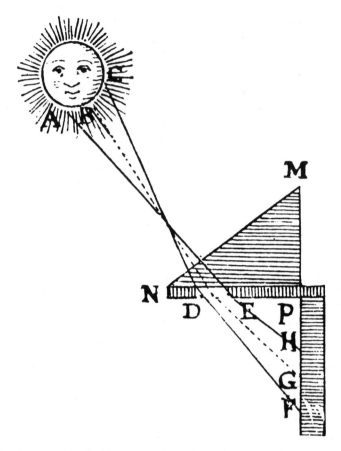

The prismatic spectrum from Descartes (1637/1902).

can roll in various ways according to the various causes which determine them; and in particular that all the refractions that occur on the same side cause them to turn in the same direction. (1965, pp. 335–336)

Digby (1644): The last sense is of seeing; whose action we can not doubt, is performed by the reflexion of light unto our eye, from the bodies which we see: and this light, cometh impregnated with a tincture drawne from the superficies of the object it is reflected from; that is, it bringeth along with it, severall of the litle atomes, which of themselves do streame, and it cutteth from the body it struke upon and reboundeth from; and they, mingling themselves with the light, do in company of it gett into the eye: whose fabrike, is fitt to gather and unite those species, as you may see by the anatomy of it: and from the eye, their iourney is but a short one to the braine: in which, we can not suspect that they should loose their force; considering, how others that come from organes further off, do conserve theirs: and likewise considering the nature of the optike spirits, which are conceived to be the most refined of all that are in a mans body. (p. 280)

Hobbes (1655): *Whiteness* is light, but light perturbed by the reflections of many beams of light coming to the eye together within a little space.... As whiteness is light, so *blackness* is the privation of light, or darkness. (1839, pp. 463–464)

Gassendi (1658): And even that range of colors which is produced by interposing a prism, or triangle of glass, which anyone can procure, conveys quite clearly what the case is. Since the colors displayed in this phenomenon are not in the things, or in the glass itself, or in the medium of the air, it is likely that how they appear to the senses is created in the eye alone and that the eye receives the impression of the colors because the rays of light reflected from objects in front of us undergo the same refraction as in the glass, and bearing this refraction as they strike the eye, imprint such-and-such an appearance upon it. (Brush, 1972, p. 343)

Pierre Gassendi (1592–1655) after an engraving in Charles Perrault (1696).

Boyle (1663): ... after all I have said of colour, as it is modified light, and immediately affects the sensory, I shall now remind you, that I did not deny, but that colour might in some sense be considered a quality residing in the body that is said to be coloured. (1966, p. 674)

Descartes (1664): ... green is the most universally agreeable [color]. (Hall, 1972, p. 59)

Hooke (1665): From the consideration of the proprieties of which impressions, we may collect these short definitions of Colours: *Blue is an impression on the Retina of an oblique and confus'd pulse of light, whose weakest part precedes, and whose strongest follows.* And, that *Red is an impression on the Retina of an oblique and confus'd pulse of light, whose strongest part precedes, and whose weakest follows.* (p. 64)

Newton (1672): ... *Light* consists of *Rays differently refrangible....* Colours are not *Qualifications of Light*, derived from Refractions, or Reflections of natural Bodies (as 'tis generally believed,) but *Original* and *connate properties*, which in divers Rays are divers.... There are therefore two sorts of Colours. The one original and simple, the other compounded of these. The Original or primary colours are, *Red, Yellow, Green, Blew,* and a *Violet-purple*, together with Orange, Indico, and an indefinite variety of Intermediate gradations. (pp. 3079, 3081, and 3082)

Mariotte (1681): There are five principals; white, black, red, yellow, and blue: all the others can be made up from mixtures of these; yellow and blue mix together to give green, red and blue give violet. (1717, p. 282)

Locke (1690): ... it being one thing to perceive and know the *Idea* of White or Black, and quite another to examine what kind of particles they must be, and how ranged in the Superficies, to make any Object appear white or black. A Painter, or Dyer, who never enquired into their causes, hath *Ideas* of White and Black, and other Colours as clearly, perfectly, and distinctly, in his Understanding, and perhaps more distinctly than the Philosopher, who hath busied himself in considering their Natures, and thinks he knows how far either of them is in its cause positive or privative; and the *Idea of Black* is no less *positive* in his Mind, than that of White, *however the cause* of that Colour in the external Object, may *be only a privation.* (pp. 54–55)

Malebranche (1699): But if the body M is such, that the subtile matter being reflected has its vibrations less quick in certain degrees, which I do not think we can determine exactly; we shall have some one of the colours which are called *primitive*, yellow, red, or blue, if all the parts of the body M equally dimish the vibrations caused by the same subtile matter. And we shall see all other colours which are formed by the mixture of the primitive, according as the parts of the body M shall unequally diminish the quickness of the vibrations of the light.... The *force* or brightness of the colours comes therefore also from the more or less *force* of the vibrations, not of the air, but of the subtile matter, and the *different species of colours* from the more or less *Quickness* of the same vibrations. (1742, pp. 34–35 and 37)

Newton (1704): But its further to be noted, that the most luminous of the prismatic Colours are the Yellow and Orange. These affect the Senses more strongly than all the rest together, and next to these in strength are Red and Green. The Blue compared with these is a faint and dark Colour, and the Indigo and Violet are much darker and fainter, so that these compared with the stronger Colours are little to be regarded. (p. 71)

Le Cat (1744): Therefore, when we talk of a red Ray, we do not mean that this Ray is actually coloured with red; but only, that this kind of Globule is made in a manner proper to excite in our Eyes the Sensation of a red Colour. In a Word, this Ray is not red, but the Agent or Cause of the Sensation of Red. (1750, p. 119)

Condillac (1754): It is then with the help of memory that the eye comes to notice up to two or three colors presented together. (1982, p. 216)

Melvill (1756): Bodies of all the principal colours, *viz.* red, yellow, green and blue, are very little altered when seen by the light of burning spirits. (p. 32)

Michail Vasilievich Lomonosov (1711–1765) after an engraving in Augé (1898).

Lomonosov (1757): I have made observations, and after many years by many surmises and also after definite experiments I have confirmed with sufficient probability that the three sorts of ether particles agree with the three sorts of true primary particles which make up sensible bodies, namely: the first size of ether corresponds to salt; the second size with mercury; and the third size with sulfur or inflammable material; and with pure earth, with water, and with air are joined all that is blunt, weak, and perfect. Finally, I find that from the first type of ether arises the color red, from the second, yellow, and from the third, blue. Other colors are generated from mixtures of the first ones.... First, that a triple number of colors is required is confirmed by all previously known opinions of the professionals from numerous optical experiments by the famous physicist and industrious student of the nature of color, Mariotte, who did not refute Newton, as some thought, but tried to correct his theory about the separation of light by refraction of the rays into colors and only confirmed that in nature there are three and not seven chief colors. (1970, pp. 260 and 261–262)

Palmer (1777): The superficies of the retina is compounded of particles of three different kinds, analagous to the three rays of light; and each of these particles is moved by his own ray.... These particles may be moved by the rays which are not analogous to them, when the intenseness of the rays exceeds their proportion. (Wasserman, 1978, p. 25)

Elliott (1780): That the rays of light could not conveniently be made to communicate their vibrations immediately to the nerves, but that the interposition of those shewn to exist in the retina was necessary to that end. That therefore there are in the retina different times of vibration liable to be excited, answerable to the times of vibration of the several sorts of rays. That any sort of rays, falling on the eye, excite those vibrations, and those only which are in unison with them, not at all affecting the others, and therefore cause only their proper colour. And that in a mixture of several sorts of rays, falling on the eye, each sort excites only its unison vibrations, whence the proper compound colour results from a mixture of the whole. (pp. 12–13)

Elliott (1782): We are therefore perhaps to consider each of these vibrations, or colours in the retina, as connected with a fibril of the optic nerve. That the vibration being excited, the pulses thereof are communicated to the nervous fibril, and by that conveyed to the sensory, or mind, where it occasions, by its action, the respective colour to be perceived. (Mollon, 1987, p. 19)

Elliot (1786): I. By *Colour*, I mean the sensations which we perceive by means of vision; as a red, yellow, white, blue, or any other colour. II. By *Light*, I mean the rays issuing from bodies without us, by which those colours are excited in our eyes; whether these rays be real emannations, as Newton supposes, vibrations excited in a medium according to Euler, or otherwise. III. By *the cause of Colour*, I mean that state, or action, or disposition, or property, of a body, (whatever it be) by which it emits or sends forth those rays in consequence of heat. (p. 40)

Brown (1796): ... when we look at a black object the fibres of the retina must be in a certain position; because, in any other position they convey the idea of colour. (Letter to Erasmus Darwin in Welsh, 1825, p. 55)

Young (1802a): Now, as it is almost impossible to conceive each sensitive point of the retina to contain an infinite number of particles, each capable of vibrating in perfect unison with every possible undulation, it becomes necessary

Thomas Young (1773–1829) after an engraving in Taylor (1846).

to suppose the number limited, for instance, to the three principal colours, red, yellow, and blue. (p. 20)

Young (1802b): In consequence of Dr. Wollaston's correction of the description of the prismatic spectrum … it becomes necessary to modify the supposition that I advanced in the last Bakerian lecture, respecting the proportions of the sympathetic fibres of the retina; substituting red, green, and violet, for red, yellow, and blue. (p. 395)

Goethe (1810): That all the colours mixed together produce white, is an absurdity which people have credulously been accustomed to repeat for a century, in opposition to the evidence of their senses. (1840, p. 225)

Brown (1820): It is as impossible for us, indeed, at present, to perceive colour without extension, as it is impossible for us to read or hear our own language, without seeming, at the same time, to perceive the very meanings of which the characters or sounds are representative, and to perceive them as if they were immediately visible and audible, like the characters and sounds themselves. (p. 158)

Brewster (1831b): 1. *Red*, *yellow*, and *blue* light exist at every point of the solar spectrum. 2. As a certain position of *red*, *yellow*, and *blue* constitute *white* light, the colour of every point of the spectrum may be considered as consisting of the predominating colour at any point mixed with white light.... The existence of three primary colours in the spectrum, and the mode in which they produce by their combination the seven secondary or compound colours which are developed by the prism, will be understood from *fig.* 51. where MN is the prismatic spectrum, consisting of three primary spectra of the same lengths, MN, viz. a *red*, a *yellow*, and a *blue* spectrum. The *red* spectrum has its maximum intensity at R; and this intensity may be represented by the distance of the point R from MN. The intensity declines rapidly to M and slowly to N, at both of which points it vanishes. The *yellow* spectrum has its maximum intensity at Y, the intensity declining to *zero* at M and N; and the *blue* has its maximum intensity at B, declining to nothing at M and N. (pp. 73–74)

Müller (1838): Light and colour are actions of the retina, and of its nervous prolongations to the brain. The kind of colour and luminous image perceived depends on the kind of external impression. With this property of the retina, by virtue of which it becomes when irritated the seat of the sensations of colour and light we are so well acquainted, that we found upon it all inquiries concerning vision. The vibrations of a fluid existing in all space, the ether, when of a certain rapidity, produce in the retina the sensation of a certain colour; when of a different degree of rapidity, that of another colour; these colours being modes of reaction of the retina. The simultaneous impression of undulations of different rapidity upon the same points of the retina excites

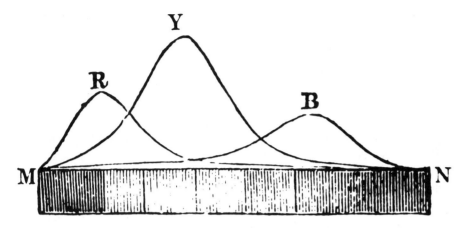

Brewster's triple spectrum.

the sensation of white light. These same sensations of colours and light may, however, be produced, without the agency of the vibrations of an ether, by mere irritation of the retina by means of electricity or mechanical pressure. (1843, pp. 1162–1163)

3.2   Color Mixing

Newton shifted the analysis of color toward the physical dimension, while not excluding the subjectivity of color perception. He also used the analysis of white light into its spectral components to synthesize new colors. For Greek scholars color seemed to be a property of objects: it could be extracted from some plants and ores to produce pigments that could be mixed with one another. Rough-and-ready rules for such mixtures were described by Plato, who recognized that the same pigments mixed in different proportions yielded different appearances. Newton confounded Plato's prediction that the precise separation and combination of colors would never be achieved by "a child of man sufficient for either of these tasks." Some colors, however, could not be produced by mixing, and Aristotle noted that these corresponded to those seen in the rainbow. Experimental investigations of color mixing started very early. Ptolemy devised a wheel on which different colors could be painted either as dots or in sectors; when the wheel was rotated the dots appeared as colored circles or the sectors combined to produced a single color. Ptolemy appreciated that the disc required to rotate rapidly, so that the persisting influence of one sector could combine with those of others. Ibn al-Haytham described a similar color wheel, and a novel form was invented by Talbot, in the nineteenth century. Another seemingly modern method of color mixing was described by Digby: small samples of color (in his case threads of fabric) could be mixed with others when they were viewed from such a distance that the individual threads could not be discriminated. Rohault related a similar observation to individual or joint stimulation of elements (capillaments) in the retina. Belief in the indivisibility of white light did not prevent Goethe discussing both primary colors and color mixture. In his *Theory of Colours*, the phenomena were divided into physiological, physical, and chemical colors. Color mixture was described in the last of these, and he also noted that adjacent colored elements that were too small to be resolved appeared as if combined.

Painting is much older than science, and so it is natural that the practical skills that the painters had learned by trial and error should have had an influence on ideas about the nature of color vision. The debate about primary colors was frequently based upon the practice of painting. Artists, in their turn, were not greatly affected by the studies by scientists in this area, as is evident from the statement by Alberti! Newton did distinguish between the mixture of pigments and that of light, noting that pigments reflected the incident light selectively. This could have proved useful to artists and scientists alike, but it was not pursued, perhaps because subsequent artists did not adopt the Newtonian primaries. For example, both le Blon and Hayter recommended that artists use the same primaries (red, yellow, and blue) that scientists had proposed. Only the black-and-white frontispiece from Hayter's book is illustrated here. The book also contains hand-colored plates demonstrating his principles of color mixture, as well as a "painter's compass" and a graduated series of squares of color. The stimulus to differentiating light from pigment mixtures was Young's (1802b, 1807) specification of a different set of primaries for light. The resolution was to await Helmholtz's (1867) clarification of the rules governing additive and subtractive color mixing. Nonetheless, more systematic studies of color mixing were undertaken with color wheels of the type described by Ptolemy. Erasmus Darwin, Brewster, and Talbot made such experiments, and manipulated the proportions of different colors exposed.

Plato (427–347 B.C.) after a frontispiece engraving in Plato (1896).

Plato (ca. 350 B.C.): Concerning colours, then, the following explanation will be the most probable and worthy of a judicious account. Of the particles which fly off from the rest and strike into the visual stream some are smaller, some larger, and some equal to the particles of the stream itself; those, then, that are equal are imperceptible, and we term them "transparent"; while the larger and smaller particles—of which the one kind contracts, the other dilates the visual stream—are akin to the particles of heat and cold . . . with these "white" and "black" are really identical affections, occurring in a separate class of sensation, although they appear different for the causes stated. These, therefore, are the names we must assign to them: that which dilates the visual stream is "white"; and the opposite thereof "black". . . . And "bright" colour when blended with red and white becomes "yellow." But in what proportions the colours are blended it would be foolish to declare, even if one knew, seeing that in such matters one could not properly adduce any necessary ground or probable reason. Red blended with black and white makes "purple"; but when these colours are mixed and more completely burned, and black is blended therewith, the result is "violet." "Chestnut" comes from the blending of yellow and grey; and "grey" from white and black; and "ochre" from white mixed with yellow. And when white is combined with "bright" and is steeped in deep black it turns into a "dark blue" colour; and dark blue mixed with white becomes "light blue"; and chestnut with black becomes "green." As to the rest, it is fairly clear from these examples what are the mixtures with which we ought to identify them if we would preserve probability in our account. But should any inquirer make an experimental test of these facts, he would evince his ignorance of the difference between man's nature and God's—how that, whereas God is sufficiently wise and powerful to blend the many into one and to dissolve again the one into many, there exists not now, nor ever will exist hereafter, a child of man sufficient for either of these tasks. (1946, pp. 173–177)

Aristotle (ca. 330 B.C.): . . . there are colours which they [painters] create by mixing, but no mixing will give red, green, or purple. These are the colours of the rainbow, though between the red and the green an orange colour is often seen. (Ross, 1931, p. 372a)

Ptolemy (ca. 150): A similar phenomenon [of colour mixing] occurs from very fast motion: for example, from the motion of a rotating disk of many colours, because one and the same visual ray does not linger upon one and the same colour, since the colour recedes from it [the visual ray] on account of the speed of rotation. And thus the same ray, falling on all colours, cannot distinguish between the first and the most recent, nor between those that are in diverse locations. For all the colours appear throughout the whole disk at the same time as though they were one colour—which would be a similar colour to the one that would actually occur from colour mixtures. For the same reason, if points that are on the disk, although not on the centre of rotation, were marked in a different colour from that of the disk, they would appear like circles of uniform colour when in rapid rotation. But if they were marked out on a line set on the disk and going through its axis, the surface of the disk will appear to have a uniform colour throughout the whole rotation. For when colour rotates about a distance perceptible to sight in the same perceived temporal moment, it is deemed to spread itself over all places through which it travels. For the phenomemon that occurs in the first rotation is always followed later by repetitions of the same sort. (Lejeune, 1956, pp. 60–61)

Ibn al-Haytham (ca. 1040): If the top is painted in different colours forming lines that extend from the middle of its visible surface, close to its neck, to the limit of its circumference, then forcefully made to revolve, it will turn round at great speed. Looking at it the observer will now see one colour that differs from all the colours in it, as if this colour were composed of all the colours of those lines. (Sabra, 1989, p. 145)

Pecham (ca. 1280): ... if the eye after being directed with fixed gaze on a bright color illuminated by an intense light is turned aside to a color illuminated more weakly, it will find that the first and second colors appear intermixed because traces of the aforeseen have been left in the eye.... many small colors appear from a distance to be one color. (Lindberg, 1970, pp. 63 and 153)

Alberti (1435): *Let us omit the debate of philosophers where the original source of colours is investigated, for what help is it for a painter to know in what mixture of rare and dense, warm and dry, cold and moist colour exists? However, I do not despise those philosophers who thus dispute about colours and establish the kinds of colours at seven. White and black [are] the two extremes of colour. Another [is established] between them. Then between each extreme and the middle they place a pair of colours as though undecided about the boundary, because one philosopher allegedly knows more about the extreme than the other. It is enough for the painter to know what the colours are and how to use them in painting. I do not wish to be contradicted by the experts, who, while they follow the philosophers, assert that there are only two colours in nature, white and black, and there are others created from mixtures of these two. As a painter I think thus about colours. From a mixture of colours almost infinite others are created.* (1966, p. 49)

Kenelm Digby (1603–1665) after an engraving in Bullart (1682).

Digby (1644): The like whereof happeneth in clothes, or stuffes, or stockings, that are woven of divers coloured but very small thriddes: for if you stand so farre of from such a piece of stuffe, that the litle thriddes of different colours which lye immediate to one an other may come together as in one line to your eye; it will appear of a middling colour, different from both those that it resulteth from: but if you stand so neere that each thridde sendeth rays enough to your eye, and that the basis of the triangle which cometh from each thridde to your eye, be long enough to make att the vertex of it (which is in your eye) an angle bigg enough to be seene singly by itselfe; then each colour will appeare apart as it truly is. (p. 264)

Hooke (1665): Whence I experimentally found what I had before imagin'd, that all the varieties of colours imaginable are produc'd from several degrees of two colours, namely Yellow and Blue, or the mixture of them with light and darkness, that is, white and black. And all those almost infinite varieties which Limners and Painters are able to make by compounding those several colours they lay on their Shels or *Palads*, are nothing else, but some *compositum*, made up of some one or more, or all of these four. (pp. 74–75)

Rohault (1671): Further, if this distant Object be composed of a great many different Parts which are of different Colours, it is evident, that if several of these Parts act together upon the same Capillament, that which is of the brightest Colour is the only one that will be seen, because the Capillament will receive the Impression only of this Part. And thus we see in a Meadow where there are a great many white Flowers mixed with a vast Number of green Spires of Grass, at a Distance it looks all White. (1723, p. 250)

Newton (1672): The same colours in *Specie* with these Primary ones may be also produced by composition: For, a mixture of *Yellow* and *Blew* makes green; of *Red* and *Yellow* makes *Orange*; of *Orange* and *Yellowish green* makes *yellow*. And in general, if any two Colours be mixed, which in the series of those, generated by the Prisme, are not too far distant from one another, they by their mutual alloy compound that colour, which in the said series appeareth in the mid-way between them. (pp. 3082–3083)

Huygens (1673): As for the composition of *White* made by all the Colors together, it may possibly be, that *Yellow* and *Blew* might be sufficient for that: Which is worth while to try; and it may be done by the Experiment, which Mr. *Newton* proposeth, by receiving against a wall of a darkn'd room the Colours of the Prisme, and to cast their reflected light upon white paper. Here you must hinder the Colors of the extremities; *viz.* the Red and Purple, from striking against the wall, and leave only the intermediate Colors, yellow, green and blew, to see, whether the light of these alone would not make the paper appear white, as well as when they all give light. I even doubt, whether the lightest place of the yellow color may not all alone produce that effect, and I mean to try it at the first conveniency; for this thought never came to my mind but just now. Mean time you may see, that if these Experiments do succeed, it can no more be said, that all the Colors are necessary to compound White, and that 'tis very probable, that all the rest are nothing but degrees of *Yellow* and *Blew*, more or less charged. (pp. 6086–6087)

Newton (1673): Concerning the business of Colors; in my saying that Monsieur *N.* hath shewn how *White* may be produced out of two uncompounded colors, I will tell him, why he can conclude nothing from *that*; my meaning was that such a White, (were there any such,) would have different properties from the White, which I had respect to, when I described my Theory, that is from the White of the Sun's immediate light, of the ordinary objects of our senses, and of all white *Phænomena* that have hitherto faln under my observation. (p. 6087)

Newton (1704): ... in attempting to compound a white, by mixing the coloured Powders which Painters use, I considered that all coloured Powders do suppress and stop in them a very considerable Part of the Light by which they are illuminated. For they become coloured by reflecting the Light of their own Colours more copiously, and that of all other Colours more sparingly, and yet they do not reflect the Light of their own Colours so copiously as white Bodies do.... With the Center O and Radius OD describe a Circle ADF, and distinguish its circumference into seven parts DE, EF, FG, GA, AB, BC, CD, proportional to the seven musical Tones or Intervals of the eight Sounds, *Sol, la, fa, sol, la, mi, fa, sol,* contained in an Eight, that is proportional to the numbers 1/9, 1/16, 1/10, 1/9, 1/10, 1/16, 1/9. Let the first part DE represent a red Colour, the second EF orange, the third FG yellow, the fourth GH green, the fifth AB blue, the sixth BC indico, and the seventh CD violet. And conceive that these are all the Colours of uncompounded Light gradually passing into one another, as they do when made by Prisms; the circumference D E F G A B C D, representing the whole series of Colours from one end of the Sun's coloured Image to the other, so that from D to E be all degrees of red, at E the mean Colour between red and orange, from E to F all degrees of orange, at F the mean between orange and yellow, and so on. . . . Also if only two of the primary Colours which in the Circle are opposite to one another be mixed in an equal proportion, the point Z shall fall upon the center O, and yet the Colour compounded of those two shall not be perfectly white, but some faint anonymous Colour. For I could never yet by mixing only two primary Colours produce a perfect white. (pp. 110–111, 114–115, and 116)

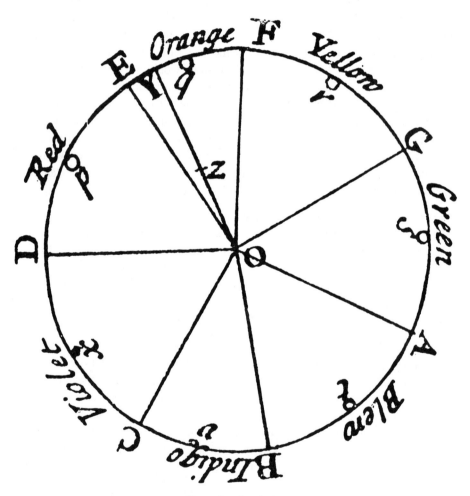

Newton's color circle.

le Blon (1720): PAINTING can represent all *visible* Objects, with three Colours, *Yellow*, *Red*, and *Blue*; for all other Colours can be compos'd of these *Three*, which I call *Primitive*. (p. 28)

Erasmus Darwin (1794): ... when the prismatic colours are painted on a wheel, they appear white as they revolve. (p. 573)

Young (1807): It is certain that the perfect sensations of yellow and of blue are produced respectively, by mixtures of red and green, and of green and violet light, and there is reason to suspect that those sensations are always compounded of the separate sensations combined: at least this supposition simplifies the theory of colours: it may, therefore, be adopted with advantage, until it be found inconsistent with any of the phenomena; and we may consider white light as composed of a mixture of red, green, and violet, only, in the proportion of about two parts red, four green, and one violet, with respect to the quantity or intensity of the sensations produced.... The sensations of various kinds of light may also be combined in a still more satisfactory manner by painting the surface of a circle with different colours, in any way that may be desired, and causing it to revolve with such rapidity, that the whole may assume the appearance of a single tint, or of a combination of tints, resulting from the mixture of colours. From three simple sensations, with their combinations, we obtain seven primitive distinctions of colours; but the different proportions, in which they may be combined, afford a variety of tints beyond all calculation. The three simple sensations being red, green, and violet, the three binary combinations are yellow, consisting of red and green; crimson, of red and violet; and blue, of green and violet; and the seventh in order is white light, composed by all the three united. (pp. 439–440)

Goethe (1810): Yellow, blue, and red, may be assumed as pure elementary colours, already existing; from these, violet, orange, and green, are the simplest combined results.... Yellow and blue powders mingled together appear green to the naked eye, but through a magnifying glass we can still perceive yellow and blue distinct from each other. Thus yellow and blue stripes seen at a distance, present a green mass; the same observation is applicable with regard to other specific colours. (1840, pp. 224 and 226)

Gall (1825): I pass over in silence all that Newton, Buffon, Goethe, and the modern natural philosophers, have said on the proportion of colors and on their mixture. I likewise abstain from examining the question, whether there exist seven primitive colors, or three only. I have no other end but to convince the reader, that there really exist, out of ourselves, determinate laws for the proportions of colors. Thus, for example, the three fundamental colors, supposing them to be only three, when placed side by side are always inharmonious. Blue, yellow, and red are inharmonious. If two of these colors are mixed, a mean color ensues. Blue and yellow compose green; blue and red, violet; red and yellow, orange. To obtain harmony, we must place by the side of a primitive color, a mixed color, in which the primitive color enters as part of the mixture; the mixed color will always be in harmony with the two primitive colors from which it results. (1835b, p. 50)

Hayter (1826): *First*—That Yellow, Red, and Blue, are entire colours of themselves, and cannot be produced by the mixtures of any other colours ... *Secondly*—Yellow, Red, and Blue, contain the sole properties of producing all other colours whatsoever, as to colour, by mixtures arising entirely among themselves, without the aid of a fourth; *Thirdly*—Because, by mixing proper portions of the Three Primitives together, *Black* is obtained, providing for every possible degree of shadow. *Fourthly*—And every practical degree of *light* is obtained by diluting any of the colours, as above producible; or in oil-painting, by the mixture of white paint. *Fifthly*—All transient or prismatic effects can be imitated with the Three Primitive Colours, as permanently considered, but only in the same degree of compensation as white bears to LIGHT. *Sixthly*— There are no other materials, in which colour is found, that are possessed of any of the foregoing perfections. (p. 12)

Brewster (1830a): To illustrate this in a more simple manner, let us suppose, that a colour wheel has a circumference divided into sectors ... and that each sector is painted of its proper colour ... then if this wheel be whirled briskly round its axis, its colour will be white. But if the red sector is taken out, or painted black, and the wheel again put in motion, the colour of the wheel will then be green. (p. 89)

Talbot (1834): This spiral figure was painted black, the rest of the circle remaining white, and the wheel was then made to revolve. The result was a gray surface varying gradually and regularly, in concentric shades, from perfect blackness in the centre to perfect whiteness in the circumference. I then chose a number of coloured papers and cut them into circles of the same size, each of which had this spiral traced upon it, which I afterwards cut out from the rest of the circle. By this means, being all of a size, they could be made to replace each other in a great variety of ways; for instance, the blue spiral was put on the yellow circle, and the yellow spiral on the blue circle. When in motion, I obtained in each case a surface varying gradually from perfect blue to perfect yellow, the central part being blue in the first case, and yellow in the latter. But the most curious part of the experiment was to observe the neutral tint, through which the colour passed at a certain point. When this neutral tint offered any singularity that made it desirable to examine it more particularly, it was easy, by measuring its distance from the centre, to compute how much of each colour entered into its composition; for instance, if the diameter of the wheel, formed of yellow and blue, was six inches, and the tint in question was two inches from the centre (supposed blue), it contained two thirds of blue

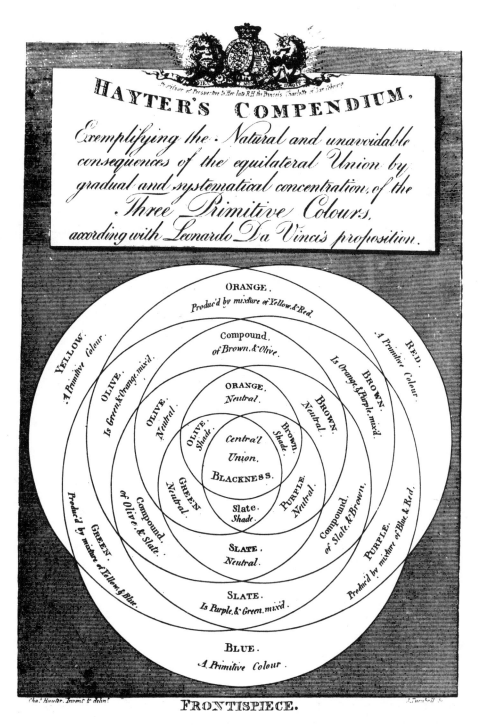

Color mixing according to Hayter.

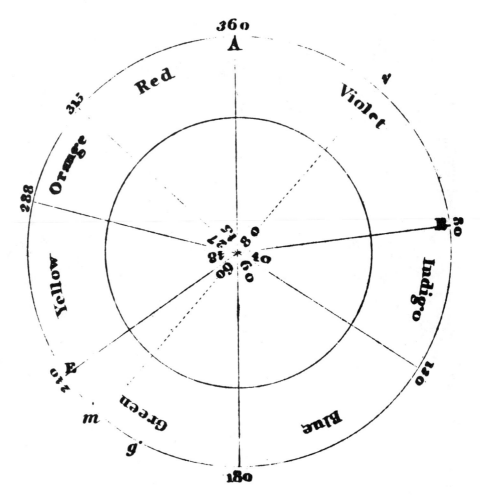

Brewster's color wheel.

and one third of yellow, and therefore I had only to make a blue circle, and place on it a yellow sector of 120°, to obtain infallibly the same tint over all its surface. (pp. 330–331)

3.3   Color Blindness

Complete color blindness is a very rare condition, and yet the term is both widely known and widely applied. Very few instances of *blindness* to all colors have been encountered, unlike the numerous cases of color defects and anomalies, in which one color is either not discriminated from others or is not seen as intensely as for the majority of people. The term "color blindness" was introduced by Brewster in 1844 in an attempt to prevent the condition

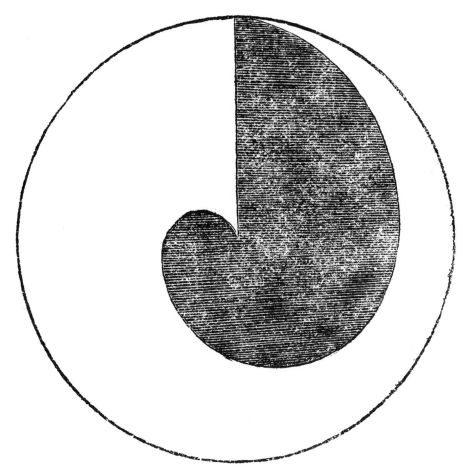

Talbot's color wheel.

from being associated with a scientist of note. Although descriptions of color defects were recorded in the seventeenth century, John Dalton's description of his own color vision had the greatest impact on the scientific community, presumably because of the high esteem in which Dalton was held for his atomic theory of matter. Indeed, such was the interest that the condition itself became called *Daltonism* in the early nineteenth century, following reference to those so afflicted as Daltonians by Prevost in 1827 (Wilson, 1855). Brewster (1844a) preferred the term "color blindness" because "no person wishes to be immortalized by his imperfection. We cannot but regard it as degrading to the venerable name which it misapplies, as one of the worst examples of vicious nomenclature" (pp. 139–140). Dalton described his own condition in a paper presented to the Manchester Literary and Philosophical Society in 1794, and it was published four years later; he believed that his inability to distinguish red from black was because the vitreous humor of his eyes was tinged with blue, thereby absorbing the longer wavelengths of the spectrum. When the hypothesis was tested after his death (on his written instructions), the humors were found to be colorless. Two centuries after Dalton's description of his own color defect, retinal pigment from one of his preserved eyes was examined and it was found to be lacking the photopigment that absorbs middle-wavelength light (Hunt, Dulai, Bowmaker, and Mollon, 1995). This is in contrast to Young's (1807) conclusion that Dalton was unable to detect the red end of the spectrum.

It is remarkable that color blindness came so late on the scientific scene. It must have been present in human societies for millenia, and yet the first descriptions of it go back barely four centuries. The likely reason for this is given in the astute comment by Malebranche that color names can mask differences in color vision. The statements by both Platter and Malebranche clearly refer to variations in color perception between individuals, but it is not so evident that they are referring to what later became called color blindness. Differences in color vision between the eyes, as described by Malebranche, are exceedingly rare. There remains no doubt, however, that Turbervile was describing a defect of color vision in the maid from Banbury; she could not distinguish any colors other than black and white, unlike the case detailed with greater precision by Boyle. She made red-green confusions which could not be attributed to poor eyesight, since she was able to discern the shapes of flowers. A similar difficulty with seeing red objects was noted in cases published by La Hire and Huddart; in both instances the subjects had good spatial acuity, but could not discriminate by color alone red fruit from the green leaves surrounding them. Huddart also commented on the familial links of those so afflicted: two brothers of the man examined had similar color defects, whereas two other brothers and a sister had normal color vision. The case reported by the Reverend Whisson to Joseph Priestley noted the links across three generations: the father was color defective, unlike the mother, but her brother had abnormal color vision; one daughter was normal, a son and a second daughter were color defective; the second daughter bore two color-defective sons and a color-normal daughter, while the son had a boy and girl who were color normal. Thus, the essential clues to the hereditary nature of certain color deficiencies were available at this early stage. The remarkable aspect of this report is that the brother of the mother was color defective, too, rendering the occurrence of a defective daughter comprehensible. Despite Whisson's description of a color-defective female, Dalton said that he had never heard of such a case. Palmer related the characteristics of anomalous color vision to the absence of one or more of the three putative receptors in the retina.

Dalton's attention was drawn to the peculiarities of his own color vision in the context of botanical specimens viewed by day and by candlelight; the geranium flower "was pink, but it appeared to me almost an exact sky-blue by day; in candle-light, however, it was astonishingly changed, not having then any blue in it, but being what I called red, a colour which forms a striking contrast to blue" (1798, pp. 29–30). When he asked his brother to look at the flower under similar circumstances, the description was like his own. His friends, on the other hand, remarked that there was little color change to be seen. This startling discovery of his (and his brother's) defective color vision started his more systematic investigations of the phenomenon. His paper was divided into three sections. The first gave an account of his own vision (which he referred to as anomalous). The second part described the vision of others with similar color anomalies; he noted that once, in a class of twenty-five pupils he was teaching, two had color vision like his own. A speculative interpretation of the cause of anomalous color vision was given in the final section. He made an analogy of color changes using filters with those that occurred in his vision, and "was led to the conjecture that one of the humours of my eye must be a transparent, but *coloured*, medium so constituted as to absorb *red* and *green* rays principally, because I obtain no proper ideas of these in the solar spectrum; and to transmit blue and other colours more perfectly" (p. 42).

In the early nineteenth century more cases were collected and described (see Wilson, 1855; Mollon, 1989), and the classification of the condition became more refined. Seebeck was foremost in this regard. He distinguished color anomalies from specific color blindness, and he categorized both into those who have difficulties with red and green, as opposed to those who confuse red and blue. Historical accounts of color blindness can be found in Wartmann (1844), Wilson (1855), Helmholtz (1896), Collins (1925), and Boring (1942).

Platter (1614): . . . occasionally the colours, in the absence of any injury to the spirits, appear in full light differently to normal vision, sometimes red, sometimes yellow, sometimes blue. (Koelbing, 1967, pp. 86–87)

Malebranche (1674): There is then great probability that all Men do not see the same Colours in the same Objects. Nevertheless I am of Opinion that it never happens, at least very rarely, that any Persons see Black and White of a different Colour from what our selves see them, though they do not see them equally Black and White. But as to the middle Colours, such as Red, Yellow, Blue, and especially those that are compounded of these three, I am persuaded there are very few Men that have exactly the same Sensations. For there are Men sometimes to be met with, who see some sort of Bodies of a yellow Colour, for instance, when they view them with one Eye, and of a Green or Blue when they behold them with the other. And yet supposing these Men to be born with one Eye only, or with two Eyes so dispos'd as to see that of a Red or Yellow Colour, which others call Green or Blue, they would believe they saw Objects of the same Colours as others do, because they would always have heard the Name *Green* given to that which they see as Yellow, and *Blue* to that which to them seems Red. (1700, p. 34)

Nicolas Malebranche (1638–1715) after an engraving in Landon (1805).

Turbervile (1684): A Maid, two or three and twenty years old, came to me from *Banbury*, who could see very well, but no colour beside *Black* and *White*. She had such Scintillations by night, (with the appearances of Bulls, Bears, &c.) as terrified her very much; she could see to read sometimes in the greatest darkness for almost a quarter of an hour. (p. 736)

Boyle (1688): ... she can distinguish some colours, as black and white; but is not able to distinguish others, especially red and green; and when I brought her a bag of a fine and glossy red, with tufts of sky-coloured silk, she looked attentively upon it, but told me, that to her it did not seem red, but another colour, which one would guess by her discription to be a dark or dirty one.... when she had a mind to gather violets, though she kneeled in that place where they grew, she was not able to distinguish them by the colour from the neighbouring grass, but only by the shape and by feeling them. (Birch, 1966, pp. 450–451)

La Hire (1711): This is also what we observe in some eyes, which being otherwise very good for seeing the smallest objects very distinctly, see red only as if it was black; and other colours they see perfectly well. (1742, p. 133)

Joseph Huddart (1741–1816) after an engraving in the group portrait "Men of Science Living in 1807–8."

Huddart (1777): He observed also that, when young, other children could discern cherries on a tree by some pretended difference of colour, though he could only distinguish them from the leaves by their difference of size and shape. He observed also, that by means of this difference of colour, they could see the cherries at a greater distance than he could, though he could see other objects at as great a distance as they; that is, where the sight was not assisted by the colour. Large objects he could see as well as other persons; and even smaller ones if they were not enveloped in other things, as in the case of cherries among the leaves. I believe he could never do more than guess the name of any colour; yet he could distinguish white from black, or black from any light or bright colour. Dove or straw-colour he called white, and different colours he frequently called by the same name: yet he could discern a difference between them when placed together. In general, colours of an equal degree of brightness, however they might otherwise differ, he frequently confounded together. Yet a striped ribbon he could distinguish from a plain one; but he could not tell what the colours were with any tolerable exactness. Dark colours in general he often mistook for black, but never imagined white to be a dark colour, nor a dark to be a white colour.... He had two brothers in the same circumstances as to sight; and two other brothers and sisters who, as well as their parents, had nothing of this defect. (pp. 262–263)

Whisson (1778): I am willing to inform you ... of my inability concerning colours, as far as I am able from my own common observation. It is a family failing: my father has exactly the same impediment: my mother and one of my sisters were perfect in all colours: my other sister and myself are alike imperfect: my last mentioned sister has two sons both imperfect: but she has a daughter who is very perfect: I have a son and a daughter, who both know all colours without exception: my mother's own brother had the like impediment with me. (p. 612)

Palmer (1786): It is quite evident that the retina must be composed of three sorts of fibres, or membranes each analogous to one of the three primary rays and susceptible of being stimulated by it alone.... There exist persons who, although they can see very clearly, cannot distinguish colours, and have the sensations of only more or less light, that is to say, of light and shade, because each class of fibre is stimulated by all of these rays.... It is impossible to remedy this structural defect. (Walls, 1956, p. 87)

Dalton (1798): I found that persons in general distinguish six kinds of colour in the solar image; namely, *red, orange, yellow, green, blue,* and *purple.* Newton, indeed, divides the purple into *indigo* and *violet;* but the difference between him and others is merely nominal. To me it is quite otherwise:—I see only *two* or at most *three* distinctions. These I should call *yellow* and *blue;* or *yellow, blue,* and *purple.* My yellow comprehends the *red, orange,* and *green* of others; and my *blue* and *purple* coincide with theirs.... It is remarkable that I have not heard of one female subject to this peculiarity. (pp. 31 and 40)

John Dalton (1766–1844) after an engraving in Muspratt (1853).

Goethe (1810): We will here first advert to a very remarkable state in which the vision of many persons is found to be. As it presents a deviation from the ordinary mode of seeing colours, it might be fairly classed under morbid impressions; but as it is consistent in itself, as it often occurs, may extend to several members of a family, and probably does not admit of cure, we may consider it as bordering only on nosological cases, and therefore place it first. I was acquainted with two individuals not more than twenty years of age, who were thus affected: both had bluish-grey eyes, an acute sight for near and distant objects, by day-light and candle-light, and their mode of seeing colours was in the main quite similar. They agreed with the rest of the world in denominating white, black, and grey.... They appeared to see yellow, red-yellow, and yellow-red, like others.... They confounded rose-colour, blue, and violet on all occasions: these colours only appeared to them to be distinguished from each other by delicate shades of lighter, darker, intenser, or fainter appearance. (1840, pp. 45–47)

Gall (1825): There are persons who are incapable of perceiving a very marked difference between one color and another. Dr. Unzer, of Altona, never was able to distinguish the difference between green and blue. A boy, who wished to learn the trade of a tailor, was obliged to renounce that design, in consequence of his incapacity to distinguish certain colors. Spurzheim cites the case of a man whom he saw at Dublin, who loved the mechanic arts and drawing, especially that of landscapes, but who was obliged to abandon painting, because he could not distinguish red from green. At Edinburgh, in Scotland, he also saw three brothers and a cousin german of theirs, who cannot discern green from brown. (1835b, p. 47)

Nicholl (1818): The colour I am most at a loss with is green; and in attempting to distinguish it from red, it is nearly guesswork.... Though I see different shades in looking at a rainbow, I should say it was a mixture of yellow and blue—yellow in the centre and blue towards the edge. (Jeffries, 1883, pp. 5–6)

Butter (1822): The colours of the rainbow or of the Moon, appear nearly the same, being twofold; at least, two distinct colours only can be seen, which he calls *yellow* and *blue*. A blue coat, however, he can distinguish from a black ... and a yellow vest is always known to him. (p. 136)

Seebeck (1837): In addition to such persons who have difficulty in the determination of colours, without however taking unequal colours for equal, there are often others to be found who, to a greater or lesser degree, confuse quite unequal colours with one another. (p. 220)

Müller (1838): Besides those persons who find difficulty in determining the colours of objects generally, without however mistaking one colour for another, there are many individuals who are in a more or less complete degree incapable of distinguishing perfectly dissimilar colours from each other. But these persons differ, again, not only in the degree of their defect, but also in its character. Hence Seebeck has divided them, smaller differences being disregarded, into two classes.... The individuals in this [first] class have a very imperfect power of distinguishing the impression of colours generally; but the defect is greatest with regard to red, and the complementary colour green; these colours being to them scarcely or not at all distinguishable from grey: the colour of which the perception is next in degree defective is blue, which is imperfectly distinguished from grey. The perception of yellow is the most perfect, though objects of this colour are also distinguished from a colourless surface much less easily than by the eye in its normal state. Individuals belonging to the second class likewise recognise yellow best; red is distinguished better, blue less perfectly, from the absence of colour, than in persons of the first class: but the distinction of red from blue is above all more imperfect here. (1843, pp. 1213–1214)

3.4   Color Zones

It is evident that by the early nineteenth century variations in color vision between individuals was of longstanding interest. In fact, Newton asked his assistant to corroborate many of his own observations, because the assistant was said to have better color discrimination than Newton (Campbell, 1986). In contrast to this realization of individual differences in the ability to differentiate colors, it was not until Purkinje's specification of color zones that attention was directed to variations of color perception within individuals. Color vision was assumed to be uniform throughout the retina. Purkinje demonstrated that it was otherwise. He constructed a device that we would now call a perimeter (see section 7.8) with which he could move colored squares systematically toward the point of fixation. At the edges of the visual field, over 90 degrees from the fixation point, objects were barely visible, and rapidly disappeared from view. Moving to more central locations, the object first became visible and only later appeared colored. He found that yellow and blue colors were visible at slightly greater peripheral angles than red and green, and that all colors were visible more peripherally in the temporal than the nasal fields.

Purkinje (1825): When one moves a coloured square slowly at the edge of the graduated curve [a perimeter] from the temporal side towards the point of direct vision, so one has initially only an impression of the shape and colour of an undefined something that is moving forwards. Every effort to distinguish the shape makes the object disappear completely; it disappears just as quickly when one fixes the attention on the point of direct vision. This occurs at an angle (to the temporal side) of 110–90 degrees. Below 90 degrees the colour quality and the shape begin to become noticable, but still very uncertainly. However, in the case of the colour it appears more generally light or dark, as happens in the gradually approaching darkness of evening when the colour of the object becomes uncertain. Cinnabar is visible at an angle of 90–70 degrees (temporally) as a very pale yellow, then orange, and then transforms gradually to its true colour quality; this is found at a visual angle from 60 degrees inwards; a beautiful pure purple appears black at an outer angle of 90 degrees, blue at 80 degrees, violet at 70 degrees, and first begins to take on its true colour at 50; light blue looks white at 90 degrees, but assumes its own colour at 80 degrees and less; a saturated blue appears as such at its first presentation in the visual field; violet appears black at 90 degrees, blue at 80 and 70 degrees, and first at 60 degrees and less as such in different shades; a saturated green looks black between 90 and 80 degrees after which it starts to develop its own colour; light yellow appears as such on its outermost presentation, as does orange; rosy red was initially white, its colour emerged first at 70 degrees; a leaf of Origanum majus appears initially dull until 40 degrees, then increasingly brighter yellow, and from there through yellow-green to its own colour. In a nasal direction and also upwards and downwards these colour changes take place even earlier, as the visual field itself is more restricted. (pp. 15–16)

Volkmann (1836): I repeated Purkinje's curious investigations concerning the change of colours in the peripheral parts of the visual field.... 1) When a coloured body moves in an arc from outside towards the centre, as it enters the visual field it is recognized earlier or later depending on the nature of the colour. At the earliest white is noticed, at an angle of more than 90°, then yellow, black, blue, vermilion, poppy-red, grass-green, in that order at angles which are difficult to specify, because the sensitivity of the eye, the strength of the illumination, and the tone of the colour introduce considerable variations. 2) I recognized the sequence of colours with certainty initially at angles from 30 to 10°. At greater angles the decision demands contemplation, and if the coloured stripes are finally removed the impression was in no way clear, but one felt the possibility of an error; this was the case with several repetitions of the experiment. 3) Frequently mistakes about the nature of the colour are made with too great an angle of incidence of the light; cinnabar appears black etc., in the manner described by Purkinje. (pp. 84 and 86)

3.5 Color Contrast

Descriptions of color contrast have a much longer history than color zones. Aristotle made a general statement about it, in the context of the rainbow: the yellow was said to be due to contrast. However, in this regard, the colors assigned to the rainbow could well have been a servant to the color theory that Aristotle maintained. His comments on color contrasts in dyed fabrics are more compelling, and this was precisely the problem that confronted Chevreul in the nineteenth century. Ptolemy, on the other hand, described the case of color assimilation: viewing through a colored veil resulted in objects taking on the color of the veil. Ibn al-Haytham gave directions on how to see contrast effects with paints. He appreciated that the perception of colors was always comparative. The example he gave was of green paint on different colored backgrounds, and it was painters, like Leonardo, who recognized the significance of color contrasts. Contrast could be experienced with black and white, and he hinted at the general principle that complementaries produce the greatest contrasts. There is a close correspondence between color contrast and color afterimages (see section 4.1), and Boyle's statement could belong to the latter. It was an instance of successive color contrast rather than the simultaneous contrast to which Leonardo referred. In his *Lectures on Natural Philosophy* Young treated color contrast immediately after colored afterimages; the examples he gave—of a gray spot modified by the color of its surround—were illustrated with hand-colored figures. In restating this demonstration, Müller interpreted it in terms of antagonistic processes operating in the retina.

   The phenomenon of color contrast is particularly associated with Chevreul, who was dyemaster of the Gobelin factory in Paris. He was meticulous in manufacturing his dyes, and incensed when his tapestries were criticized for the inconsistency of their colors. This spurred him to examine color contrast in more detail, and he published a book under the title *De la loi du Contraste Simultané des Couleurs* (On the Law of Simultaneous Color Contrast) in 1839. Most of the book is concerned with the application of colors in the arts, but he did enunciate his law with respect to vision.

Aristotle (ca. 330 B.C.): Hence also the rainbow appears with three colours ... The outer band of the primary rainbow is red: for the largest band reflects most sight to the sun, and the outer band is largest.... The appearance of yellow is due to contrast, for the red is whitened by its juxtaposition with green. We can see this from the fact that the rainbow is purest when the cloud is blackest; and then the red shows most yellow. (Yellow in the rainbow comes between red and green.). . . . Bright dyes too show the effect of contrast. In woven and embroidered stuffs the appearance of colours is profoundly affected by their juxtaposition with one another (purple, for instance, appears different on white and on black wool), and also by differences of illumination. Thus embroiderers say that they often make mistakes in their colours when they work by lamp-light, and use the wrong ones. (Ross, 1931, pp. 374b–375a)

Ptolemy (ca. 150): A surface colour is produced when we look through thin linen coloured red or purple. The visual rays pass the weft of the linen without any deviation and carry with then some of their colour. (Lejeune, 1956, p. 66)

Ibn al-Haytham (ca. 1040): . . . if designs are made with fresh-green paint on a dark-blue body, the paint will look [red] and of a clear colour; but if designs are made with the same paint on a clear-yellow body, the paint will look [green] and of a dark colour. . . . For the qualities of lights and colours are perceived by the eye only by comparing them with one another. (Sabra, 1989, pp. 99–100)

Leonardo da Vinci (ca. 1500): Of colours of equal whiteness that will seem most dazzling which is on the darkest background, and black will seem most intense when it is against a background of greater whiteness. Red also will seem most vivid when against a yellow background, and so in like manner with all the colours when set against those which present the sharpest contrasts. (MacCurdy, 1938b, p. 295)

Digby (1644): And we see painters heighten their colours, and make them appear lighter by placing deepe shadowes by them: even so much, that they will make objects appear neerer and further of, meerly by their mixtion of their colours. (p. 258)

Boyle (1663): . . . it is sometimes possible that the colour that would otherwise be produced by an outward object, may be changed by some motion, or new texture already produced in the sensory, as long as that unusual motion, or new disposition lasts; for I have divers times tried, that after I have through a telescope looked upon the sun, though thorough a thick, red, or blue glass, to make its splendour supportable to the eye, the impression upon the retina would not only be so vivid, but so permanent, that if afterwards I turned my eye towards a flame, it would appear to me of a colour very differing from its usual one. (pp. 672–673)

Young (1807): A similar effect [to colored afterimages] is often produced, when a white, or grey object is viewed on a coloured ground, even without altering the position of the eye: the whole retina being affected by sympathy nearly in the same manner as a part of it was affected in the former case. . . . The [gray] spot, which is tinted with [thin] black lines only, appears, upon the yellow ground, of a purple hue . . . A grey spot on a purple ground appears of a greenish yellow or olive hue. (pp. 455 and 787)

Goethe (1810): In taking off green spectacles, we see all objects in a red light.... If a green paper is seen through striped or flowered muslin, the stripes or flowers will appear reddish. A grey building seen through green pallisades appears in like manner reddish. (1840, pp. 25 and 26)

Müller (1838): A very small dull-grey strip of paper, lying upon an extensive surface of any bright colour, does not appear grey, but has a faint tint of the colour which is the contrast of that of the surrounding surface. Thus, for example:—A strip of grey paper upon a green field often appears to have a tint of red, and when lying upon a red surface, a greenish tint.... The colour excited thus, as a contrast to the exciting colour, being wholly independent of any rays of the corresponding colour acting from without upon the retina, must arise as an opposite or antagonistic condition of that membrane. (1843, p. 1188)

Michel-Eugène Chevreul (1786–1889) after an engraving in Muspratt (1853).

Chevreul (1839): *In the case where the eye sees at the same time two contiguous colours, they will appear as dissimilar as possible, both in their optical composition and in the height of their tone....* In the *simultaneous contrast of colours* is included all the phenomena of modification which differently coloured objects appear to undergo in their physical composition and in the height of tone of their respective colours, when seen simultaneously. The *successive contrast of colours* includes all the phenomena which are observed when the eyes, having looked at one or more coloured objects for a certain length of time, perceive, upon turning them away, images of these objects, having the complementary to that which belongs to each of them. (1854, pp. 15 and 34)

3.6   Color Shadows

Color shadows are also closely related to color contrasts and have been commented on by artists, like Alberti and Leonardo, as well as scientists, like Buffon and Rumford. Count Rumford (Benjamin Thompson) was alerted to the phenomenon in the course of some experiments on light, which led to his design of a photometer. He let sunlight into a room through a small hole and placed a small wooden cylinder in its path so that its shadow was cast on a screen a few inches away. The cylinder was also illuminated by a candle, casting an adjacent shadow. Rumford (1794) was surprised to see that the shadow illuminated by the candle was yellow, whereas that illuminated by the sun was "of the most beautiful *blue* it is possible to imagine" (p. 108). His subsequent experiments, using colored glasses and Argand lamps, convinced him that he was dealing with color contrast. For example, by placing a dense yellow glass in the path of the sunlight he could reverse the colors of the shadows. Young succinctly summarized the situation: the shadows are opposite, or complementary, to the source, a view affirmed by both Goethe and Müller.

Alberti (1435): As the shadow deepens the colours empty out, and as the light increases the colours become more open and clear. For this reason the painter ought to be persuaded that white and black are not true colours but are alterations of other colours. (1966, p. 50)

Leonardo da Vinci (ca. 1500): If you see a woman clad in white, in an open landscape, she will be of such brilliance on that side towards the sun as to dazzle the eyes almost as much as the sun itself. That side of her, however, which is illuminated by the light from the sky will have a bluish hue. (Minnaert, 1940, p. 109)

Buffon (1774): Frequent observations have led me to recognize that shadows never look green at sunset and sunrise unless the horizon is filled with much reddish vapour; in all other cases the shadows are always blue, the more bluish as the sky is more serene. This blue colour of the shadows is nothing other than the colour of the air. (Binet and Roger, 1977, p. 147)

Rumford (1794): Reflecting upon the great variety of colours observed in these last experiments, many of which did not appear to have the least relation to the apparent colours of the light by which they were produced, I began to suspect that the colours of the shadows might, in many cases, notwithstanding their apparent brilliancy, be merely an optical deception, owing to contrast, or to some effect of the other neighbouring colours upon the eye. (p. 115)

Benjamin Thompson, Count Rumford (1753–1814), after an engraving in the *Philosophical Magazine* 9:plate 6, 1803.

Young (1807): These appearances [of color contrast] are most conveniently exhibited by means of the shadows of objects placed in coloured light: the shadow appearing of a colour opposite to that of the stronger light, even when it is in reality illuminated by a fainter light of the same colour. (pp. 455–456)

Goethe (1810): The colour of the shadow may be considered as a chromatoscope of the illumined surface, for the spectator may always assume the colour of the light to be the opposite of that of the shadow. (1840, p. 32)

Müller (1838): The coloured shadows are usually ascribed to the physiological influence of contrast; the complementary colour presented by the shadow being regarded as the effect of internal causes acting on that part of the retina, and not of the impression of coloured rays from without. (1843, p. 1190)

3.7 Color Constancy

The phenomena of color contrasts and shadows emphasize the variability of color vision, whereas color constancy refers to the stability of color perception under varying conditions: objects appear to retain their color under a wide range of illuminations. Color constancy also entered the scene rather late; it was alluded to in La Hire's comparison between the colors of objects illuminated by the sun or candles; he also realized that colors are seen in the context of all others present. This point was amplified by Monge: he enlisted members of the Royal Academy of Paris to view white paper through a piece of red glass, and they were surprised that it continued to look white. However, when they viewed the paper down a long tube, to exclude the visibility of the surrounding objects, it appeared red (see Mollon, 1985). This elegant demonstration enabled Monge to conclude that colors of objects are based not only on the characteristics of the light they transmit to the eye but also on those of the surrounding objects. Unlike the case of color contrast, in which the relationship between the color and its background are changed, color constancy occurs when the ratio of an object's color to that of its background is unchanged, even if the total illumination varies. Young gave a clear example of such color constancy.

Philippe de La Hire (1640–1718) after an engraving in Augé (1898).

La Hire (1694): One cannot easily persuade onself that one sees all objects of different colour by day and by candlelight, because one compares all colours together. It is nevertheless true that blue appears green under this condition, and if we never saw blue except by candlelight we would not distinguish this colour from green. In order to know what difference there is between objects illuminated by candlelight and those illuminated by sunlight, one must close all the windows of a room during the day and light enough candles to illuminate all the objects there. Then, moving into another room that is well illuminated by sunlight, one looks through the door of the room at the objects illuminated by the candle. They will appear reddish yellow in comparison to those that are illuminated by sunlight and can be seen at the same time, which cannot be observed when one is in the room with the candle. (p. 239)

Gaspard Monge (1746–1818) after an engraving in Augé (1898).

Monge (1789): So the judgements that we hold about the colours of objects seem not to depend uniquely on the absolute nature of the rays of light that paint the picture of the objects on the retina; our judgements can be changed by the surroundings, and it is probable that we are influenced more by the ratio of some of the properties of the light rays than by the properties themselves, considered in an absolute manner. (Mollon, 1985, p. 6)

Young (1807): … when a considerable part of the field of vision is occupied by coloured light, it appears to the eye either white, or less coloured than it is in reality: so that when a room is illuminated either by the yellow light of a candle, or by the red light of a fire, a sheet of writing paper still appears to retain its whiteness. (p. 456)

3.8   Binocular Color Combination

The combination of different colors presented to corresponding regions of each retina became an issue of theoretical importance following Newton's experiments on color mixing. Indeed, it was a supporter of Newton, Desaguliers, who was among the first to draw attention to the phenomenon. Desaguliers was a Frenchman who gave lectures on many branches of natural philosophy at Oxford and London (Gunther, 1937). He was an ardent advocate of Newtonian optics, on which he gave lectures and demonstrations, some of which were attended by royalty. His demonstrations of phenomena to students was a great innovation, about which there was much debate at the time, and he wrote textbooks based on his lectures (Desaguliers, 1719, 1745). He applied a method of binocular combination that became widely employed in other studies of binocular vision (see chapter 6), namely, placing an aperture in such a position that two adjacent objects were in the optical axes of each eye (as shown in the illustration). Under these circumstances red and green patches of silk did not mix after the manner of combining prismatic lights. Taylor, on the other hand, reported that red and blue combined binocularly to yield purple. Similarly, Le Cat stated that yellow and blue resulted in green.

Du Tour achieved dichoptic combination by other means: he placed a board between his eyes and attached blue and yellow fabric in equivalent positions on each side, or the fabric was placed in front of the fixation point, as illustrated. When he converged his eyes to look at them they did not mix but alternated in color. This is a clear description of binocular color rivalry. Du Tour also applied the method of observing the colors through an aperture, as adopted by Desaguliers, and obtained similar results. Yet another technique was to view different colored objects through two long tubes, one in each optic axis. This was used by Reid, and he saw the colors combined. In these initial reports we have the start of a dispute that was to extend beyond the period of the present inquiry, and was the source of much acrimony between two towering theoretical opponents in the second half of the nineteenth century— Helmholtz and Hering: do colors fuse when presented to corresponding regions of each eye, or do they undergo rivalry? Reid's description was not without its ambiguity: the colors were not only said to be combined but also one "spread over the other, without hiding it." That is, a single color was not seen, which was also the conclusion of Bell and Weber. Wells amplified this by noting that one or other color was dominant during this "transparent" phase. The complexity of the binocular percepts is evident from Müller's account: sometimes one or the other color will predominate, whereas at other times "nebulous spots" of one color are visible on the other.

Wheatstone dispensed with the difficulties of dissociating accommodation from convergence, which obtained in the methods previously adopted, by viewing different colors with the stereoscope. The outcome was binocular rivalry, but this was either of the whole monocular stimulus or local parts of the two colored discs.

Desaguliers (1716): But if instead of the Candles, ρ be a piece of red Silk, and γ a piece of green Silk, the same Position of the Eyes will make the Image at

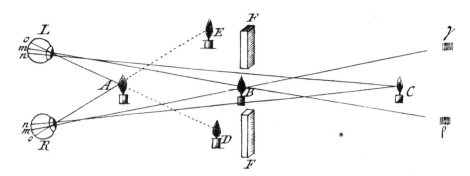

Desaguliers's method for binocular stimulation of corresponding retinal areas with different colors.

Jean Théophile Desaguliers (1683–1744) after an engraving in Landon (1805).

B, appearing like a red and green Spot together without a Mixture of Colours. If ρ be a red hot Iron, and γ a Candle of Sulphur, the Phænomenon will be more distinct. (p. 451)

Taylor (1738): ... one takes a piece of blue glass in front the candle E and a piece of red glass in front of the candle D with the intention of distinguishing them one from the other, without changing the quality of the respective images; if then the eyes approach the position that is necessary to look at a distant object, directing the axes of the eyes in the line Gg and Ff one can see a blue candle with the left eye B, and a red candle with the right eye C; and if one looks through the aperture with attention, directing the optic axes in the line dDeE, one sees the blue and red candles together in the aperture E, where they will have the appearance of a candle of the colour purple. (pp. 170–171).

Le Cat (1744): ... if you put a yellow Glass before one of the Candles, and a blue Glass before the other [presented to different eyes], the single Candle which you see will be green; that is to say, composed of the yellow of the first Candle, and the blue of the second. (1750, p. 200)

Etienne-François Du Tour (1711–1784) after a mezzotint kindly supplied by the Académie des Sciences, Paris.

Du Tour (1760): I glued a round patch of blue taffeta of about an inch in diameter onto one side of a sheet of cardboard, and on the opposite side, another patch of yellow taffeta of the same size, so that the two were exactly back to back. I placed the cardboard against my nose in a vertical plane and perpendicular to my face. Through my right eye I saw the blue patch and not the yellow, and vice versa for my left eye. Thus each one of the two patches formed separate images; blue in my right eye, yellow in my left eye. However, I was aware of only one patch. If that awareness was the result of the simultaneous combination of two images, should not the patch have seemed to be green? Now I was unable to discern the least tint of green. That single patch I saw sometimes appeared blue, sometimes yellow, apparently according to the rays of light from one or the other patches striking my eyes with more energy. Also, sometimes the patch appeared partly blue and yellow.... If one looks with both eyes at a point *A*, four or six inches away, and places on the optical

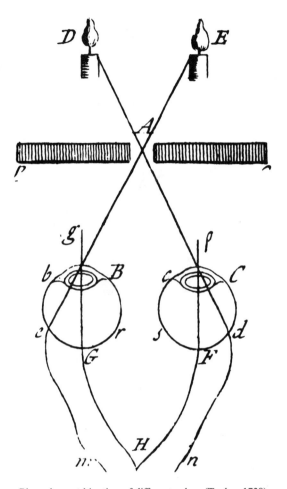

Binocular combination of different colors (Taylor, 1738).

axes *EA*, *GA*, short of the point *A* and their intersection, two small pieces of taffeta, one blue at *D*, the other yellow at *C*, one sees only a single spot of blue or yellow or a combination of colours, and never green. (pp. 514–515)

Reid (1764): And if two shillings are placed at the extremities of the two tubes, one exactly in the axis of one eye, and the other at the axis of the other eye, we shall see but one shilling. If two pieces of coin, or other bodies, of different colour, and of different figure, be properly placed, in the two axes of the eyes, and at the extremities of the tubes, we shall see both bodies in one and the same place, each as it were spread over the other, without hiding it; and the colour will be that which is compounded of the two colours. (pp. 325–326)

Wells (1792): But in all my experiments upon this subject I have remarked, that, when the two objects appeared united, each was seen, notwithstanding, in its proper colour; the red, for example, appearing as it were through a

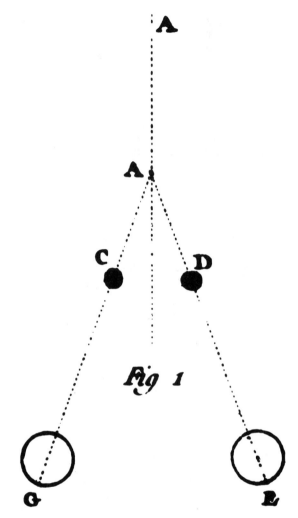

One of Du Tour's methods for stimulating the eyes with different colors.

transparent green, and the green, in the same experiment, as through a transparent red. (p. 45)

Bell (1803): ... if we look at a halfpenny and a shilling, placed each at the extremity of two tubes, one exactly in the visual axis of one eye, and the other, in the axis of the other eye, we shall see but one piece of coin, and of a colour neither like the shilling nor like the halfpenny, but intermediate, as if the one were spread over the other. (p. 354)

Weber (1834): ... the impressions of the colours merge into one if we look through differently coloured pieces of glass placed simultaneously in front of the two eyes.... Objects are also perceived by the two eyes simultaneously rather than alternately; and they appear to have a different colour from the one seen through either coloured glass alone. (Ross and Murray, 1978, p. 85)

Volkmann (1836): If one disentangles the the complicated appearances [of binocular color combination] one can scarcely doubt that different colours at identical retinal locations appear different, and that the colour appearance in the eye receiving a brighter light weakens or inhibits the appearance in the other. (pp. 93–94)

Müller (1838): The experiment of looking upon a sheet of white paper through two differently coloured glasses at the same time, may serve as an illustration for the present. The impressions of blue and yellow, for example, are found in such an experiment not to mingle readily; at one moment the blue, at another the yellow is predominant. Sometimes blue nebulous spots are seen upon the yellow field; at other times, yellow spots of varying magnitude upon the blue field: sometimes one colour alone prevails, and has absorbed the other; sometimes the reverse is seen. The appearance of one colour in spots upon a ground of the other colour, shows indeed that the attention can be directed at the same time to one part of one retina, and to the other parts of the other retina. (1843, pp. 1085–1086)

Wheatstone (1838): If a blue disc be presented to the right eye and a yellow disc to the corresponding part of the left eye, instead of a green disc which would appear if these two colours had mingled before their arrival at a single eye, the mind will perceive the two colours distinctly one or the other alternately predominating either partially or wholly over the disc. (pp. 386–387)

3.9 Subjective Colors

All color perception is, of course, subjective, as Newton noted. Nonetheless, the term has been, and still is, used for the perception of color from black and white stimuli. That is, the stimulus does not have the variations in wavelength that would be expected from the colors perceived. The history of subjective colors is a fascinating one, although most of it relates to the period after which this survey ends. According to Erb and Dallenbach (1939) and Cohen and Gordon (1949) the phenomenon was discovered independently many times, largely because of the wide range of stimuli that can induce the colors. They are often referred to as Fechner colors, following his description of the colors seen in a rotating black-and-white disc. However, Fechner's was not the earliest report, and those given by Brewster and (anonymously) by Charles Wheatstone have been overlooked in the reviews of the early history of the phenomenon (see Wade, 1977a). Cohen and Gordon did mention Prévost's account of the colors seen when a beam of sunlight entering a dark room is interrupted by a moving card. Indeed they referred to the phenomenon as Prevost-Fechner-Benham subjective colors; Benham's name was added because of the attention he drew to the colors seen with rotating black-and-white patterns on a child's spinning top (Benham, 1894). Prévost attributed the colors to differences in the times they required to act on the retina. There remains doubt concerning the origin of the colors observed with Prévost's method and on their interpretation (see Wade, 1983). Wheatstone considered that the perceived colors were due to diffraction at the edge of the moving card, and the sequence of colors does have many of the characteristics of the "flight of colors" seen in brief after-images.

One of the features of subjective colors is that they can be seen in both stationary and moving black-and-white patterns. Brewster observed pastel colors when maintaining fixation on high-contrast, fine parallel lines in an engraving. Not only were colors visible but the lines themselves became wavy (see section 4.5). It is likely that the slight involuntary eye movements over the pattern were the basis for the subjective colors (Wade, 1977a). Surprisingly, Wheatstone seems to have been unaware of Brewster's observation, since it was not commented upon in his brief and anonymous paper. That "C.W." is Charles Wheatstone is suggested by reference in the article to the kaleidophone, which was an instrument of his own invention (see Wheatstone, 1827), and from interchanges with Brewster at a meeting of the British Association for the Advancement of Science in 1832 (see Wade, 1983). Wheatstone did refer to other pertinent aspects of Brewster's writings on pattern disappearances. His attention was drawn to the phenomenon as a consequence of translating Prévost's account, and in disputing its interpretation. He went on to state: "There is another case of coloration which, I believe, has not yet been noticed" (C.W., 1831, p. 537). This involved the oscillating motion of black and white lines. When these were placed on the rods of the kaleidophone, colors could be seen. He appreciated that the lines should be perpendicular to the motion for the effect to be visible, and that the colors could be seen when printed letters were oscillated.

Fechner himself was most surprised to see the colors, since he was using the disc, with different sectors of black, as a means of producing gradations of gray during rotation. Having experienced the colors, he confirmed their appearance by asking other people to look at the rotating disc; they described the colors in similar terms, although there were considerable differences in their apparent intensities. This led Fechner to realize that the phenomenon must have a lawful basis, and he studied the effect of angular velocity on the colors seen, as well as variations in the number of sectors on the disc. He remarked, ruefully, that Goethe would have liked the phenomenon, since colors could be produced from black and white! As a final footnote to his paper, Fechner expressed surprise that Talbot had not noticed the phenomenon during his photometric experiments, which involved rotating similar discs to produce a gradation of grays.

Brewster (1825): When the eye is stedfastly fixed, for some time, upon the parallel lines which are generally used to represent the sea in maps, the lines will all break into serpentine lines, and *red*, *yellow*, *green*, and *blue* tints will appear in the interstices of them. (p. 292)

Prévost (1826): In a chamber sufficiently dark, into which a ray of the sun penetrates, move a rectangular piece of white card, about two inches in breadth, backwards and forwards, as if you would cut this ray nearly perpendicular to its axis. At the moment the white card traverses this axis, the eye which regards it evidently receives from this object a white light, as if the card remained stationary at this place. But it happens, however, that the disc, illuminated by the ray, the section of which it represents, appears coloured; it is

David Brewster (1781–1868) after an engraving in *The Illustrated London News*, 1868, vol. 52, p. 189.

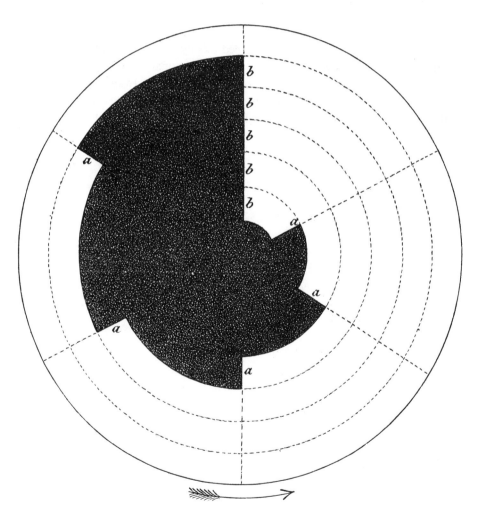

Fechner's black and white patterned disc for producing subjective colors.

white only in the centre. The very small white space which surrounds the centre, changes to violet, deepening as it recedes. The violet spot is surrounded by a zone of a deep indigo colour, very distinct and well defined, and exactly resembling the colour of the heart's-ease (*viola tricolor*). Around this indigo zone is a zone of greenish-yellow, equally well defined; then, surrounding it externally, a red tint. (C.W., 1831, p. 535)

C.W. (1831): If a sheet of paper, with black characters, either printed or written, be moved rapidly backwards and forwards at the ordinary distance of distinct vision, the lines described by the motion will appear accompanied by very evident colours, the green and red obviously predominating. The experiment succeeds better if the lines are far apart, and perpendicular to the direction of motion. (p. 537)

Fechner (1838): Rapidly rotate a disc covered with variations in black and white and one sees colours. I first noticed this by accident. . . . when the disc rotates in the direction of the arrow, at *a* the stationary eye is exposed to white followed by black. . . . The colors more or less die away with continued rotation of the edge *b* until the eye is exposed to another white area *a*. However, the impressions of the different colours disappear at different rates. (pp. 227–228)

4   Subjective Visual Phenomena

Jan Evangelista Purkyně or Purkinje (1787–1869) after an illustration in Zaunick (1961).

Purkinje (1823a): *Generally there are sensations which do not correspond to anything outside the body. In so far as they imitate the qualities and forms of external things, they thereby often give rise to illusions, phantoms, or appearances with no corresponding reality. These can be referred to as subjective sensory phenomena.... It is an imperative belief of the natural scientist that each and every modification of a subjective state in the sphere of the senses corresponds to an objective state. (pp. 3–4 and 92)*

The name of Purkinje is inextricably linked with subjective visual phenomena; he gave them that title, and he added greatly to the detail of their description, as well as defining various subcategories within which the phenomena could be considered. Purkinje was in the phenomenological line of Goethe, to whom the second volume of his *Beobachtungen und Versuche zur Physiologie der Sinne* (Observations and Experiments on the Physiology of the Senses) was dedicated.

The subtitle of both volumes is *Beiträge zur Kenntniss des Sehens in subjectiver Hinsicht* (Contributions to the Understanding of Vision in its Subjective Aspect). He commenced his work on subjective visual phenomena because, as a medical student in Prague, he had little access to equipment which would have allowed him to conduct experimental inquiries into other aspects of physiology. The first volume formed the basis of his doctoral thesis; it was published in 1819, but I have not been able to consult a copy of it. It was reprinted in 1823, and it is this slim volume that has been used for the translations included here. It is likely that Goethe was instrumental in Purkinje's appointment to the post of professor of physiology at the University of Breslau in 1823. Goethe had been impressed by reading the doctoral dissertation, as it both employed his method and provided support for his theoretical position at a time when it was generally shunned by the German scientific community (see Hykes, 1936; John, 1959). On accepting the appointment Purkinje took up Prussian citizenship, and continued to publish under the German spelling of his name. He remained in Breslau for 18 years, and returned to the German-language (Charles) university at Prague in the final years of his academic life.

Most of the subjective phenomena had been described before Purkinje, although his references to earlier observations were scant. His contribution was to classify them, and to describe them with precision. The first volume has sections on stroboscopic patterns, pressure figures, effects of galvanic stimulation, blind spot, pattern disappearances and distortions, visibility of retinal blood vessels and blood flow, afterimages, aftereffects, entoptic phenomena, single and double vision, and eye movements. The *Neue Beiträge* (New Contributions), which was published in 1825, amplified several of the earlier topics but added some new ones too, such as indirect vision, real and apparent motion, accommodation, near and far vision, and voluntary squinting. His preliminary estimations of color zones were included in the section on indirect vision, and what has become known as the Purkinje shift was described in a section concerned with color blending of afterimages and real images.

The significance of Purkinje's observations was quickly appreciated both in Germany and in other European countries. Goethe was not alone in his admiration of Purkinje's powers of observation. In 1826 the experimentalist Johannes Müller extended the work in a book entitled *Über die phantastischen Gesichtserscheinungen* (On Fantastic Visual Appearances); in that same year he published a book on comparative physiology and eye movements. In Britain, a summary and translation of Purkinje's first volume was provided (anonymously) by Wheatstone in 1830 (see Wade, 1983). Wheatstone gave his article the title "Contributions to the Physiology of Vision"; he was to use the same title for his own article describing the stereoscope in 1838. Wheatstone (C.W., 1830) commenced by noting that "this little volume has excited considerable interest in Germany" (p. 102), but he took issue with the use of the term "subjective":

*To distinguish these phenomena from those which arise on the presence of their appropriate
external objects, the author employs the term subjective, which, as denoting this class of
phenomena better than any other we are acquainted with, and, to avoid circumlocution,
we have purposely retained; it will, however, on consideration, be perceived, that the term
is not strictly proper, as, correctly speaking, all phenomena, as such, are subjective, i.e. in
the mind; and were we, without qualification, to admit the classification of phenomena into
objective and subjective, we should be unable to determine, with any degree of accuracy,
where the objective ends or the subjective begins." (p. 102)*

Wheatstone not only gave a summary of selected sections from Purkinje's book
but he also added novel methods for observing some phenomena—particularly
the visibility of retinal blood vessels (see section 4.8).

The subheadings used in this chapter do not include all the categories given
by Purkinje, since there was considerable overlap in the ones he used. Moreover, some of the topics are discussed in other chapters: blind spot (section 2.8),
color zones (section 3.4), motion aftereffects (section 5.4), eye movements
(section 5.6), strabismus (section 6.7), distinct vision (section 7.7), and peripheral vision (section 7.8). As was noted above, Purkinje did not cite many earlier
sources in his two books, even though some of them had a long history. This
was particularly the case with afterimages. They were readily observable in
the natural environment, and they provided a puzzle insofar as they could be
seen in the absence of an inducing (primary) stimulus.

4.1   Afterimages (Spectra, Accidental Colors)

Afterimages can be seen following either brief, intense illumination of the eye
or prolonged fixation on an illuminated stimulus. Thus, looking briefly at a
bright light can result in its continued visibility when the eyes are closed, or
directed away from the light to a uniform surface. Historically, they have been
given several names, like ocular spectra and accidental colors, both of which
refer to the colored characteristics of the phenomenon. In fact, the term
"afterimage" (*Nachbild*) itself was used by Purkinje (1823a) in the final section
of his book, and it was taken up by Fechner (1840) in his more detailed study
of afterimages; it has since superceded the other names given to them. They
are the phenomena that receive the greatest attention in this chapter, not only
because there are many records of their visibility but also because of the light
they throw on the understanding of vision. Franz (1899) summarized this
succinctly: "In the history of after-images we seem to have an epitome of the
interrelations of physics, physiology and psychology; and probably no other
single phenomenon is so good an example of the growth of experiment and
measurement in psychology" (p. 1).

As with many of the other phenomena in this book, the survey of afterimages
commences with an account given by Aristotle; it is not without its ambiguities,
but it does raise some recurring themes. Aristotle said that the color of the

afterimage was the same as that of the stimulus previously fixated; in modern parlance this would be called a positive afterimage. These are less often seen following fixation on colored than on black-and-white stimuli. The consequences of looking at an intense light can be more readily interpreted: the afterimage was visible with closed eyes, and it appeared in whatever direction the eyes were pointing. Moreover, there is a suggestion that the sequence of colors is rapid, and this has been corroborated by many others (see Plateau, 1878b; Berry, 1922; Scheerer, 1984). Indeed, it was called the "flight of colors" by Helmholtz (1867). Ptolemy was concerned with a more specific instance of afterimages: fixation on a bright color resulted in subsequently viewed objects taking on that color, which is an example of a positive afterimage. Ibn al-Haytham's description was more precise insofar as it noted that the shape of the primary stimulus was visible in the afterimage, and that the lingering effect was a negative afterimage. Leonardo produced multiple afterimages of the sun by moving his eyes rapidly in its vicinity, and then looking into a dark area.

A technique that was rapidly adopted, once it had been described by Aguilonius, was to form a patterned afterimage following fixation on a paned window: a negative afterimage resulted from such observation. That is, the parts of the afterimage that corresponded to the glass panes looked dark, whereas the bars appeared brighter. Castelli pursued this in a more systematic manner in his remarkable study of afterimages which also set out a number of their critical features. He used a stimulus (a lead-framed window) that afforded good fixation: with the eyes closed, the afterimage retained the shape of the window and it passed through a sequence of colors. The visibility of the afterimage was itself cyclical: it faded from view only to reappear and fade again. With the eyes open, the apparent size of the afterimage varied with the distance of the surface onto which it was projected. This last observation was made with greater precision in the late nineteenth century (Emmert, 1881), and it is commonly referred to as Emmert's law. Robert Darwin also commented on this aspect of the apparent size of afterimages (see Wade, 1978a).

The dangers of looking at the sun are evident from the description by Boyle's acquaintance, who amplified the problem by using a telescope! The fact that he was able to see the effects of this unwise procedure a decade later suggests strongly that he destroyed the retinal cells in the observing eye; what he noted subsequently was the scotoma so formed. Nonetheless, both Newton and Malebranche followed a similar procedure. Newton viewed the sun's image reflected in a mirror, as he described in a letter to John Locke. His interest in the phenomenon was a consequence of reading Boyle's account, which he repeated "with the hazard of my eyes." He might well have damaged his eyesight since the "phantasm" remained visible long after the initial observation of the sun. Having formed the afterimage, Newton noted that it could be retained for longer by winking, and that it seemed to be influenced by attention. In fact, he wrote that "I could make the phantasm return without looking any more upon the sun"; he was so disturbed by the phenomenon that

he shut himself in a dark room for three days "to recover the use of my eyes."
He equated the afterimages with aspects of imagination, and so he remarked
to Locke: "Your question about the cause of this phantasm, involves another
about the power of fancy, which I must confess is too hard a knot for me to
untie." What is of particular significance is that he conducted the first experiment on interocular transfer; the afterimage could be seen when using his left
eye even though it had been induced in the right eye. The transferred afterimage
was said to be "almost as plain" as the monocular one. Newton's experiment
had little impact in his own day, and it was only with the publication of Locke's
correspondence by Lord King in 1829 that its impact was made. Brewster
appreciated the significance of the observation, and reprinted Newton's letter
in the *Edinburgh Journal of Science*, which he edited (Brewster, 1831a). He
also carried out a similar experiment on himself, and confirmed the visibility
of an afterimage in the previously unstimulated eye. Brewster's theory of
vision was essentially retinal; indeed, he referred to the retina as the seat of
vision, and sought to account for phenomena in projective terms. Thus, his
interpretation of interocular transfer was peripheral—the impressions were
transferred from one retina to the other.

Newton did see colors but did not dwell upon them or their sequence, unlike
Malebranche; the sequence was similar to that given by Castelli, and they
were accorded a physiological interpretation. Malebranche considered that
vision was a consequence of vibrations in the optic nerve, and different colors
followed from different frequencies of vibration. Because yellow was the first
color seen in the afterimage, he concluded that it must be associated with the
highest frequency of vibration.

General characteristics of afterimages were adumbrated by both Jurin
and Buffon. Jurin's contribution was made in his "Essay upon distinct and
indistinct vision" which was appended to Smith's *Opticks* (1738). Jurin was
prominent in British science, and held the post of Secretary to the Royal
Society from 1721 to 1727; he contributed extensively to the *Philosophical
Transactions* but rarely on optics. He brought a keen eye and a sharp intellect
to a broader range of topics in vision than the title of his essay might imply.
In the case of afterimages he proposed the principle of reciprocal action. The
afterimage of the window was in the opposite contrast to that seen under
normal viewing, and so the processes in the retina that produced the afterimage
were the reverse of those generated by the primary stimulus. Buffon stated a
similar relation with respect to colored afterimages: objects of one color produce afterimages that are complementary. Buffon introduced the term "accidental colors" to describe them, and to distinguish them from natural colors.
An interpretation of such accidental colors was advanced by Scherffer. Any
colored stimulus will produce a strong effect for that color and a weaker one
at other colors; following intense stimulation, there is a loss of sensibility, and
viewing a white surface will result in a more powerful influence of the previously more weakly stimulated color. The afterimages were said to be more

intense when the natural color was viewed against a black background and then projected onto a white surface. In addition, Scherffer indicated that accidental colors mixed in a similar way to natural colors: he placed a red and a yellow square next to one another, and alternately fixated them; when a white surface was subsequently viewed, three afterimages were evident—a greenish blue in the center, flanked by green and violet squares.

Franklin added a quantitative note by indicating the duration for which afterimages remained visible: the colors faded faster than the form, which could last for many seconds. Purkinje examined the relationship between the duration of an afterimage and the prior fixation. The general principle that obtained was that the afterimage of a candle lasted for one twentieth of the period of prior observation. He also found that the negative afterimage of the candle was visible long after the colors had faded from view.

A clear distinction between direct (positive) and reverse (negative) afterimages was made by Robert Darwin, the son of Erasmus and the father of Charles. Robert was a medical man, like his father, and his paper on ocular spectra, published in the *Philosophical Transactions of the Royal Society*, was one of his few forays into science. He made a series of observations on this and other phenomena, and his article was reprinted in Erasmus's *Zoonomia*. The simple "tadpole" demonstration was applied to particularly good effect: a negative afterimage could be seen on the white paper, and it appeared whiter than the paper itself. His interpretation of this effect, essentially in terms of differential retinal adaptation, has a modern ring to it. The visibility of afterimages could be revived by intermittent stimulation (moving the hand in front of the eyes), and it was noted that there can be a latency before the afterimage is initially visible. The stimulus that he used for generating an afterimage of a word was illustrated in black and white for the *Philosophical Transactions*, but it was hand-colored in *Zoonomia*. Erasmus was evidently intrigued by his son's investigations, and he elaborated a little on them himself, by inducing a sequence of concentric colored afterimages.

With the establishment of the basic features of afterimages, they began to be used as a tool to investigate other aspects of vision—particularly those involving eye movements (see sections 5.5 and 5.8). For example, Wells generated afterimages in each eye and noted that the single impression moved with voluntary movements of the eyes, and remained single even with passive movements of one eye. When the afterimage was projected onto a white surface and one eye was displaced, the singleness of the afterimage remained even though the paper onto which it was projected appeared double. This was one of the many subtle demonstrations Wells made with afterimages, and he pursued their implications with penetrating logic. In like manner, Young studied aspects of color vision as well as central and peripheral vision by means of afterimages. Wheatstone realized that afterimages were stabilized retinal images, so that the effects of eye movements could be excluded in their observation. To this end, he generated binocular afterimages for figures with retinal disparities, and noted that depth was still seen with them. His conclusion that stereoscopic depth perception was not dependent on eye movements was confirmed by Dove (1841), who illuminated paired images with an electric spark, and saw them in relief.

The two nineteenth-century researchers who conducted the most extensive investigations into afterimages were Plateau and Fechner, both of whom suffered sorely for their science, as did Newton and Brewster before them. Fechner was a professor of physics at Leipzig University when he conducted his afterimage experiments. However, the frequency with which he had observed the sun damaged his eyesight, and might have played a part in inducing a crisis that resulted in resignation from his chair and in his shunning of company and comforts for some years. When he did emerge from this crisis his concerns shifted first toward philosophy and then to psychophysics (Fechner, 1860), which transformed the study of the senses.

The most complete historical survey of afterimages was presented by Plateau (1878b and c); he gave a brief summary of accounts before 1800, and an annotated bibliography is provided for those between 1800 and 1876. It is a sad irony that Plateau himself was blind when he compiled his history. Although Plateau considered that his blindness was a consequence of forming afterimages by observing the sun, this is unlikely to have been the case. In the course of conducting his doctoral research on visual persistence (Plateau, 1829) he often looked at the sun for as long as half a minute, following which he suffered from retinitis. However, his blindness, which became total in 1843, would seem to have been due to uveitis (see Verriest, 1990). He prefaced his survey with the following caution: "At the outset, and this point is of extreme importance, the experiments which are cited in this section are dangerous: they are of a type that has resulted in my developing the source of an affection which has completely deprived me of vision; I cannot advise physicians and physiologists too strongly to refrain from making such experiments, which are of slight importance in comparison to the problems they might cause" (1878a, p. iii). Sadly, this cautionary note has often been ignored in more recent studies of afterimages.

Aristotle (ca. 330 B.C.): . . . when we have looked steadily for a long while at one colour, e.g. at white or green, that to which we next transfer our gaze appears to be of the same colour. Again if, after having looked at the sun or some other brilliant object, we close the eyes, then, if we watch carefully, it appears in a right line with the direction of vision (whatever this may be), at first in its own colour; then it changes to crimson, next to purple, until it becomes black and disappears. (Ross, 1931, p. 459b)

Ptolemy (ca. 150): When we have looked at a very bright colour for a long time and then we look elsewhere, objects seem to have that colour, because the impression of the bright colour persists for some time. (Lejeune, 1956, p. 66)

Ibn al-Haytham (ca. 1040): . . . if an observer looks at a pure white body irradiated by daylight, so that the light on this body is strong although it is not sunlight, and he continues to look at the body for some time, then turns his eyes to a dark place, he will find the form of that light in the dark place, and with the same shape. When, subsequently, he closes his eyes and stares for a while, he will experience in his sight the form and shape of that light; then all that fades away and sight returns to its own condition. . . . For if the eye looks at length at a strong light, the light will dim its sight, thus producing a certain darkness that lingers in the eye for a while before it clears up. (Sabra, 1989, pp. 51 and 267)

Leonardo da Vinci (ca. 1500): A dark place will seem sown with spots of light and a shining place with dark round spots, when seen by the eye which has recently gazed many times and rapidly at the body of the sun. (MacCurdy, 1938a, p. 255)

Aguilonius (1613): When one looks fixedly at a window for some time, and then closes the eyes, one retains the image of the window, but in a contrary manner: it is as if the parts that allowed the light through appear obscure, while the opaque bars seem luminous. (p. 56)

Benedetto Castelli (1557–1643) after an engraving kindly supplied by The Wellcome Institute Library,

Castelli (1639): I got all those present to fix their eyes on a glassed window illuminated by the sun with the following precaution: that they would not let the eye wander over the window but would hold their sight on a determinate spot of one of the panes and keep the eye steady for as much time as it takes to say, for instance, the Miserere psalm. Once this was done, I did so that all those who had done the operation closed their eyes and on questioning them as to what they were seeing with the eyes thus closed all answered that they were seeing the very same window with its glass panes differentiated one from the other by their lead [frames] and by small details (particolari minuzie). [But] what was amazing to all was that the window appeared as painted in very diverse hues, now with yellow colours, then green, now red, then peacock blue. Then it would vanish, reappear, and fade away again. The wonder of everyone increased even more when the same operation was repeated with open eyes and these eyes were turned now toward one part, now toward another part of a white wall, all saw the image of the same window but with the following additional wonder: when they looked at a wall further away from their eyes than the window, they saw an image larger than the real window; when they looked at a wall nearer and nearer, the image of that same window appeared smaller, and smaller in such a way that when looking at a sheet of paper placed three spans away from the eyes the same image appeared very small on the paper. (Ariotti, 1973, p. 7)

Boyle (1663): . . . having upon a time looked too fixedly upon the sun, thorough a telescope, without any coloured glass, to take off from the dazzling splendor of the object, the excess of light did so strongly affect his eye, that ever since, when he turns it towards a window, or any white object, he fancies he seeth a globe of light, of about the bigness of the sun then appeared of to him, to pass before his eyes: and having inquired of him, how long he had been troubled by this indisposition, he replied, that it was already nine or ten years since the accident, that occasioned it, first befel him. (Birch, 1966, p. 674)

Newton (1691): I looked a very little while upon the sun in the looking-glass with my right eye, and then turned my eyes into a dark corner of my chamber, and winked, to observe the impression made, and the circles of colours which encompassed it, and how they decayed by degrees, and at last vanished. This I repeated a second and a third time. At the third time, when the phantasm of light and colours about it were almost vanished, intending my fancy upon them to see their last appearance, I found to my amazement, that they began to return, and by little and little to become as lively and vivid as when I had newly looked upon the sun. But when I ceased to intende my fancy upon them, they vanished again. After this, I found that as often as I went into the dark, and intended my mind upon them, as when a man looks earnestly to see any thing which is difficult to be seen, I could make the phantasm return without looking any more upon the sun. . . . and, which is still stranger, though I looked upon the sun with my right eye only, and not with my left, yet my fancy began to make the impression upon my left eye, as well as upon my right. For if I shut my right eye, or looked upon a book or the clouds with my left eye, I could see the spectrum of the sun almost as plain as with my right eye, if I did but intende my fancy a little while upon it; for at first, if I shut my right eye, and looked with my left, the spectrum of the sun did not appear till I intended my fancy upon it; but by repeating, this appeared every time more easily. (1829, pp. 217–218)

Malebranche (1699): When one has looked at the sun, and the optic nerve has been very much shaken by the brightness of its light, because the fibres of this nerve are situated in the *focus* of the transparent humours of the eye; then if one shuts the eyes, or enters into a dark place, the shaking of the optic nerve will alter only as the more or less. However, we shall see different colours, white at first, yellow, red, blue, and some of those which are formed by the mixture of the primitives, and in the last place black. Whence we may conclude, that the vibrations of the *retina*, which are very quick at first, become slower by degrees. For once more, it is not the greatness or *force* of these vibrations, but their *quickness* that changes the species of the colours; since red, for example, appears red to a weak as to a strong light. We might therefore judge perhaps by the series of colours, if it was very constant, that the vibrations of the yellow are more quick than those of the red, and of the red than of the blue, and so of the other colours that succeed. (1742, p. 37)

James Jurin (1684–1750)
after an illustration in
Robinson (1980).

Comte de Buffon
(George Louis Leclerc)
(1707–1788) after a
frontispiece engraving in
Le Clerc (1866).

Jurin (1738): A person sitting to be shaved against a light sash window, fixed his eyes intently upon the window for some time, and afterwards shutting them, he had now the appearance of a window similar to that he saw before: only the glass panes were dark, and the wood between them was luminous.... These, and many more *phænomena* of like kind, seem to depend upon the principle, that when we have been for some time affected with one sensation, as soon as we cease to be so affected, a contrary sensation is apt to arise in us, sometimes of itself, and sometimes from such causes, as at another time would not produce the sensation at all, or at least not to the same degree. (p. 169)

Buffon (1743a): Thus there is a range of accidental colours which corresponds to the range of natural colours; red produces an accidental colour of green, yellow produces blue, green produces purple, blue produces red, black produces white, and white produces black. (p. 153)

Scherffer (1761): If a sense is stimulated by two impressions, one of which is strong and the other is weak, we do not experience the weak stimulus. This occurs principally when the two stimuli are of the same kind, or when a strong action of an object is followed by another of the same kind, but weaker and less violent. Because the sense organ is tired and relaxed, it needs more time to be able to transmit to the nerves the weak sensations, or because the motion and the violent vibration of all the parts of this organ do not stop functioning immediately with the action of the outside object. (Plateau, 1878b, p. 13)

Franklin (1765): For the impression made on the visual nerves by a luminous object will continue for twenty to thirty seconds. Sitting in a room, look earnestly at the middle of a window a little while when the day is bright, and then shut your eyes; the figure of the window will still remain in the eye, and so distinct that you might count the panes. A remarkable circumstance attending this experiment, is, that the impression of forms is better retained than that of colors; for after the eyes are shut, when you first discern the image of the window, the panes appear dark, and the cross bars of the sashes, with the window frames and walls, appear white or bright; but if you still add to the darkness in the eyes by covering them with your hand, the reverse instantly takes place, the panes appear luminous and the cross bars dark. And by removing the hand they are again reversed. This I know not how to account for. Nor for the following; that, after looking long through green spectacles, the white paper of a book will on first taking them off appear to have a blush of red; and after long looking through red glasses, a greenish cast; this seems to intimate a relation between green and red not yet explained. (1970, pp. 379–380)

Robert Darwin (1786): These ocular spectra are of four kinds: 1st, Such as are owing to a less sensibility of a defined part of the retina; or *spectra from a defect of sensibility*. 2d, Such as are owing to a greater sensibility of a defined part of the retina; or *spectra from excess of sensibility*. 3d, Such as resemble their object in its colour as well as form; which may be termed *direct ocular spectra*. 4th, Such as are of a colour contrary to that of their object; which may be termed *reverse ocular spectra....* Make with ink on white paper a very black spot, about half an inch in diameter, with a tail about an inch in length, so as to represent a tadpole; look steadily for a minute on this spot, and, on moving the eye a little, the figure of the tadpole will be seen on the white part of the paper, which figure of the tadpole will appear whiter or more luminous than the other parts of the white paper; for the part of the retina on which the tadpole was delineated, is now more sensible to light than the other parts of it, which were exposed to the white paper.... I covered a paper about four inches square with yellow, and with a pen filled with a blue colour wrote upon the middle of it the word BANKS in capitals, and, sitting with my back to the sun, fixed my eyes for a minute exactly on the center of the letter N in the middle of the word; after closing my eyes, and shading them somewhat with my hand, the word was distinctly seen in the spectrum in yellow letters on a blue field; and then, on opening my eyes on a yellowish wall at twenty feet distance, the magnified name of BANKS appeared written on the wall in golden characters. (pp. 313, 321, and 347)

Robert Waring Darwin (1766–1848) after a mezzotint kindly supplied by The Wellcome Institute Library, London.

Wells (1792): When we have looked steadily for some time at the flame of a candle, or any other luminous body, a coloured spot will appear upon every object, to which we shortly after direct our eyes.... The spot not only appears

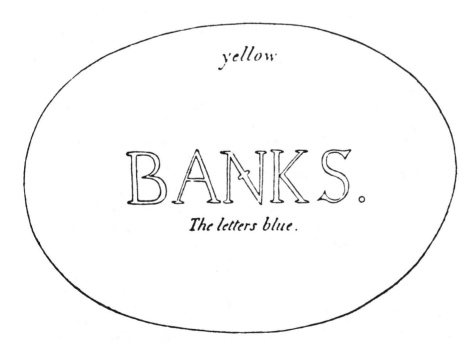

Robert Darwin's stimulus used to demonstrate the variation in the apparent size of an afterimage with the distance of the surface onto which it is projected.

single in every ordinary position of the optic axes, but cannot even be made to appear double, by any means whatever. If it be projected, for example, upon a piece of white paper, whoever makes the trial will find, that, although, on pressing one eye upward or downward, or to either side, the paper will be seen double, yet the spot will always appear single, and possess its former place on the paper, as seen by the eye, which is not disturbed. (pp. 65 and 67–68)

Erasmus Darwin (1794): Place a circular piece of white paper, about four inches in diameter, in the sunshine, cover the center of this with a circular piece of black silk, about three inches in diameter; and the center of the black silk with a circle of pink silk, about two inches in diameter; and the center of the pink silk with a circle of yellow silk, about one inch in diameter; and the center of this with a circle of blue silk, about half an inch in diameter; make a small spot with ink in the center of the blue silk ... look steadily for a minute at this central spot, and then closing your eyes, and applying your hand at about an inch distance before them, so as to prevent too much or too little light from passing through the eye-lids, and you will see the most beautiful circles of colours that imagination can conceive; which are most resembled by the colours occasioned by pouring a drop or two of oil on a still lake in a bright day. But these circular irises of colour are not only different from the colours of the silks above mentioned, but are at the same time perpetually changing as long as they exist. (pp. 20–21)

Young (1801): ... if the image of the sun itself be received on a part of the retina remote from the axis, the impression will not be sufficiently strong to form a permanent spectrum, although an object of very moderate brightness will produce this effect when directly viewed. (p. 45)

Young (1807): When the eye has been fixed on a small object of a bright colour, and is then turned away to a white surface, a faint spot, resembling in form and magnitude the object first viewed, appears on the surface, of a colour opposite to the first, that is, of such a colour as would be produced by withdrawing it from white light; thus a red object produces a bluish green spot; and a bluish green object a red spot. The reason of this appearance is probably that the portion of the retina, or of the sensorium, that is affected, has lost part of its sensibility to the light of that colour, with which it has been impressed, and is more strongly affected by the colour of the other constituent parts of white light. (p. 455)

Goethe (1810): If in the morning, on waking, when the eye is very susceptible, we look intently at the bars relieved against the dawning sky, and then shut our eyes or look towards a totally dark place, we shall see a dark cross on a light ground before us for some time.... If, on the other hand, we turn the open eye towards the side of a room, and consider the visionary image in relation to other objects, we shall always see it larger in proportion to the distance of the surface on which it is thrown.... If, when the eye is impressed with visionary images that last for a while, we look on coloured surfaces, and intermixture also takes place; the spectrum is determined to a new colour which is composed of the two. (1840, pp. 7–8 and 227)

Purkinje (1823a): I have varied the period of observation [of the candle flame] from twelve seconds to a minute, and the duration of the afterimage is always the same proportion of the initial observation (1 : 20), as is the sequence of images of the flame, except that in this experiment the intensity and duration of the colours predominate. In order to imagine the whole experience more easily and clearly, think about a dazzling white, a yellow, a red, a blue, a mild white and a black flame image of the same size, and like leaves placed on top of and completely obscuring one another. In the first moment after observation of the light source, with the eyes covered, one sees the dazzling image of the flame, but it disappears very quickly from the outside inwards, leaving the yellow; this lasts longer than the previous one, and disappears in a similar way; the same applies for each of the following until the black one, which lasts the longest. (pp. 100–101)

Müller (1826b): The afterimages . . . change their relative position to our own bodies with movement of the eyes; the imaginary images maintain a fixed position with respect to our own space with every movement of the closed eyes. (p. 72)

Brewster (1830a): My right eye being tied up, I viewed this luminous disc [the sun] with the left through a tube, which prevented any extraneous light from falling on the retina. . . . I was surprised to find, upon uncovering my right eye, and turning it to a white ground, that it also gave a coloured spectrum exactly the reverse of the first spectrum [in the left eye], which was *pink* surrounded by *green*. This result was so extraordinary, that I repeated the experiment twice, in order to secure against deception, and always with the same result. The spectrum in the left eye was uniformly invigorated by closing the eye-lids, because the images of external objects efface the impression upon the retina; and when I refreshed the spectrum in the left eye, the spectrum in the right was also strengthened. On repeating the experiment a third time, the spectrum appeared in both eyes, which seems to prove, *that the impression of the solar image was conveyed by the optic nerve from the left to the right eye*; for the right eye being shut, could not be affected by the luminous image. (pp. 91–92)

Müller (1838): If we look for a long time at a black square on a white ground, and then divert our eyes slightly, so as not entirely to leave the square, but rather to look more directly at its border, a portion of the spectrum which it has produced, a', c', d', will appear free upon the white ground, as a bright margin to one part of the dark image a, b, c, d. (1843, p. 1182)

Wheatstone (1838): Another and a beautiful proof that the appearance of relief in binocular vision is an effect independent of motions of the eyes, may

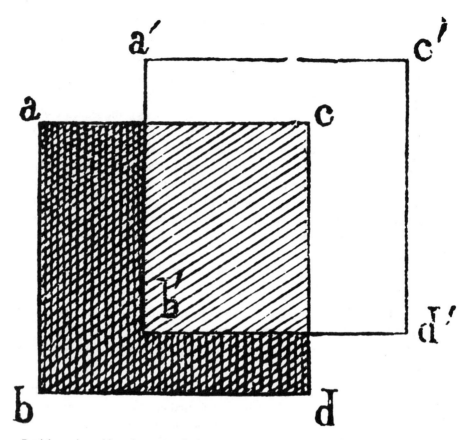

Partial superimposition of a square afterimage on the stimulus that generated it (Müller, 1838).

be obtained by impressing on the retinæ ocular spectra of the component figures. For this purpose the drawings should be formed of broad coloured lines on a ground of complementary colour, for instance red lines on a green ground, and be viewed ... in a stereoscope ... taking care however to fix the eyes only to a single point of the compound figure; the drawings must be strongly illuminated, and after a sufficient time has elapsed to impress the spectra on the retinæ, the eyes must be carefully covered to exclude all external light. A spectrum of the object in relief will then appear before the closed eyes. It is well known, that a spectrum impressed on a single eye and seen in the dark, frequently alternately appears and disappears: these alternations do not correspond in the spectra impressed on the two retinæ, and hence a curious effect arises; sometimes the right eye spectrum will be seen alone, sometimes that of the left eye, and at those moments when the two appear together, the binocular spectrum will present itself in bold relief. (pp. 392–393)

Plateau (1839): *When the retina is submitted to the action of rays of any colour whatever, it resists that action, and tends to recover its normal state, with a* *gradually increasing force. Then, if the organ be suddenly withdrawn from the exciting cause, it returns to the normal state by a sort of oscillatory motion, the more intense as the action has been further prolonged, a motion in virtue of which the impression passes first from the positive to the negative state, and then continues generally to oscillate, in a manner more or less regular, by becoming weaker, until it has entirely vanished.* (p. 332)

Fechner (1840): When one is observing an afterimage, the appearance jumps to an earlier or a later phase, depending upon the sudden increase or decrease of light entering the eye.... The longer one has observed an object, the more intense the whole appearance of the afterimage and the longer it lasts, before it becomes indistinct.... This result appeared to me so distinctive and important that I have repeated my observations several times. The experiments with measurements of duration in diffuse daylight and with the cross-bar have yielded very constant results.... The conclusion is that if the afterimage disappears quickly, only the first phase can be seen clearly.... Movement of the eye, or the rest of the body, leads to the disappearance of complementary afterimages. (pp. 215, 217–218, and 221)

Gustav Theodor Fechner (1801–1887) after a photogravure in Kuntze (1892).

4.2 Dark and Light Adaptation

The phenomenon of momentary blindness to weak light after exposure to intense illumination seems so compelling that it must have been recorded from the earliest times. Yet I have found relatively few reports of the phenomenon in the early literature. Aristotle did discuss it briefly in the context of persisting images, and it was associated with afterimages, visual persistence, and motion aftereffects (see sections 4.1, 4.11, and 5.4). The descriptions by Ibn al-Haytham, Pecham, and Leonardo all have an element of ambiguity about them, unlike that by Francis Bacon, which describes both dark and light adaptation. Jurin judiciously excluded pupil dilation as the cause of dark adaptation on the basis of differences in time course; while the pupil adapts quickly to changes in illumination, the recovery from exposure to intense light requires much longer. Despite this dissociation, Porterfield retained the belief in the link between pupil size and perception. Both Haller and Robert Darwin noted the changes in pupil size, but they refrained from ascribing the recovery from light adaptation to that cause alone. The theories that had been proposed to account for afterimages were often applied to dark and light adaptation, too. For example, Scherffer's speculations about retinal fatigue were taken to explain dark adaptation. Goethe suggested that there are differences in the duration of dark adaptation according to the strength of the eyes, and he did provide some values for the recovery period.

Aristotle (ca. 330 B.C.): ... the affection continues in the sensory organs, both in their deeper and in their more superficial parts, not merely while they actually engaged in perceiving, but even after they have ceased to do so. That they do this, indeed, is obvious in cases where we continue for some time engaged in a certain form of perception, for then, when we shift the scene of our perceptive activity, the previous affection remains; for instance, when we have turned our gaze from sunlight into darkness. For the result of this is that one sees nothing, owing to the motion excited by the light still subsisting in our eyes. (Ross, 1931, p. 459b)

Ibn al-Haytham (ca. 1040): Faint light can produce a sensible effect in the eye only after a fairly long interval of time. (Sabra, 1989, p. 266)

Pecham (ca. 1280): ... after a glance [at bright lights], images of intense brightness remain in the eye, and they cause a less illuminated place to appear dark until the traces of the brighter light have disappeared from the eye. (Lindberg, 1970, p. 63)

John Pecham (ca. 1230–1292) after a wooden effigy in Canterbury Cathedral.

Leonardo da Vinci (ca. 1500): Every concave place will appear darker if seen from outside than from within. And this comes about because the eye that is outside in the air has the pupil much diminished, and that which is situated in the dark place has the pupil enlarged. (MacCurdy, 1938a, p. 240)

Francis Bacon (1627): ... if men come out of a great light into a dark room; and contrariwise, if they come out of a dark room into a light room; they seem to have a mist before the eyes, and they see worse than they shall do after they have stayed a little while either in the light or in the dark. (1857a, p. 629)

Jurin (1738): In coming out of a strong light, into a room with the window-shutters almost closed, we have an immediate sensation of darkness; and this continues much longer than the pupil requires to dilate and accomodate itself to that weak degree of light, which is almost instantly done. But after staying some time in a much darker place, the same room, which appeared dark before, will be sufficiently light. (pp. 169–170)

Hartley (1749): ... being kept much in the dark should enable the Persons to see with a very obscure Light. (p. 199)

Porterfield (1759b): From this we may see, why it becomes advisable that such as, from a very light House, are obliged to go home in the dark, should keep their Eyes shut for some Time before they go out; for by this Means the Pupil will dilate, so as to enable them to see the Road, and avoid the Precipices and Dangers that may be in their Way. (p. 191)

Bouguer (1760): All our readers have noticed that one can only see with the greatest difficulty what happens in a dark room when one looks into it from a brightly illuminated place.... Certain places in a darkened room are more strongly lighted than others; there is often an infinity of different nuances in the shadows, and it is very certain that shadows serve to render other parts more prominent; but if the impression which remains to us from the light outside is still too strong, not only shall we not discover the dark parts of objects, but we shall not even distinguish those which are the most strongly illuminated. (1961, p. 52)

Scherffer (1761): ... when we come from a very light place to a dark place, it seems at first to be like night, because the weak light cannot stimulate the eyes which have just been exposed to an intense light. (Plateau, 1878b, p. 13)

Haller (1767): Why are we blind when brought out of a strong light into a weak one? Because the optic nerve, having suffered the action of stronger causes, is incapable of being moved by weaker ones. Whence have we a pain, by passing suddenly from a dark place into the light? Because the pupil, being widely dilated in the dark, suddenly admits too great a quantity of light before it can contract; whence the tender retina, which is easily affected by a small light, feels, for a time, an impression too sharp and strong. (1786, p. 32)

Robert Darwin (1786): On emerging from a dark cavern, where we have long continued, the light of a bright day becomes intolerable to the eye for a considerable time, owing to the excess of sensibility existing in the eye, after having been long exposed to little or no stimulus. This occasions us immediately to contract the iris to its smallest aperture, which becomes gradually dilated, as the retina becomes accustomed to the greater stimulus of the daylight. (p. 322)

Goethe (1810): In passing from bright daylight to a dusky place we distinguish nothing at first: by degrees the eye recovers its susceptibility; strong eyes sooner than weak ones; the former in a minute, while the latter may require seven or eight minutes. (1840, p. 3)

4.3   Purkinje Shift

Variations in color vision within an individual were seldom considered before the time of Purkinje. His brief description of the changes in the brightness of different colors at daybreak appeared in his second book on subjective visual phenomena. Another translation of this passage can be found in Brožek (1989); the term "Purkinje phenomenon" was coined by de Lepinay and Nicati in 1882 (see Brožek and Kuthan, 1990; Brožek, Kuthan, and Arens, 1991), and it is also referred to as the Purkinje shift. Purkinje himself did not attribute great importance to the phenomenon as it was the ninth entry in the section on color combination. Earlier reports of a similar phenomenon can be found in Aristotle and Leonardo, although they are by no means as explicit as the clear phenomenological description given by Purkinje. The statement by Aristotle on color contrast (section 3.6) is perhaps more indicative of the phenomenon: embroiderers made color confusions by lamplight that were not made by daylight. However, the color of the lamplight itself would have been a factor in such confusions.

A shadowy figure in the history of this phenomenon is Mathias Klotz (1748–1821), a Bavarian artist who wrote two books on color theory (Klotz, 1806, 1816). Brožek et al. (1991) discovered, to their surprise, a reference in the seventeenth edition of the *Brockhaus Enzyklopädie* (1972) to a description of the phenomenon by Klotz; the author of the entry was anonymous, but they did trace (with difficulty) the appropriate publication. Klotz commenced *Gründliche Farbenlehre* (A Comprehensive Theory of Colors) (1816) by regretting the fact that it followed so closely on Goethe's tome, and hence was less likely to make an impact. He was precise in this prediction, since the book has rarely been cited, despite insightful passages on brightness contrast (see section 8.2), as well as color changes under varying illumination. As was the case with Purkinje, the color change was noticed by chance; Klotz had been painting a military portrait and noted that some features of the uniform appeared to change color toward dusk. This led him to conduct more controlled observations with patches of colored paper on a neutral background: red appeared brighter than blue by daylight, with the reverse by twilight. Unlike Purkinje, Klotz did not base his conclusions on his own observations alone as they were confirmed by a friend.

Aristotle (ca. 330 B.C.): We never see a colour in absolute purity: it is always blent, if not with another colour, then with rays of light or with shadows, and so it assumes a new tint. That is why objects assume different tints when seen in light and sunshine, and according as the rays of light are strong or weak. (Ross, 1913, p. 793b)

Leonardo da Vinci (ca. 1500): Green and blue are invariably accentuated in the half-shadows, yellow and red and white in the light parts. (Minnaert, 1940, p. 111)

Klotz (1816): If one places two discs (each of about three inches diameter), one of blue and the other of red, in the middle of a sheet of grey paper, and compares them by daylight, then the red will appear much brighter than the blue. Thereafter, with approaching twilight, one observes the same discs every two minutes and one will note that the blue loses its colour earlier than the red; then—what is still more remarkable is that the blue appears to lose its darkness—whereas the red gets darker. (p. 19)

Purkinje (1825): The degree of objective illumination has a great influence on the intensity of colour quality. In order to prove this most vividly, take some colours before daybreak, when it begins slowly to get lighter. Initially one sees only black and grey. Then the brightest colours, red and green, appear darkest. Yellow cannot be distinguished from a rosy red. Blue looks to me the most noticeable. Nuances of red, which otherwise burn brightest in daylight, namely carmine, cinnabar and orange show themselves as darkest, in contrast to their average brightness. Green appears more bluish, and its yellow tint develops with the increasing daylight. (pp. 109–110)

4.4 Pattern Disappearances

The disappearance of peripheral patterns when maintaining fixation on a central one is often referred to as the Troxler effect, after Troxler's description at the beginning of the nineteenth century. He arranged a sequence of color patches on a wall so that they extended into the periphery of vision. When he fixated on one, after some time the most peripheral ones disappeared, followed by those closer to the fixation point. In all cases, the lost patterns were replaced by the blue wall, which was the background. A decade earlier, Erasmus Darwin observed a similar fading and then disappearance of centrally fixated targets. He made this discovery in the context of his experiments on afterimages, which demanded prolonged fixation on color patches in order to render them visible. The intensity of the color declined until the whole patch ceased to be visible. Brewster, on the other hand, maintained that fixated objects never disappear, whereas peripheral ones did, even when viewed binocularly. For Purkinje the patterns were replaced by cloudy streaks, rather like some of the effects seen with stroboscopic patterns (section 4.9). How he was able to conduct the experiment described is a puzzle, since the act of blowing the pieces of paper away would have resulted in some eye or head movement, and such movements are known to restore faded images, as was detailed in the case of seeing the retinal blood vessels (section 4.8).

Erasmus Darwin (1794): On looking long on an area of scarlet silk of about an inch in diameter laid on white paper ... the scarlet colour becomes fainter, till at length it entirely vanishes, though the eye is kept uniformly and steadily upon it. (p. 19)

Erasmus Darwin (1731–1802) after an engraving in *The European Magazine and London Review* 27:75, 1795.

Troxler (1804): I stuck a long row of all types of brightly coloured stripes and patches on a light blue surface, and always arranged them at a particular distance and direction from my eyes and the row disappeared to be replaced by the blue surface. (p. 7)

Brewster (1818): When the eye is steadily fixed upon any object, this object will never cease to become visible; but if the eye is steadily directed to another object in its vicinity while it sees the first object indirectly, the first object will, after a certain time, entirely disappear, whether it is seen with one or both eyes, whatever be its form or colour, or its position with respect to the axis of vision. When the object is such as to produce its accidental colour before it vanishes, the accidental colour disappears, also, along with the object. (p. 151)

Ignaz Paul Vitalis Troxler (1780–1866) after an illustration in Attinger, Godet, and Türler (1932).

Purkinje (1823a): I distributed several pieces of paper evenly on a dark ground, removed any lighting from the side of the eye, and fixated on the central piece.

After a short time the pieces of paper started to disappear. I waited until the effect reached its maximum, blew the papers away, and saw the well-known cloudy streaks appear in motion, wander around and disappear in their usual way. This is related to the disappearance and reappearance of letters seen when one becomes sleepy while reading, because that is the best time to experience the cloudy circles. (p. 78)

4.5 Pattern Distortions

There are some similarities between pattern disappearances and distortions: both require steady fixation, but the latter are more readily visible with regular geometrical patterns (see Wade, 1977b). Patterns of finely engraved parallel lines lose their straightness with prolonged viewing, as noted by both Purkinje and Brewster. It is not clear from Purkinje's description whether observation was monocular or binocular; in his case it might not have made too much difference because he had a marked difference in the acuity of his eyes, with the right being much stronger than the left. He attributed the apparent waviness of the straight lines to the overlap of the real image of the lines with a displaced afterimage, after the manner of moiré fringes. Brewster did state that he viewed the patterns with one eye, and he commented on the transfer of the unease experienced in the viewing eye to the closed one. He adopted a similar interpretation to Purkinje, attributing the distortions to interferences fringes on the retina. It was when observing such engravings that Brewster reported seeing subjective colors (section 3.9), although Purkinje's account gave no hint of seeing colors.

Concentric circles provide a different type of distortion: cloudy streaks appear to radiate from the center, and they can themselves rotate. Purkinje associated these effects with variations in accommodation, although he was not precise as to the mechanism of the relation between them. The distortions were radial with concentric circles and circular with a pattern of radiating lines; that is, they were perpendicular to the lines in the pattern. Such effects can be attributed to transient forms of astigmatism that are a consequence of instabilities in the ciliary muscles (Wade, 1977b). Although regular astigmatism had been described by Young (section 2.5), the transient form was not measured until a century and a half later; its intensity varies with accommodation. Accordingly, Purkinje was not far from the mark when linking the phenomenon to accommodation.

Purkinje (1823a): During intense viewing of the parallel lines of an engraving one observes an oscillation of the lines which on closer inspection involves some being closer together and others farther apart, so that the lines appear in the form of waves. (p. 122)

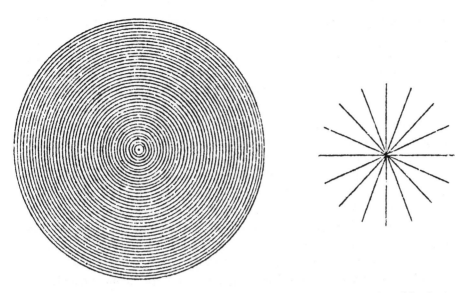

Concentric circles and radiating lines for producing pattern distortions with prolonged fixation.

Purkinje (1825): One draws a series of closely spaced concentric circles, constructed as neatly as possible. . . . These are displaced outside the distance of distinct vision, and there appear in all directions bands of clearly distinguishable parallel lines, over which the multitude of lines slide and entwine as cloudy streaks and points; they all radiate from the centre to the periphery, and their number, width and direction differ with different individuals, but remain constant for any one. . . . One can produce the same appearance very clearly with a figure comprised of 16 or more radii. (pp. 144–145)

Brewster (1832a): If the eye looks at any of these objects [parallel black lines drawn on white paper] steadily and continuously, the black lines soon lose their straightness and parallelism, and inclose luminous spaces somewhat like the links of a number of parallel chains. When this change takes place, the eye which sees it experiences a good deal of uneasiness, an effect which is communicated to the eye which is shut. (p. 170)

4.6   Pressure Figures (Phosphenes)

Grüsser and Hagner (1990) trace the history of theories of vision from the initial description of pressure figures by Alcmaeon; it introduced the concept of emission of light from the eye, and subsequent writers sought to reconcile this aspect of seeing with the phenomena of optics. That is, the emission theories could be taken as having some empirical support, and no competing theory could adequately account for the the light seen in darkness as a consequence of a blow to the eye or pressure applied to it. The phenomenon has been reported many times, with occasional novelties of observation or interpretation.

The first set of reports given here essentially redescribed the phenomenon and the conditions for its observation. A major refinement in its phenomenology was provided by Scheiner. The phosphene was located opposite the point of pressure, and it could take on a variety of configurations; it could also be seen by day as well as in darkness, although the figures so generated were different. Scheiner believed that pressure generated some internal light that was reflected from the lens back to the retina. Descartes, shortly afterward, provided a mechanistic interpretation of the phenomenon that was adopted by others like Hobbes: the force on the optic nerve results in exciting the fibers in a similar way to light. For Newton, the pressure figures were intensely colored, and they were transient, unless variations in the pressure exerted by the finger were introduced, an aspect confirmed by Young. Hartley was noted for applying Newtonian mechanistic principles to psychological phenomena, particularly memory; small vibrations (vibratiuncles) in the nerves of the brain could be associated with particular events, and could provide a mechanistic model of memory (see Wilkes and Wade, 1997). In the context of pressure figures, he echoed Newton's interpretation of them, as did Harris.

Descartes and Newton assigned the phenomenon to internal causes, but an empirical vindication of this was provided by Morgagni. He generated pressure figures in his own eye in darkness, and asked an assistant if any light was observed to issue from his eye. In the absence of such light, Morgagni assumed that the figures were subjective, and an index of healthy retinal function. His observations were confirmed and elaborated by Langguth, who examined the interocular transfer of the phosphenes, and by Müller. Morgagni also extended the detail of description of the phenomenon: the characteristics of the phosphenes were dependent upon the area and pressure applied, and paired phosphenes could be produced. Further refinement was supplied by Elliott, Goethe, and Purkinje. The checkered and streaky appearance of the pressure figures was commented on by Elliott, and Goethe drew attention to the light and dark circles that can be seen. It was, however, Purkinje who gave the most detailed account of the checkerboard figures, together with diagrams illustrating them.

Alcmaeon (ca. 500 B.C.): And the eye obviously has fire within, for when one is struck [this fire] flashes out. Vision is due to the gleaming,—that is to say, the transparent—character of that which [in the eye] reflects the object; and sight is the more perfect, the greater the purity of this substance. (Stratton, 1917, p. 89)

Lucretius (ca. 56 B.C.): . . . the pupil of the eye receives one kind of blow when it is said to perceive a white colour, and quite another when it perceives black. (1975, p. 159)

Averroes (ca. 1180): Furthermore, that for seeing it is necessary that light be transmitted to these coats [of the retina], can be proved by the fact that, when a person is struck on the eyelid, his eye suddenly becomes dim and the light which was in his eye is extinguished just as a candle is, and then he can see nothing. (1961, p. 9)

Averroes (1126–1198) after an illustration in Runes (1959).

Albertus Magnus (ca. 1250): A certain sensation appears in the eyes in darkness, either when they are compressed from outside or suddenly beaten. (Grüsser and Hagner, 1990, p. 65)

Scheiner (1619): The appearance [of the phosphene] is round. But if the pressure is stronger the perceived light extends and finally takes an elliptical form. One always perceives it on the opposite side [in the visual field] to where the eye is deformed. The spark is mainly to be seen in the darkness, independent of whether the eyelids are closed or open. But one perceives it even by day and with open eyes, especially near the angle of the eye. The whole appearance consists of a gleaming border, it is then in the middle dark and nearly black. (Grüsser and Hagner, 1990, p. 67)

Descartes (1637): ... it seems to those who receive some injury in the eye that they see an infinity of fireworks and lightning flashes before them, even though they shut their eyes or else are in a very dark place; so that this sensation can be attributed only to the force of the blow which moves the small fibres of the optic nerve, as a strong light would do. (1965, pp. 101–102)

Hobbes (1640): ... upon every *great agitation* or *concussion* of the *brain* (as it happeneth from a stroke, especially if the stroke be upon the eye) whereby the optic nerve suffereth any great violence, there *appeareth* before the *eyes* a certain light, which is *nothing without*, but an apparition only, all that is real being the concussion or motion of parts of that nerve. (1840, p. 5)

Thomas Hobbes (1588–1679) after an engraving in Molesworth (1839).

Boyle (1663): ... when a man receives a great stroke upon his eye, or a very great one upon some other part of his head, he is wont to see, as it were, flashes of lightening, and little vivid, but vanishing flames, though perhaps his eyes be shut. (p. 671)

Malebranche (1674): ...there is no necessity there should be Light in my Hand, when I see a Flash, upon giving my Eye a blow. (1700, p. 30)

Mariotte (1681): ... push a finger against a corner of your eyes during the night, and on the opposite side you will see appear a luminous circle. (Weale, 1957, p. 650)

Newton (1717): When a Man in the dark presses either corner of his Eye with his Finger, and turns his Eye away from his Finger, he will see a Circle of Colours like those in the Feather of a Peacock's Tail. If the Eye and the Finger remain quiet these Colours vanish in a second Minute of Time, but if the Finger be moved with a quavering motion they appear again. Do not these Colours arise from such Motions excited in the bottom of the Eye by the Pressure and Motion of the Finger, as at other times are excited there by Light for causing Vision? (p. 137)

Giovanni Battista Morgagni (1682–1771) after a frontispiece engraving in Morgagni (1761).

Morgagni (1719): As I observed repeatedly it is certain that no light appears when the cornea is pressed. When, however, the region near the cornea is deformed, light appears immediately in the shape of half an annulus. Deformation a little further away produces a light annulus. If during the same time two regions of the eyeball are deformed, two light figures appear immediately. When, however, the pressure is exerted not only with the finger tips but with the whole finger, the light takes on an elliptical form. Applying instead of the finger a much smaller round body, such as the head of a needle, the light ring is much smaller. It always appears on the side [of the visual field] opposite that indented. . . . even when he observed extremely carefully and very bright light appeared to me [Morgagni], he could never observe any light by himself. (Grüsser and Hagner, 1990, pp. 72–73)

Langguth (1742): A friend, who became curious about these phenomena . . . visited me in the dark room. I briefly explained to him what I was doing. The doors were closed and I asked him to observe my eyes very closely. While I was perceiving the small lights [i.e., the deformation phosphenes] he was not able to observe any small flashes or oscillating lights. Thereafter he performed the same experiment by himself according to the same rules, experimental procedure and design. He asked me to confirm whether light appears to him immediately by pressure on the eye. I, however, could not discover any light leaving his eyes. Later I performed this experiment repeatedly, always with the same result; in the middle of the night or when I was lying horizontally, early in the morning when I awoke from sleep before sunrise. Even when I covered one eye for the whole day and occluded [during the experiment] the other with the hand, the described pressure on the eye led to the same effect. Similarly, when I opened one eye, observed the normally illuminated objects, closed the other and pressed on it, the same sensations appeared, but somewhat weaker. (Grüsser and Hagner, 1990, p. 74)

Hartley (1749): Flashes of Light, and other luminous Appearances, are occasioned by Strokes upon the Eye, rubbing it, Faintings, etc. Now it is very easy to conceive, that violent Agitations in the small Particles of the Optic Nerve should arise from such Causes; and consequently that such Deceptions of Sight, as one may call them, should be produced, if we admit the Doctrine of Vibrations. (p. 198)

Harris (1775): There seems to be little room to doubt, but that like sensations do always excite like ideas, although the causes of these similar sensations may be different. The flashes of lightning sometimes perceivable from a sudden blow upon the eye, or other part of the head, are probably owing to the same kinds of motions being excited in the optic nerve by the blow, as by the impressions of real lightning.... And these phænomena do somewhat corroborate the hypothesis, of vision being caused by some motion excited in the optic nerve, by the impulses of light upon the bottom of the eye. (p. 103)

Elliott (1780): The luminous appearance which ariseth on pressing the centres of the eyes ... is chequered; that is, some parts of it are darker than others, and sometimes there appear spots, and streaks which are much brighter than the other parts of the lumination. (pp. 6–7)

Young (1793): It must be observed, that the sensation of light from pressure of the eye subsides almost instantly after the motion of the pressure has ceased, so that the cause of the irritation of the retina is a change, and not a difference, of form; and therefore the sensation of light appears to depend immediately on a minute motion of some part of the optic nerve. (p. 180)

Erasmus Darwin (1794): If any one in the dark presses the ball of his eye, by applying his finger to the external corner of it, a luminous appearance is observed; and by a smart stroke on the eye great flashes of fire are perceived ... So that when the arteries, that are near the auditory nerve, make stronger pulsations than usual, as in some fevers, an undulating sound is excited in the ears. Hence it is not the presence of the light and sound, but the motions of the organ, that are immediately necessary to constitute the perception or idea of light and sound. (p. 21)

Bell (1803): There are also corruscations seen before the eyes in consequence of a blow upon the eye ball, and accompanying violent headach, vertigo, phrenitis, epilepsy, &c. Whatever forces the blood with great violence to the head, as coughing, vomiting, sneezing, will cause, for the instant, such corruscations, by means of the disturbed circulation through the retina. (p. 295)

Goethe (1810): If the eye is pressed only in a slight degree from the inner corner, darker or lighter circles appear. At night, even without pressure, we can sometimes perceive a succession of such circles emerging from, or spreading over, each other. (1840, pp. 42–43)

Gall (1825): Hence it is, that the nerves of the senses are able to perform their special functions, in consequence of inward irritations only, and without the concurrence of the external world.... the flow of blood to the eye, makes us see sparks and brilliant objects.... a blow on the eye, and the contact of two different metals, one of which is applied upon the upper lip and the other placed under the tongue, occasion light. (1835c, p. 200)

Purkinje (1823a): If I apply gentle pressure with my fingertips to the cornea of my closed eye there first appears a broad luminous ring, which becomes increasingly visible, and is composed of small light and dark rectangles, (Fig. 5) which run obliquely from below left to above right. The outermost boundary of the ring approximates a rounded rhomb standing upright on a corner. The circular shaped gap in the centre initially appears as dark as the outermost

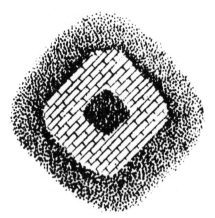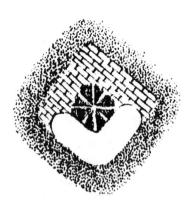

Purkinje's representation of pressure figures.

region. Gradually eight pale radiating lines become visible, (Fig. 6.) in which the rectangles in the area of the ring itself become even lighter, so that soon all the shaded parts disappear. (pp. 22–23)

Müller (1826a): In both eyes there are specific locations which, if pressed present one and the same pressure figure, which lies in the same position of the subjective visual field. (p. 73)

Müller (1838): The mechanical influence of a blow, concussion, or pressure excites, for example, in the eye the sensation of light and colours. It is well known that by exerting pressure upon the eye, when the eyelids are closed, we can give rise to the appearance of a luminous circle; by more gentle pressure the appearance of colours may be produced, and one colour may be made to change to another. Children, waking from sleep before daylight, frequently amuse themselves with these phenomena. The light thus produced has no existence external to the optic nerve, it is merely a sensation excited in it. However strongly we press upon the eye in the dark, so as to give rise to the appearance of luminous flashes, these flashes, being merely sensations, are incapable of illuminating objects. Of this any one may easily convince himself by experiment. I have in repeated trials never been able, by means of these luminous flashes in the eye, to recognise in the dark the nearest objects, or to see them better than before; nor could any other person, while I produced by pressure on my eye the appearance of brilliant flashes, perceive in it the slightest trace of real light. (1843, pp. 1061–1062)

4.7 Entoptic Phenomena

Entoptic phenomena are taken to refer to the visible expression of structures in the eye. Retinal blood vessels (section 4.8) should be included under this heading, but they have had a more specific observational history and are treated separately. This section is concerned with phenomena that have basked in an array of graphic appellations. They stem mostly from the Latin *muscae volitantes*, which has been translated as *mouches volantes*, *fliegende Mücken*, *flying gnats*, or, much more prosaically, *floaters*. All refer to seeing inclusions in the eye, which are most clearly visible against a bright, unpatterned background.

Muscae volitantes were clearly described by Benevenutus Grassus, who probably flourished in the twelfth century, and whose works were transcribed several times and eventually printed in 1474 (see Wood, 1929). He adopted a Galenic interpretation of the phenomenon, by attributing the obstruction to an interruption in the flow of visual spirits from the brain to the eye. He noted that the muscae were more prevalent in those of melancholic mind, and they increased with age. A "restorative electuary" was prescribed for banishing the visibility of the muscae and "reviving the spirit of vision." Platter's accidental observation of a spot before his eye was followed by a precise interpretation of its cause: debris from the ciliary processes floated freely in the aqueous humor, blocking the light entering the eye. Willis remarked on the phenomenon, which he had frequently encountered in his medical examinations, and assigned its cause to blockages in the fibers of the optic nerve. La Hire distinguished between two types of *mouches volantes*: one retained a fixed visual direction, and the other appeared to move about; they were more readily seen by presbyopes, and La Hire illustrated the form they could take, which resemble the retinal blood vessels. Taylor mentioned that the moving *mouches* only do so when the eyes move, a point made with greater clarity by Bell. Boerhaave, Robert Darwin, and Müller commented on a much more subtle entoptic feature—the circulation of the blood in the retinal blood vessels.

Wheatstone (C.W.) applied his optical knowledge to enhance the visibility of ocular inclusions. Under normal circumstances, the light entering the eye from many directions will prevent the visibility of any obstructions in the ocular media, whereas a narrow pencil of light will illuminate regions of the retina not obscured by floaters, and provide a potential means of locating them. In this way he was able to discern both moving and stationary *mouches* in his own eyes. The technique was so sensitive that the secretions from the tear ducts could be rendered visible.

Grassus (1474): When an excessive amount of black bile is carried to the brain, the latter is thereby disordered, the optic nerve is obstructed, the spirit of vision is unable to find its way to the eye, and this obstruction shows itself as flies flying through the air, and when the patient looks at a candle the light seems to him to be divided into four parts. The human face also looks abnormal. (Wood, 1929, p. 59)

Platter (1614): One morning, when I was in conversation with someone, suddenly something like a black, round spot the size of a lens appeared in front of my left eye. It appeared to dance around.... If I consider the reason for this disturbance I reach the opinion that some filament of a ciliary process had severed its connection with the uvea and now swims freely in the aqueous humor, wandering back and forth in front of the pupil, obscuring the view. (Koelbing, 1967, p. 115)

Willis (1664): As often as in the disempers of the Eyes blackish pricks or concaternated pieces of any thing seem to be rolled before the Eyes, it is likely that this apparition is so made, because certain filaments or small strings of the Optick Nerve are shut up, which when the light cannot pass through rightly, as through the rest, so many as it were shadowy spaces appear in the middle of the clearness. (1681, p. 139)

Thomas Willis (1621–1675) after an engraving in Birch (1743).

La Hire (1694): Presbyopes are subject to see spots and threads, like the mouches volantes that are always before the eyes, principally when looking at a white or bright object. These spots are not all of the same nature; there are those that I call permanent because they remain in the same position with respect to the visual axis.... others float about and continually change their location. (p. 260)

Boerhaave (1703): The tiny shapes which are imagined to float before our eye are treated by those people who have no knowledge of mechanics as being the beginnings of an impending cataract in the watery humour; and the delicate eye is often corroded by their sharp remedies and faulty art. Now that these

Hermann Boerhaave (1668–1738) after an engraving in Pettigrew (1840a).

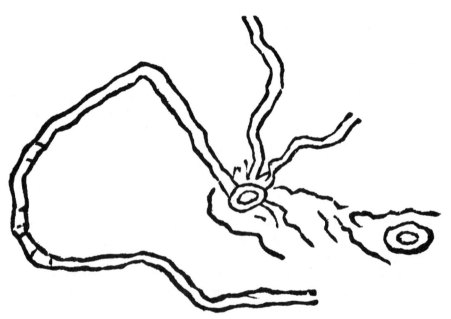

Mouches volantes as depicted by La Hire.

phenomena have been localized in the retina, and their causes are the arteries
... how changed is the form of the treatment. (1983, p. 111)

Taylor (1738): The mouches volantes are occasioned by malady of the organ
of vision itself, in which one believes one sees little foreign bodies; they have
some resemblance with the gnats or other objects which fly through the air; they
vary in their shape, number, opacity and location; they seem to be in motion
when the eyes move, and remain stationary when the axis of the eye retains a
fixed direction. (p. 98)

Smith (1738): People in growing old, are often troubled by the appearance of
dark, irregular spots continually flying before their eyes, like flies; especially in
looking at bright objects, such as white paper or the sky-light. (*Remarks*, p. 5)

Robert Darwin (1786): By being accustomed to observe such small sensations
in the eye, it is easy to see the circulation of the blood in this organ. I have
attended to this frequently, when I have observed my eyes more than commonly sensible to other spectra. The circulation may be seen either in both eyes
at a time, or only in one of them. (p. 344)

Bell (1803): Old people are often troubled with the appearance of dark irregular
spots flying before their eyes. In fever, also, it is very common to see the patient
picking the bed-clothes, or catching at the empty air. This proceeds from an
appearance of motes or flies passing before the eyes, and is occasioned by an
affection of the retina, producing in it a similar sensation to that produced by
the impression of images; and what is deficient in the sensation, the imagination
supplies; for, although the resemblance betwixt those diseased affections of
the retina and the idea conveyed to the brain may be very remote, yet, by that
slight resemblance, the idea, usually associated with the sensation, will be
excited in the mind.... when my eyes are fatigued, and, sitting in my room, I
look towards the window, I see before me small lucid circles which seem to
descend in quick succession; upon attending more particularly to my eyes, I find
them in perpetual motion; my eye is turned gradually downward, which gives
the spectrum the appearance of descending; but it regains its former elevation
with quick and imperceptible motion. During the slow inclination of the eye
downward, the motes or little rings seem to descend; but in lifting the eye again,
the motion is so quick, that they are not perceived. (pp. 292 and 294)

Purkinje (1823a): If during inflammation of the vascular system (either through
physical exertion or fever) one stares fixedly at a bright white surface (like a
uniformly covered sky or a snowfield) then many individual light spots suddenly appear in the visual field, which look like shooting stars in different
locations, and moving rapidly in curved and straight lines. (pp. 128–129)

C.W. (1830): In the ordinary circumstances of vision, particles floating in the
humours of the eye, or specks in the cornea and crystalline lens, are invisible,
because their shadows are projected by different rays of light on every part of

the retina, thus permitting no distinct image of them to be formed; but they may be rendered visible by allowing only a single ray of light to fall on the eye through a hole made in a card by a very fine needle, and placing the light and aperture so that the object within the eye may be in a right line with them and the centre of the retina: they may be projected on any part of the retina, but they will be most distinctly seen at this point of most perfect vision. I have thus observed, in my own eye, collections of transparent globules which, from their free motions, evidently exist in the humours; and one remarkable spot (in my left eye), which, from its permanence, must be either in the cornea or the lens; after winking, the secretion from the lachrymal ducts is also very obvious. (p. 112)

Müller (1838): There are many circumstances under which a general expression of the circulation is perceptible. It is seen, however, more particularly in looking at surfaces which are brightly illuminated, but not to a dazzling degree, as the sky; or after fixing the eyes for some time upon a surface of snow or white paper. The appearance consists in an indistinct confused movement, as of points crossing each other in all directions, or like the motion of vapours. The appearance is so undefined, that the direction of motion cannot be determined. It evidently is due to the motion of the blood. (1843, p. 1211)

4.8 Retinal Blood Vessels

Interest in the visibility of retinal blood vessels arose from conflicts concerning the blind spot (see section 2.8). Mariotte, in locating the blind spot, concluded that the choroid rather than the retina was the surface receptive to light. Pecquet disputed this conclusion, but marveled at the fact that such a compelling phenomenon as the blind spot had not previously been recorded. The dispute introduced another novel and compelling phenomenon, the previous ignorance of which is similarly remarkable. Pecquet argued that the blood vessels of the retina prevent light passing to both the retina and the choroid, and therefore should be visible. Moreover, the vessels themselves were larger at their trunk (the optic disc) than in the axis of vision. They were not considered to pose a problem for central vision because of their small size. That is, they were considered to be below the threshold for visual acuity. Mariotte, in defending himself against Pecquet, actually used the smallness of the central vessels as support for his position.

Bell realized that the visible ramifications were shadows on the retina cast by the vessels, and subsequent researches specified their characteristics in greater detail. Indeed, the pattern is often called the Purkinje tree after Purkinje's description and illustration of the ramifications. Wheatstone (C.W.), in his précis and translation of Purkinje's first book, devised an even better method for rendering the vessels visible (see Wade, 1983, for the full account of

Wheatstone's report). Directing a narrow, moving beam of light into the corner of the eye, he noted that the shadows moved in the opposite direction to the light source; if the motion stopped the ramifications fragmented and disappeared. Of greater significance was the crescent-shaped shadow cast by the foramen centrale (fovea). He demonstrated that this region did correspond to the center of the visual axis by generating a foveal afterimage and then locating it with respect to the crescent shadow. Thus he was using the afterimage as a stabilized retinal image in order to define the relative location of other retinal structures. He suggested a slightly different procedure for observing the finer vessels: a card with a large pinhole was moved by the side of the eye, so that diffuse light incident on its back surface could enter the eye. He could not see any vessels around the fovea, and his reference to the "differences of colour observed by anatomists" was an allusion to Soemmerring's yellow spot.

More detailed illustrations of the retinal vessels were provided by Horner, who was unable to observe the crescent shadow around the fovea, but did note a fine granulated texture in the central region. A sound summary was given by Müller, who also drew attention to individual differences in seeing them.

Pecquet (1668): These Vessels, which are no other but the ramifications of the Veins and Arteries, are derived from the Heart, and having no communication with the brain, they cannot carry thither the Species of the Object. If therefore the Visual rays, issuing from an Object fall on these Vessels at the place of their Trunk or main Body, 'tis certain that the Impression made thereby will produce no Vision, and that the picture of that Object will be deficient; as when on a white paper in an obscure Chamber, there is some black spot, or in it some hole considerably bigg: for the more sensible this blackness or hole is, the more of the image of the object it intercepts from our Eyes. It is not so in respect of the *small* ramifications, that issue from those trunks and shoot into the *Retina*. For if they be met with at the place of the bottom of the Eye, where Vision is made distinct, they will not render the image of the Object deficient, because they are so small, as not to be sensible. Thus it is, that in Looking-glasses, where they want lead or tin in any place big enough to be perceived, the image, we there see, appears to have a hole; which happens not, when there is but so small a one, as might be made by the point of a needle. (p. 670)

Mariotte (1683): ... these *Vessels* that are nearest the *Axis Opticus*, are no bigger than a *Hair*, or the 240th part of an Inch; and being in the surface of the *Retina*, are at some distance from the *Choroide*, so as to let *Rays* enough pass under, for the distinguishing of *Objects* not very small. The *Vessels* also that carry the blood are clear and *Pellucid*, causing *Refraction* that is helpful to *Vision*. (p. 266)

Bell (1803): There is a kind of umbræ seen before the eyes which are occasioned by the vessels of the retina ... the person sees umbrageous ramifications which strike across the sphere of vision, and are synchronous with the pulse,

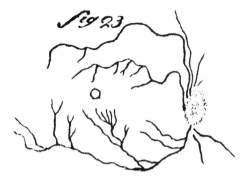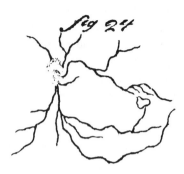

The Purkinje tree.

showing its dependance on the full and throbbing pulsation of the head.
(p. 295)

Purkinje (1823a): If I move a candle flame, held several inches away from my
right eye, in different directions and even in a circle, there appears, through
the diffusely illuminated light, a dark pattern of vessels which originates from
the optic nerve and has two principal branches towards the top and bottom;
they ramify and bend towards the centre of the visual field. (pp. 89–90)

C.W. (1830): This is an easy experiment to repeat, and is certainly a singularly
beautiful one; the blood-vessels of the retina, with all their ramifications, are
distinctly seen projected, as it were, on a plane without the eye, and greatly
magnified. I have found the experiment to work more perfectly when, the eye
being stedfastly directed forwards, the light is made to move right and left
below the eye, or upwards and downwards at the side of the eye; for when the
flame is in the field of view the image is indistinct: the eyelids of the unemployed
eye should not be closed, but the light should be obstructed by the hand or
any other covering. It is indispensable that the light be in motion, for directly
it becomes stationary the image breaks into fragments and disappears: during
the motion of the light the image also moves, and in a contrary direction to that
of the light. . . . One of the most remarkable circumstances of this phenomenon
is that the point corresponding to the foramen centrale, a crescent-image is
occasionally observed. . . . That the variable mark just mentioned is in the centre
of distinct vision I ascertained by the following experiment: I impressed on my
eye the spectrum of a coloured wafer, by looking intently at a black dot at its
centre; on causing then the vascular image to appear, I saw the centre of the
spectrum coincide with that of the mark. . . . The mark in the middle of the
field of vision is most probably a shadow, occasioned by a slight convexity or
concavity of the retina at that point. . . . The absence of vessels at the centre
of the retina will probably account for the greater distinctness with which
small objects are there seen, and also for the difference of colour observed by
anatomists in that part of the nervous expansion. (pp. 111–112)

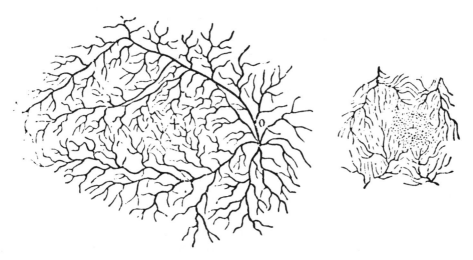

Ramifications of the retinal blood vessels illustrated by Horner.

Horner (1834b): In all experiments upon one eye, the comfort of the other contributes materially to success. A case should be bound over it, so as completely to darken it, without touching the eyelids. It has, by all observers, been experienced, that a distinct view is not to be maintained, unless the light is kept in motion. The lens or the cardboard must be moved slowly backwards and forwards edgewise, so that the light may traverse the interval between the cornea and the angle or lid of the eye. The drawing (fig. 1) exhibits, with as much accuracy as my slender graphic skill admits, the result of numerous and varied observations of the vessels of my *right eye*. The cross (+) indicates the *centre* of the field of view, or the point of direct vision. Beneath 0 is the *origin* of the larger vessels.... To exhibit the more minute vessels, which either from my perception improving from habit, or possibly from continued excitement, appeared much more numerous in my later than my earlier trials, fig. 2 is an enlarged figure of the more central vessels, and of the peculiar appearance of the central portion of the ground of the picture. At the centre (+) of the ground of the picture, which "corresponds to the projection of the *foramen centrale*," C.W. observed a crescent-like appearance, indicating in his opinion "a slight convexity or concavity in the retina at that point." In my own eye, whether the right or the left, no trace of such a crescent is found, but the appearance of a granulated texture in the level surface, like a number of exceedingly minute polished spherules collected within an obscurely defined circular space, as represented in fig. 2. (pp. 263–264)

Müller (1838): By the movement of the candle to and fro, the light is made to act on the whole extent of the retina, and all parts of the membrane which are not covered by the vasa centralia are feebly illuminated; those parts, on the contrary, which are covered by those vessels cannot be acted on by the light, and are perceived, therefore, as dark arborescent figures. In most persons this experiment succeeds readily; but in some individuals the phenomenon is produced with difficulty, or not at all. The figures of the vessels appear to lie before the eyes, and to be suspended in the field of vision. We have here a distinct demonstration of the axiom, that in vision we perceive merely certain states of the retina, and that the retina is itself the field of vision. (1843, p. 1163)

4.9   Stroboscopic Patterns

What have become called stroboscopic patterns (Smythies, 1957) were described before the stroboscope was invented. When the eye is stimulated by an unpatterned, flickering light, patterns of bewildering complexity become visible. They are called stroboscopic patterns after the modern form of the instrument, which can deliver pulses of light at very high frequencies. The original stroboscopic disc, devised by Stampfer in 1833, produced intermittent stimulation by means of a sequence of slits passing in front of the eye. It was used to study visual persistence and apparent motion (see sections 4.11 and 5.2), and Purkinje made a variant of one in 1840; he called it the phorolyt, and it was sold commercially as a magic disc (Matousek, 1961). Purkinje produced flicker by waving his fingers in front of one eye, while looking at the bright sky, and reported seeing checkerboards, zigzags, spirals, and ray patterns. Brewster produced the intermittency another way: he walked alongside evenly spaced, vertical railings and directed his closed eyes toward the sun which was behind them. The patterns were similar to some of those described by Purkinje, with the addition of color. In true Scottish style, Brewster likened the colored checkerboard patterns to "the brightest tartan."

Purkinje (1823a): I stand in bright sunshine with closed eyes and face the sun. Then I move the outstretched, somewhat separated, fingers up and down in front of the eyes, so that they are alternately shaded and illuminated. In addition to the uniform yellow-red that one expects with closed eyes, there appear beautiful regular figures which are initially difficult to define but become clearer. With continuation of the finger movement the figure changes from simple to complex and fills the whole visual field. (pp. 11–12)

Brewster (1834): If, when we are walking beside a high iron railing, we direct the closed eye to the sun so that his light shall be successively interrupted by the iron rails, a structure resembling a kaleidoscopic pattern, having the *foramen centrale* in its centre, will be rudely seen. The pattern is not formed in distinct lines, but by patches of reddish light of different degrees of intensity. When the sun's rays are powerful, and when their successive action has been kept up for a short time, the whole field of vision is filled with a brilliant pattern, as if it consisted of the brightest tartan, composed of red and green squares of dazzling

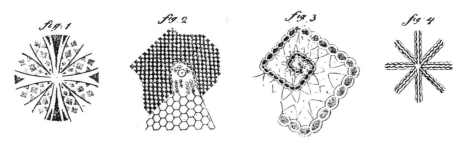

Purkinje's stroboscopic patterns.

brightness.... The brilliancy of the spectrum thus produced, and the beauty of its colours, exceed any optical phænomenon which I have witnessed, and so dazzling is its effect that the eye is soon obliged to withdraw itself from its overpowering influence. (pp. 241–242)

4.10   Talbot-Plateau Law

Intermittent stimulation could be controlled by means of the stroboscopic disc or a white disc with a black sector, and its systematic study provided the basis for determining one of the first lawful relationships in visual science—the Talbot-Plateau law. Talbot is best known for his invention of the negative-positive process in photography, in the late 1830s, but he had developed an interest in optics before that time. The persisting visibility of a moving light stimulated his concern with the apparent brightness of intermittent stimuli: if the repetition was rapid, then the apparent brightness was reduced by the ratio of the period light and dark to the period of dark. Thus, the baseline was taken to be the apparent brightness of the source when continuously presented, and this was diminished by intermittent periods of darkness. Plateau had been conducting experiments in this area for longer than Talbot, and was a little aggrieved that Talbot had established priority in stating the relationship. Plateau's experiments had been more extensive, and the ratio that he proposed was phrased in a slightly different way: apparent brightness was a function of the period of light and dark divided by the period of light. The two formulations are essentially equivalent, and both Talbot and Plateau restricted their relationship to situations in which the light was seen as continuous; that is, for repetitions below the threshold for visual persistence.

Talbot (1834): ... a regularly intermittent luminary whose observations are too frequent and too transitory for the eye to perceive, loses so much of its apparent brightness from this cause, as is indicated by the proportion between the whole time of observation, and the time during which it disappears. This seems a very simple law, and therefore likely to be true; but such an experiment

William Henry Fox
Talbot (1800–1877)
after a photographic
reproduction in *100
Years of Photography
1839–1939*. London:
The Science Museum,
1939.

Joseph Antoine
Ferdinand Plateau
(1801–1883) after an
illustration in Nuttin
(1961).

[using a rotating glowing coal] is much too inaccurate to be relied on to establish it. But as it suggests a very important idea, namely, that time may be employed to measure the intensity of light, it becomes desirable to establish it on solid grounds. And this is easily accomplished; we have only to take a white circle, with one of its sectors painted black, and make it revolve rapidly. It will appear, as every one knows, of a uniform gray tint, without any variation from the centre to the circumference; and yet, if we look first at a point near its centre, and afterwards at a point near its circumference, a much greater portion of the white surface will pass beneath the eye in the latter case than in the former. But a greater portion of the black surface likewise passes, and causes a compensation. In every point of the circle the white and black parts meet the eye during the same proportion of time, and therefore the tint is uniform throughout. (pp. 328–329)

Plateau (1835): "M. Talbot has recently informed us of a principle of photometry that is very simple and for which there are a number of useful applications. But this principle, even if true, has not been demonstrated conclusively to my mind. I propose ... to report the results of experiments that I have conducted, without knowledge of the work of M. Talbot, which establishes the same principle in a direct manner. This principle can be described in the following way: *When a luminous object excites the eye in a regular and intermittent manner, and its successiveness is not distinguishable because of their close temporal proximity, resulting in it being perceived as continuous, then the apparent brightness of this object is found to be diminished in the ratio of the sum of the durations of each period of light and dark, to the duration of a single period of light.*" (pp. 52–53)

4.11   Visual Persistence

Under normal circumstances the moving luminous objects we observe can be resolved during their motion, either because they move slowly or because they can be pursued by the eyes. Certain naturally occurring events (like comets and lightning) do not satisfy these conditions, and they were commented on in antiquity. Aristotle discussed persisting images in general, and he likened the effects to those of a projectile moving through space. However, the examples that he gave were concerned with afterimages and aftereffects (see sections 4.1 and 5.4) rather than to the visually persisting effects of observing projectiles. Seneca's account of shooting stars is remarkable because it attributed the visible trail to the "slowness of vision" rather than to any attribute of the object itself. Ptolemy had used the same principle implicitly in his color top (section 3.2), and he made explicit reference to the visibility of rapidly rotating discs and wheels (section 5.1); there is also brief mention of a rapidly moving flame. Ibn al-Haytham also referred to a rotating top and he developed a variation on

Ptolemy's flame theme that was replayed constantly throughout the following centuries: rapid motion of a burning stick results in the visibility of the path through which it passes. Leonardo drew circles of flame in this way; he also added the subtle variation of moving the eyes with respect to a fixed flame, with a similar visual effect. Newton appreciated that the phenomenon could be employed to measure the duration of the persisting images. In the second edition of his *Opticks*, he suggested that its value was less than a second, but precise measurements were made by Chevalier D'Arcy. He constructed a machine with rotating arms onto which a glowing coal could be attached. The device must have been very large, as it was viewed in a large, dark room at a distance of twenty-eight toises—an ancient French unit of distance. A toise was divided into six feet, which in turn consisted of twelve inches; the Paris line was one twelfth of an inch, and it was contrasted with the English line, which was one tenth of an English inch. Accordingly, a toise was almost two meters, and the viewing distance of D'Arcy's rotating cross was about fifty-five meters. By measuring the velocity required to complete a visible circle of light (in an otherwise dark room), he calculated the duration of visual persistence to be "8 tierces," a tierce or a third is a unit of time we no longer use. As a second is a sixtieth of a minute, a third is a sixtieth of a second, and so D'Arcy's estimate of visual persistence was 8/60 second, or about 130 milliseconds—a value that corresponds closely to those subsequently obtained with more modern devices.

Another aspect of the phenomenon was commented on by Locke: why are we not aware of the interruptions to vision caused by blinking? He related this to the continued visibility of rotating objects, and provided an interpretation in terms of his associationist theory of mind. The absence of any visible effects of blinking was given a more physiological account by Erasmus Darwin. His interests in the phenomenon was doubtless triggered by his son's studies in this area. Robert Darwin elaborated on an aspect of visual persistence described by Leonardo: brief exposure of parts of a single figure (if the intervals between them are sufficiently short) results in the visibility of the whole figure. Darwin's observations were more acute because he noted that the brightness of the scene seen through a rotating sail was less than that without the sail. Wheatstone made a more elaborate device with rotating sectors for viewing pictures, the parts of which were successively exposed. Wheatstone's concerns with visual persistence were stimulated by his researches in acoustics, and particularly by Young's description of vibrating piano strings. The figure illustrates the device Wheatstone called the kaleidophone—after Brewster's kaleidoscope; whereas the latter created beautiful forms based on visual components, the former did so on the basis of acoustic components. Metal rods of different cross-section and shape could be set in vibration and their extremities passed through the paths depicted. While the kaleidoscope took the popular imagination by storm (see Gordon, 1869), it remained an instrument of amusement rather than science. At the beginning of the nineteenth century a bewildering variety of such philosophical toys was produced, and visual persistence was a prime phenomenon incorporated in them. There were three ways in which this was achieved. First, by rendering visible changing patterns of stimulation during continuous viewing, as in the kaleidophone. Second, by brief presentation of parts of a single figure, as in Wheatstone's rotating sectors. Third, by successive stimulation of slightly different figures. It is the last that had the greatest impact on visual science. In the second case, an aperture moved in front of a scene or picture. A variant of this was the thaumatrope, or wonder turner, a very simple arrangement invented by Paris: a circular piece of card had different drawings on each side, and its ends were connected by string. When it was whirled both designs were seen superimposed: rats could be seen caged during rotation but free when the thaumatrope was stationary; fragments of words written on each side of the disc could be rendered complete due to visibility of their persisting parts. Visual and verbal puns were intentionally incorporated into the two elements.

The third mode of exploiting visual persistence was developed by Plateau in 1829; he investigated the effects of moving a pattern behind a moving or a fixed aperture (see Plateau, 1836, 1878c), and he gave it the name anorthoscope. The aperture was typically a slit, and the shapes were rotated on a disc behind it. Distorted shapes appeared to become regular during such movement. The anorthoscope was a variant of Plateau's phenakistoscope, a diagram of which can be seen on page 208. The instrument was elaborated by Zöllner (1862), who established the importance of pursuit eye movements in the perceptual distortions. Plateau's contrivance was matched by an almost identical one made by Stampfer (see section 5.2) called a stroboscopic disc. As noted above, the term "stroboscope" has become associated with the brief illumination of rapidly moving objects in order to freeze their motion. Wheatstone appreciated this principle of momentary stimulation and described it in 1834. It became widely adopted in visual science as a means of compensating for eye movements: stimuli were illuminated by a spark in order to determine whether particular phenomena (like spatial illusions) remained visible.

Seneca (ca. 6–65) after an engraving in *The Historic Gallery of Portraits and Paintings*, Vol. 5. London: Vernor, Hood, and Sharpe, 1809.

Aristotle (ca. 330 B.C.): The objects of sense-perception corresponding to each sensory organ produce sense-perception in us, and the affection due to their operation is present in the organs of sense not only when the perceptions are actualized, but even when they have departed. (Ross, 1931, p. 459a)

Seneca (ca. 63): These so-called stars leap out and fly across and on account of their great speed seem to trail a long flame. Our sight does not discern their passing but believes the entire path is on fire wherever they fly. The speed of their transit is so great that its stages are not observable. Only the movement as a whole is grasped. We understand more where the star has gone than where it is. So, it marks the entire route as though by a continuous flame because the slowness of our vision does not follow the successive instants of its flight but sees at the same instant where it started and where it ended. The same thing happens in lightning. (1971, pp. 75–77)

Ptolemy (ca. 150): This phenomemon also occurs in shooting stars whose light seems elongated because of the speed of their course, according to the extent of perceived elongation of distance and the perceived phenomemon that strikes vision.... Thus, objects which move slowly appear to move rapidly when they disappear quickly from view, like a fire which moves during an instant. (Lejeune, 1956, pp. 61 and 77)

Ibn al-Haytham (ca. 1040): Again, if the motion is circular and extremely rapid, like the [rotary] motion of a top, sight will not perceive it, but will perceive the moving top, if the motion is very rapid, as if it were motionless.... For let someone take a stick aflame at one of its ends, and let him quickly move it right and left in a dark night: looking at such a flame will find it extended through the interval along which it moves, which interval will be many times larger than the flame's magnitude. For sight can perceive an object's size or position or motion only after a measurable interval of time. (Sabra, 1989, pp. 254 and 348)

Leonardo da Vinci (ca. 1500): For if when the eye is fixed you draw a brand of fire in a circle or from below the eye upward this brand will seem to be a line of fire which rises upwards from below, and yet this brand cannot actually be in more than one part of this line at one time. And in the same way if this brand remain fixed and the eye move downward from above it will appear to this eye that the brand is rising from below in a continuous line.... the movement of certain instruments worked by women, made for convenience of gathering their threads together ... For these in their revolving movement are so swift that through being perforated they do not obstruct to the eye anything behind them. (MacCurdy, 1938a, pp. 267 and 272)

Castelli (1639): This image [of a rapidly moving hand] does not vanish immediately but persists for such a time that the hand goes from right to left and back to right before the first and the intermediate images have vanished. Thus it is that the object appears to us to be be spread out. (Ariotti, 1973, p. 9)

Locke (1690): How frequently do we, in a day, cover our Eyes with our Eyelids, without perceiving that we are at all in the dark?.... For any thing, that moves round in a Circle, in less time than our *Ideas* are want to succeed one another in our Minds, is not perceived to move; but seems to be a perfect entire Circle of that Matter, or Colour, and not a part of a Circle in Motion. (pp. 63 and 86)

Newton (1717): ... when a Coal of Fire moved nimbly in the circumference of a Circle, makes the whole circumference appear like a Circle of Fire: Is it not because Motions excited in the bottom of the Eye by the Rays of Light are of a lasting nature, and continue till the Coal of Fire in going round returns to its former place? (p. 322)

Le Cat (1744): These Phænomena depend on the Durations of the Sensations, which an Object excites in the Nerves, and on the Quickness with which the Action is repeated.... Now if the Action of an Object is renewed on a nervous

Papilla, before its former Impression is extinguished, the Impressions will continue, as if the Object had not ceased to act. This is the Case in regard of the fiery Circles produced by the frequent and rapid Whirling of a burning Coal thro' the same Track. (1750, pp. 274–275)

Melvill (1756): If a white rod be moved rapidly backwards and forwards with an angular motion, the whole circular space which it runs over will appear whitish; but not equally so, being faintest and most dilute in the middle, and brighter towards the two sides, which seem to be distinctly terminated by two white rods intersecting each other in the center of rotation. (p. 69)

Chevalier Patrice D'Arcy (1725–1779) after a painting in the Académie des Sciences, Paris.

D'Arcy (1765): For this purpose he contrived a machine, which consisted of a cross, turning horizontally upon its center, by means of a wheel and a weight, the velocity of which he could vary at pleasure, and ascertain to the utmost exactness. To view this machine, he placed an observer at 28 toises [about 55 m] from it, in a room in which no light was admitted but what came from the object (which was a live coal fastened upon the cross) and the experiments were made in the night time.... He found that the coal seemed to make an uninterrupted circle, when it revolved in 8 thirds of a minute [8/60 second]; and that no velocity less than this would answer the purpose. It made no difference at whatever distance from the center of the cross the coal was placed, or whether the machine was viewed through a telescope, or in any other manner he could think of applying. The result was also the same, at whatever distance the machine was viewed. (Priestley, 1772, pp. 634–635)

Robert Darwin (1786): When a fiery meteor shoots across the night, it appears to leave a long lucid train behind it, part of which, and perhaps sometimes the whole, is owing to the continuance of the action of the retina after having been thus vividly excited. This is beautifully illustrated by the following experiment: fix a paper sail, three or four inches in diameter, and made like that of a smoke jack, in a tube of pasteboard; on looking through the tube at a distant prospect, some disjoined parts of it will be seen through the narrow intervals between the sails; but as the fly begins to revolve, these intervals appear larger; and when it revolves quicker, the whole prospect is seen quite as distinct as if nothing intervened, though less luminous. (p. 324)

Erasmus Darwin (1794): So we many times in an hour cover our eye-balls with our eye-lids without perceiving that we are in the dark; hence the perception or idea of light is not changed for that of darkness in so small a time as the twinkling of an eye; so that in this case the muscular motion of the eye-lid is performed quicker than the perception of light can be changed for that of darkness. (p. 24)

Young (1800): Take one of the lowest strings of a square piano forte, round which a fine silvered wire is wound in a spiral form; contract the light of a window, so that, when the eye is placed in the proper position, the image of the

The curved appearance of the spokes of a wheel moving behind vertical railings (Roget, 1825).

light may appear small, bright, and well defined, on each of the convolutions of the wire. Let the chord be now made to vibrate, and the luminous point will delineate its path like a burning coal whirled round, and will present to the eye a line of light, which, by the assistance of a microscope, may be accurately observed. (p. 135)

Roget (1825): A curious optical deception takes place when a carriage wheel, rolling along the ground, is viewed through the intervals of a series of vertical bars, such as those of a palisade, or of a Venetian window-blind. Under these circumstances the spokes of the wheel, instead of appearing straight, as they would normally do if no bars intervened, seem to have a considerable degree of curvature.... If the impressions made by these limited portions of the several spokes follow one another with sufficient rapidity, they will, as in the case of the luminous circle already alluded to, leave in the eye the trace of a continuous curve line; and the spokes will appear to be curved, instead of straight. (pp. 131 and 136–137)

Peter Mark Roget (1779–1869) after an engraving in Pettigrew (1840b).

Paris (1827): "*A Wonder-turner*," or a toy which performs wonders by turning round ... This philosophical toy is founded upon the well-known optical principle, that an impression, made on the retina of the eye, lasts for a short interval, after which the object which produced it has been withdrawn. During the rapid whirling of the card, the figures on each of its sides are presented in such quick transition that they both appear at the same instant, and thus occasion a very striking and magical effect. (p. 6)

John Ayrton Paris (1785–1856).

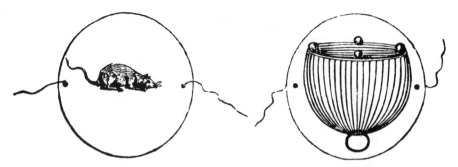

Two sides of the thaumatrope; when the strings make the disc whirl round, the rat will be seen in the cage (Paris, 1827).

Wheatstone (1827): These experiments principally consist in subjecting to ocular demonstration the orbits or paths described by the points of greatest excursion in vibrating rods, which in the most frequent cases, those of the combinations of different modes of vibration, assume the most diversified and elegant curvilinear forms. The entire track of each orbit is rendered simultaneously visible by causing it to be delineated by a brilliantly luminous point, and the figure being completed in less time than the duration of the visual impression, the whole orbit appears as a continuous line of light.... At the back of a wooden frame, about six inches in height and breadth, and from one to three inches more in depth, a circular plate of glass is placed, upon which a design is painted with transparent colours; at the front, is placed, parallel to the glass, a circle of tin, covered on its exterior surface with white paper, and having the space between two adjacent radii cut out. This circle moves freely on its centre round an axis, supported by a bar in front, and is put into rapid and regular motion.... If a light be placed behind the transparent painting, and still better if it be concentrated by a lens, on making the circle revolve with rapidity, the whole of the picture will be rendered visible at one view, although but very limited portions are successively presented to the eye. (pp. 344 and 351)

Talbot (1834): I will select the experiment known to every one of whirling round a glowing coal, which from its rapid motion conveys to the eye the impression of a continuous fiery circle. The question deserves consideration whether the eye receives from this circular ring exactly the same quantity of light which it received from the much smaller surface of the coal at rest. There can be no doubt, as it seems, that such must be the case; for if the luminous circle sent more rays to the eye, it would likewise send more in every other direction, and thus the apartment would become more luminous than before, which is not the fact. If, then, the total quantity of light remains the same, it follows that its apparent intensity must have diminished exactly in the same proportion as its apparent area has been enlarged. For the sake of more accuracy if we confine our reasoning to a very small portion of the luminous body, the enlarged area

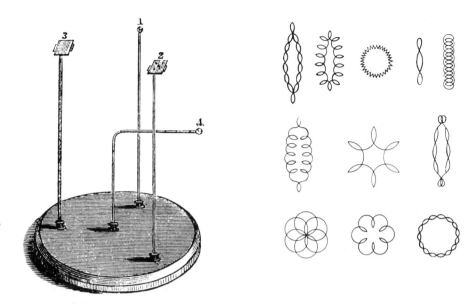

Wheatstone's kaleidophone on the left and the patterns that can be produced with it on the right.

which it seems to occupy is evidently proportional to the circumference of the circle it describes; therefore the preceding argument alleges that the intensity of light diminishes as the circumference increases. But in the same proportion in which the circumference increases, the time during which the coal is found in any particular point of the circle diminishes. The rapidity of rotation does not affect the argument. For instance, if the rotation is in a vertical plane, the time during which the coal occupies the summit of the circle, compared with the time of one whole revolution, necessarily diminishes when the circle grows larger; for the time of its being so situated bears the same proportion to the whole time, that the diameter of the moving body (which is constant) bears to the whole circumference through which it moves. Since, then, these two things— the intensity of light and the time of the body's remaining in any given part of the circle—are each inversely proportional to the circumference of the circle it describes, it follows that they must be directly proportional to each other. (p. 328)

Weber (1834): Embers burning in the dark in a brisk circular motion arouse an image of a present circle of fire. This is because the rays of light fall on adjacent parts of the eye, and the impressions made on each point do not fade immediately. Perhaps the perception of a lasting effect of light is not distinguished from the perception of light itself. Rather, this perception is unitary and continuous. (Ross and Murray, 1978, p. 103)

Wheatstone (1834): The instantaneousness of the light of electricity of high tension … affords the means of observing rapidly changing phenomena during a single instant of their continued action, and of making a variety of experiments relating to the motions of bodies when their successive positions follow each other too quickly to be seen under ordinary circumstances. A few obvious instances will at present suffice. A rapidly moving wheel, or a revolving disc on which any object is painted, seems perfectly stationary when illuminated by the explosion of a charged jar. Insects on the wing appear, by the same means, fixed in the air. Vibrating strings are seen at rest in their deflected positions. A rapid succession of drops of water, appearing to the eye a continuous stream, is seen to be what it really is, not what it ordinarily appears to be, &c. (p. 591)

Plateau (1836): I have the honour to present to the Academy an example of an instrument of my invention which is now published with the name *anorthoscope*. It is a new version of anamorphosis, the idea for which I initially gave in my university dissertation, printed in Liége in 1829. The anorthoscope is essentially composed of: first, a series of transparent discs, on which are represented distorted figures; secondly, one black cardboard disc pierced with many slits; thirdly, an instrument made with a large pulley with two grooves which move two small pulleys of different diameter, and arranged on a common axis. When one wants to use the apparatus, one attaches the black disc on the little pulley which is placed on the front of the instrument, namely on the side of the handle, and one similarly attaches to the other little pulley one of the transparent discs. Then the latter disc is very strongly illuminated from the back; one must be positioned in front of the instrument at a distance of a few feet, fixating at the height of the little pulley; the handle is turned by another person. When the transparent disc, even when really rotating at a great speed, seems to lose its movement, and the distorted figures are changed into perfectly regular designs. (p. 7)

5 Motion

Ibn al-Haytham (ca. 965–1039) after an illustration in Duke-Elder and Abrams (1970).

Ibn al-Haytham (ca. 1040): *Sight perceives motion in one of three ways: by comparing the moving object with other objects, or with a single object, or with the eye itself.... Now sight perceives the motion of an object only by perceiving the object in two different places or positions. And an object's position can vary only in time.... But a certain duration must exist between any two different moments, and therefore sight can perceive motion only in time. (Sabra, 1989, pp. 193 and 195)*

Visual persistence is a clear instance of the temporal dimension in vision, and it was increasingly linked with motion perception in the nineteenth century. Classical categories of vision, however, rarely included motion, although motion phenomena were frequently recounted. Lucretius was aware of the threshold for motion perception, and he described apparent motion, induced motion, the motion aftereffect, and vertigo. Ptolemy did include a heading for the perception of movement in Book II of his *Optics*, and they were interpreted in terms of a Euclidean emission theory and the visual pyramid. However, by assuming that the visual rays required time to travel to objects, he was able to address issues of motion thresholds. He also integrated motion perception with the distance that objects were away from the observer, and examined instances in which the eyes themselves moved through space, either as a consequence of forward head movements or lateral eye movements.

The perception of motion was also one of the visible properties listed by Ibn al-Haytham for consideration in his *Optics*, and his analysis has a much more modern flavor to it. He outlined three general ways in which motion can be detected—with reference to one or more visible objects or to the observer—all of which required a certain (but unspecified) interval for their detection. The inclusion of the eye itself as a reference was an acknowledgment that eye movements were implicated in motion perception. However, when errors of sight were described, Ibn al-Haytham included the perception of duration, but not that of motion.

General statements about motion perception tended to amplify the distinctions made by Ptolemy and Ibn al-Haytham, whereas particular phenomena were examined in specific detail. Surprisingly, although visual persistence had been commented on extensively for centuries, its connection with apparent motion is much more recent, and it was dependent on the invention of contrivances for its visibility. A sequence of discrete images could not be observed in the natural environment, unlike induced motion and vertigo, which have been consistently discussed since their description in Greek science. The motion aftereffect provides a puzzle in this regard. It can be seen in the natural environment and was reported by Aristotle, and yet it did not become a part of the Aristotelian corpus that was repeated and refined over almost two millenia. Significantly, it was not described by either Ptolemy or Ibn al-Haytham. It resurfaced in the nineteenth century, and then had a popularity that almost rivaled that of apparent motion. Vertigo has an eye movement aspect, but this was disclosed at a relatively late period in its study. The gross characteristics of eye movements, on the other hand, were a constant source of survey. While rotations about a vertical and horizontal axis were easily observed, the existence of torsional movements, which are much smaller, was often a matter of dispute.

5.1   Motion Perception

Moving objects often elicit eye movements in their pursuit. Euclid discussed this in the context of an array of objects moving at different speeds, with the eye following one of them. As would be expected from his geometrical theory of vision, the motions perceived are a consequence of the optical projections: the fixated object remains apparently stationary while the others are seen moving in the direction relative to this. Pursuit eye movements were returned to by many later writers, and were integrated with distinct vision (see section 5.6).

Many early accounts of motion perception addressed the matter of thresholds for detecting it. Motion involves change of position over time, and those objects that clearly changed their relative position (like heavenly bodies), but were not seen as moving, were thought to move very slowly with respect to the observer. That is, the motion occurred but was not detected. Both Lucretius and Ptolemy made this inference, and the latter concluded that it was a consequence of the great distances between the stars and the eye. Ptolemy added that some objects move too quickly to be perceived, like the spokes of rapidly rotating wheels. According to the emission theory he embraced, the rays emanating from the eye remained for long enough in the same location with respect to a rotating spoke for it to be replaced by its neighbor. Ptolemy introduced the concept of threshold (minimum perceptible) to account for this. Ibn al-Haytham gave a generalization of this effect on the basis of Ptolemy, as did Pecham on the authority of Ibn al-Haytham. Leonardo provided a geometrical analysis of objects moving at different speeds and at different distances from the eye; he appreciated that an infinite variety of these two factors could yield equivalent changes of visual angle over time, and these would be seen as equal.

With the advent of mechanistic interpretations of the physical world, matter in motion played a crucial conceptual role. However, relatively little attention was paid to motion perception. When it was considered by Rohault, Malebranche, and Locke, the analyses added little that was novel. Malebranche did appreciate the ambiguity of motion projections on the retina and he gave a general statement of what has become called size-distance invariance (see section 8.1). Thus, he elaborated upon Ptolemy's concern with the link between motion and distance. From the eighteenth century, attention focused on the distinction between retinal image motions caused by object and eye movements, and by the relativities of object motions. John Herschel did describe the characteristics of motion parallax, although this had been discussed earlier in the context of distance perception (see section 7.11).

Hippocrates (ca. 400 B.C.): For the senses of the soul that act through sight or hearing are quick; while those that act through touch are slower. (1923b, pp. 285–287)

Euclid (ca. 300 B.C.): *If, when several objects move at unequal speed, the eye also moves in the same direction, some objects, moving with the same speed as the eye, will seem to stand still, others, moving more slowly, will seem to move in the opposite direction, and others, moving more quickly, will seem to move ahead.* (Burton, 1945, p. 371)

Lucretius (ca. 56 B.C.): The stars all seem to be fixed and stationary in the vaults of ether, yet all are in constant motion, since they rise and return to their far distant settings when they have traversed the sky with bright body. And the sun and moon in like manner appear to remain in their places, while experience proves that they move along. (1975, p. 307)

Ptolemy (ca. 150): If the amplitude of motion is moderate and if the time during which it occurs is not perceptible, as is often produced with circular motion (for example, with a rotating disc and the wheels of an horse-drawn chariot, when they turn very rapidly), then no motion is perceived because they retain their position in a shorter time than the minimum perceptible: the visual ray stays fixed on the location of the object which intercepts it, as happens with things that do not move. If, on the other hand, an imperceptible motion is produced in a perceptible but short time, as often occurs with moving objects, when the distance between them and the observer is great (for example, the stars which reside in the sky and distant objects at sea), then they do not appear to move. (Lejeune, 1956, pp. 54–55)

Ibn al-Haytham (ca. 1040): For sight perceived motion only by perceiving the object in two different places one after the other, or in two different positions one after the other.... For if sight perceives the moving object in the second position without at that moment perceiving it in the first position where it was formerly perceived, then it will sense the difference between the two moments. And if it senses that difference, then it will sense the time between them. That being the case, the time in which sight perceives the motion must be sensible. (Sabra, 1989, p. 195)

Pecham (ca. 1280): *All things that are seen require time for their perception.* [This is so] because change becomes sensible only in a period of time, as we are taught by illusions of the senses in the swift movement of things. Furthermore, it is evident that an object can be discerned only in a period of time, since a point on a rotating body appears as a circle. (Lindberg, 1970, p. 135)

Leonardo da Vinci (ca. 1500): If the proportions of the movements of two movable things is the same as that of their distance from the eye in the same direction the movements of the movable things will always appear equal although they may be of almost infinite diversity. (MacCurdy, 1938a, p. 269)

Rohault (1671): Now it is not difficult to conceive, that we know a Body to be in Motion; first, when its Image appears successively applied to different Images of certain Objects, which we do not compare with any other, but imagine to be immoveable; or when we find that we must turn our Head or our Eyes in order to have the Object always at the End of the Line, along which we carry our principal Attention; or lastly, when, if we move neither our Eyes nor our Head, we find it is gone out of that Line. The contrary to all which makes an Object appear to us to be at rest. (1723, p. 257)

Malebranche (1674): It is certain that we know not how to judge how great the Motion of a Body is, but by the Length of the Space the Body has run over.

Thus, our Eyes not informing us of the true Length of the Space describ'd by the Motion, it follows that 'tis impossible for us to know the true Quantity of the Motion. (1700, p. 20)

Locke (1690): Motion can neither be, nor be conceived without space.... *Motions very slow*, though they are constant, *are not perceived* by us; because in their remove from one sensible part towards another, their change of distance is so slow, that it causes no new *Ideas* in us; but a good while after one another: And so not causing a constant train of new *Ideas*, to follow one another immediately in our Minds, we have no Perception of Motion.... On the contrary, *things that move* so swift, as not to affect the Senses distinctly with several distinguishable distances of their Motion, and so cause not any train of *Ideas* in the Mind, *are not* also *perceived* to move. (pp. 78, 85, and 86)

Hartley (1749): We judge of Motion by the Motion of Pictures on the *Retina*, or our Eyes in following the Objects. (p. 203)

Porterfield (1759b): ... we can never know the absolute Magnitude or Celerity of their [Bodies] Motions, but only the Proportion that these Motions bear to one another. (p. 417)

Harris (1775): In like manner, as the different situations of several bodies, or of several parts of the same body, whilst at rest, are perceived by impulses of the light proceeding from them upon correspondent parts of the retina; so also are the motions of bodies, perceived by the successive progress of these impulses over different parts of the retina. Thus, if an object moves from P along the line PQ, its image upon the retina will also move from *p* along the line *pq*.... Apparent motion, or the motion of the images of moving objects upon the retina, must have a certain limited degree of velocity to become perceptible; that is, the space described upon the retina in a given time, must be neither less than some given space, nor greater than some determinate space. (pp. 104 and 107)

Wells (1792): If the eye be at rest, we judge an object to be in motion when its picture falls in succeeding times upon different parts of the retina; and if the eye be in motion, we judge an object to be at rest, as long as the change in the place of its picture upon the retina, holds a certain correspondence with the change of the eye's position. (pp. 94–95)

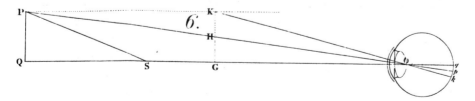

Retinal stimulation by a moving object according to Harris.

Steinbuch (1811): We can see *the movement* of objects. This perceived movement is of two kinds: *one relative* and *one absolute*. We see relative movement if many objects are present in the visual field and one or several change their position in opposition to the majority ... We see absolute movement if nothing else is visible other than the moving object. (p. 193)

John Herschel (1833): Not only do external objects at rest appear in motion generally, with respect to ourselves when we are in motion among them, but they appear to move one among the other—they shift their *relative* apparent places. Let any one travelling rapidly along a high road fix his eye steadily on any object, but at the same time not entirely withdraw his attention from the general landscape,—he will see, or think he sees, the whole landscape thrown into *rotation*, and moving round that object as a centre; all objects between it and himself appearing to move *backwards*, or the contrary way to his own motion; and all beyond it, forwards, or in the direction in which he moves: but let him withdraw his eye from that object, and fix it on another,—a nearer one, for instance,—immediately the appearance of rotation shifts also, and the apparent centre about which the illusive circulation is performed is transferred to the new object, which, for the moment, appears to rest. This apparent change of situation of objects with respect to one another, arising from a motion of the spectator, is called *parallactic* motion; and it is, therefore, evident that, before we can ascertain whether external objects are really in motion or not, or what their motions are, we must subduct, or allow for, any such *parallactic* motion which may exist. (pp. 13–14)

Müller (1838): We judge the motion of an object, partly from the motion of its image over the surface of the retina, and partly from the motion of our eye following it. (1843, p. 1178)

5.2 Apparent Motion (Stroboscopic Motion)

Apparent or stroboscopic motion refers to perceiving motion from a rapid sequence of systematically differing static images. It is contrasted to real motion, in which objects change their location over time, although the same objections to the use of the term could be raised as were for subjective colors. As was mentioned earlier (section 4.9), the subjective visual phenomena referred to as stroboscopic patterns were produced by rapidly moving the outstretched fingers in front of the closed eye, but the instruments for generating apparent motion were much more contrived, and several similar versions were invented at about the same time. Indeed, it was with the invention of instruments like the stroboscopic disc that motion perception was rendered experimentally tractable. Since the 1830s, apparent motion has been the most intensely studied area of motion perception. However, the history of the devices that could synthesize motion from a series of static pictures is more extensively documented in chronicles of film and photography than in histories of vision (see Eder, 1945).

All but one of the descriptions given under this heading cluster around the end of the period under consideration. Lucretius gave an account of what appears to be apparent motion, although it occurred in the context of discussing dreams. Plateau (1878a) remarked that Lucretius had come very close to appreciating the nature of apparent motion. No further reference to this phenomenon is encountered until the nineteenth century. Apparent motion is dependent upon visual persistence, and so it is not surprising that the impetus to examine it derived from a description of a phenomenon linking the two. Roget, more renowned for his *Thesaurus* than for his studies of vision, not only described the pattern of spokes moving behind a palisade (section 4.11) but also suggested that this could provide an accurate means for measuring the duration of visual persistence. Faraday, fueled by this "optical deception," and noting another instance of it with counterrotating cogwheels, enunciated the principle upon which all the subsequent devices were based. He constructed a simple arrangement of cut-out sectored discs to examine the effects further. Faraday's paper appeared in the new, and short-lived, *Journal of the Royal Institution*, but it was rapidly translated into German (Faraday, 1831b), thus extending its influence. Plateau's phenakistoscope or fantascope was almost identical to Stampfer's stroboscopic disc—both presented a short sequence of slightly different figures intermittently to the eye. Wheels could be seen to rotate, dancers pirouette, and soldiers march. Plateau, with his greater understanding of vision, appreciated that there were limits to the visibility of such apparent motion: if the rotation was too slow, then each individual figure was seen; if it was too fast, then they were all seen together in a confusion. These instruments could be used by just one person at a time, whereas Horner developed a variant for group viewing. This device became widely used in the latter half of the nineteenth century under the name of the zoetrope. Roget also had his interests in visual persistence rekindled by Faraday's article, and claimed to have made a similar instrument in 1831.

Lucretius (ca. 56 B.C.): ... when the first image perishes and a second is then produced in another position, the former seems to have altered its pose. Of course this must be supposed to take place very swiftly: so great is their velocity, so great the store of things, so great the store of particles in any single moment of sensation, to enable the supply to come up. (1975, p. 337)

Roget (1825): The velocity of the apparent motion of the visible portions of the spokes [moving behind a palisade] is proportionate to the velocity of the wheel itself; but it varies in different parts of the curve: and might therefore, if accurately estimated, furnish new modes of measuring the duration of the impressions of light on the retina. (pp. 139–140)

Lucretius (ca. 98–55 B.C.) after an illustration in Wood (1885).

Michael Faraday (1791–1867) after an engraving in *The World's Great Men*. London: The London Printing and Publishing Company, 1854.

Faraday (1831a): The eye has the power, as is well known, of retaining visual impressions for a sensible period of time; and in this way, recurring actions, made sufficiently near to each other, are perceptibly connected, and made to appear as a continuous impression. (p. 210)

Plateau (1833): The apparatus [phenakistoscope] ... essentially consists of a cardboard disc pierced along its circumference with a certain number of small openings and carrying painted figures on one of its sides. When the disc is rotated about its centre facing a mirror, and looking with one eye opposite the openings.... the figures are animated and execute movements. (p. 305)

Stampfer (1833): The principle on which this device [the stroboscopic disc] is based is that any act of vision which creates a conception of the image seen is

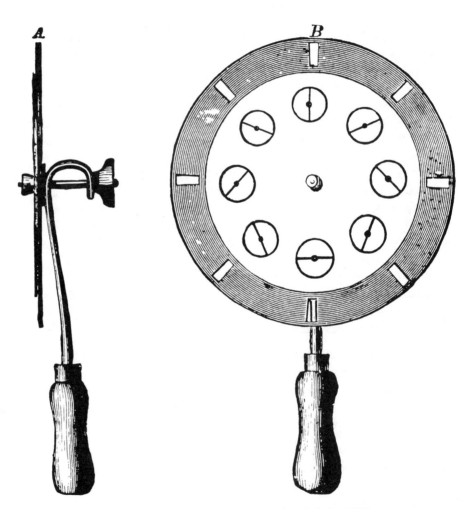

Plateau's phenakistoscope as illustrated in Helmholtz (1867).

divided into a suitable number of single moments; these present themselves to the eye in rapid succession, so that the ray of light falling on the change of the images is interrupted, and the eye receives only a momentary visual impression of each separate image when it is in the proper position. (Eder, 1945, pp. 499–500)

Simon Stampfer (1792–1864) after an illustration in Schmitz (1982).

Horner (1834a): The apparatus is merely a hollow cylinder, or a moderately high margin, with apertures at equal distances, and placed cylindrically round the edge of a revolving disk. Any drawings which are made on the interior surface in the intervals of the apertures will be visible through the opposite apertures, and if executed on the same principle of graduated action, will produce the same surprising play of relative motions as the common magic disk does when spun before a mirror. But as no necessity exists in this case for bringing the eye near the apparatus, but rather the contrary, and the machine when revolving has all the effect of transparency, the phænomenon may be displayed with full effect to a numerous audience. I have given the instrument the name of *Dædaleum*, as imitating the practice which the celebrated artist of antiquity was fabled to have invented, of creating figures of men and animals endued with motion. (p. 37)

Roget (1834): This [Faraday's observation above] again directed my attention to the subject, and led me to the invention of an instrument which has since

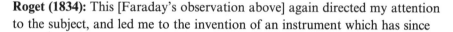

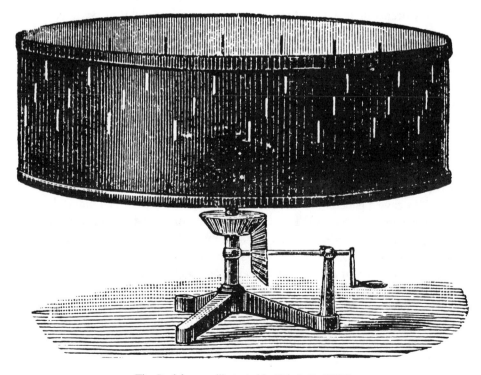

The *Dædaleum* as illustrated in Helmholtz (1896).

been introduced into notice under the name of the Phantasmascope or Phenakistoscope. I constructed several of these at that period (in the spring of 1831)
which I showed to many of my friends; but in consequence of occupations and
cares of a more serious kind, I did not publish any account of this invention,
which was reproduced on the continent in the year 1833. (p. 466)

5.3 Induced Motion

Induced motion has a more continuous history than does apparent motion. It
is a contrast effect in the motion domain: if several objects move together and
another remains stationary, then the latter will appear to move in the opposite
direction. Euclid provided what seems to be a clear description of this phenomenon, which is most puzzling, since it flies in the face of his geometrical
account of vision. It is a phenomenon that cannot be explained in projective
terms, and it is the only such instance in his *Optics*. One possible explanation
is that Euclid was assuming that the eyes were following the moving objects.
The quotation given is sandwiched between two others that are concerned with
eye movements, and both define the perceived motion in terms of the optical
geometry. Therefore if the eye was following the two moving objects, the
stationary one would move in the opposite direction with respect to the eye.

There is no such doubt about the description given by Lucretius of a
naturally occurring instance of induced motion: stars at nighttime seem to
move in the direction opposite to the passing clouds. Similarly, Ptolemy noted
the phenomenon with regard to water flowing past a stationary boat: if vision
was restricted to the water, then the boat appeared to be in motion, but when
the field of view included the land and the boat, the water was seen as moving.
A similar statement was made by Malebranche. Ptolemy's account displays
an appreciation of the relativity of motion perception with respect to the frames
of reference that are in operation. Perhaps the most common and compelling
instance of induced motion is when moving clouds partially obscure the moon,
as Ibn al-Haytham noted. He also hinted, following Ptolemy, that the clouds
provide a reference relative to which the motion of the moon is assigned.
Leonardo gave a general statement of the conditions under which induced
motion occurs, and Porterfield provided a much more specific and experimental
instance of this. He realized that it required more than one moving object for
the phenomenon to occur, but as few as two would suffice. Both Porterfield
and Harris noted that the moving objects would appear stationary; Harris
suggested that the eyes might pursue the moving objects involuntarily, so that
the stationary object would move with respect to the retina. Wells returned to
a point that was implicit in Ptolemy's account—that the total amount of
motion perceived in a given situation will be constant, no matter how it is
apportioned.

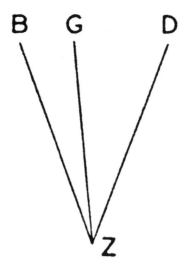

Euclid's representation of induced motion.

Euclid (ca. 300 B.C.): *If, when certain objects are moved, one is obviously not moved, the object that is not moved will seem to move backward.* For, let *B* and *D* move, and let *G* remain unmoved, and from the eye let the rays fall, *ZB*, *ZG*, and *ZD*. So *B*, as it moves, will be nearer to *G*, and *D*, receding, will be farther away; therefore, *G* will seem to move in the opposite direction. (Burton, 1945, p. 371)

Lucretius (ca. 56 B.C.): ... when the winds carry scattered clouds across the sky in the night time, then the shining stars seem to glide against the clouds and to pass above them in a very different direction from their true one. (1975, p. 311)

Ptolemy (ca. 150): ... when a boat is stationary on a calm river flowing without waves, a passenger who does not look at the distant land (but only at the water) thinks that the boat is moving quickly and that the water is still.... Since the water appears stationary, the motion must be seen as coming from the boat. However, if the water, land, and boat are observed at the same time, and if we are aware that the land is stationary, we will see the water in motion. (Lejeune, 1956, pp. 78–79)

Ibn al-Haytham (ca. 1040): Sight therefore perceives the moon as moving behind the swiftly moving cloud because it correlates a part of the moon with successive parts of the cloud. (Sabra, 1989, p. 261)

Leonardo da Vinci (ca. 1500): The movement of an object near a stationary object often causes the stationary object to seem to transform itself to the movement of the the moving object and the moving object to seem stationary and fixed. (MacCurdy, 1938a, p. 255)

Malebranche (1674): ... it very often happens that things which seem to us in Motion, are perfectly at Rest; and on the Contrary those which seem to us at Rest, notwithstanding, are in Motion. As when a Man, for instance, sits on a shipboard whilst the Vessel is under sail in a swift and steady Motion, he seems to see the Lands and Towns fly from him; they seem to be in motion, and the Vessel to stand still. (1700, p. 21)

Porterfield (1759b): If two or more Objects move with the same Velocity, and a third remain at rest, the Moveables will appear fixed, and the Quiescent in Motion the contrary Way. Thus, Clouds moving very swiftly, their Parts seem to preserve their Situation, and the Moon to move the contrary Way. (p. 424)

Harris (1775): If two or more objects, having the same apparent velocity, move all the same way; an object at rest, by which they pass, may appear to move the contrary way, whilst the objects in motion may seem at rest. For no motion among themselves is perceivable, since their images upon the retina keep the same distances from each other; and if the spectator insensibly moves his eye, so as to keep these images in the same place, the image of the object at rest will pass successively over them, after the same manner as if that object had been in motion the contrary way. (p. 106)

Wells (1794a): If from any deception of sight we attribute motion to an object at rest, we necessarily suppose all other objects which are in its neighbourhood, or are placed in the same direction from us, to move the same way, and with the same velocity, provided these be also at rest: for, no deception ever does or can increase or diminish the angle which any two objects subtend at the eye. (p. 796)

5.4   Motion Aftereffect

Induced motion is a simultaneous phenomenon: it occurs during observation of moving and stationary bodies. The motion aftereffect, as its name implies, is a successive phenomenon. Following fixation on some moving source, like a river or a waterfall, subsequently viewed stationary objects appear to move in the opposite direction. The fixation must be prolonged (many seconds or a minute) and pursuit eye movements should be suppressed. This is what Aristotle seems to have done when viewing running water in a river, and he reported that thereafter stationary objects appeared to be moving. There has been more than a little debate concerning this account, because the direction of the motion aftereffect was not specified (see Verstraten, 1996). Gregory (1981) considered that the direction was wrongly stated by Aristotle, being in the same direction as the water. Verstraten compared different translations of the appropriate section, and concluded that no direction had been specified by Aristotle.

Aristotle's account of the motion aftereffect is well known because it was cited in Wohlgemuth's (1911) excellent review of the phenomenon. That by Lucretius is less well known, in part because it was not mentioned by Wohlgemuth. It is also a more complex description because it includes reference to induced motion as well as the motion aftereffect. Lucretius was crossing a river on horseback; when surrounded by flowing water and looking down at it the horse seemed to be moving against the current. This is an instance of induced motion, other examples of which were given by Lucretius (see section 5.3). When he looked at the bank of the river it seemed to be moving in the same direction as the induced motion—that is, opposite to the flowing water. Verstraten (1996) has compared four translations of this passage, and concluded that Lucretius did specify the appropriate direction of the motion aftereffect.

It cannot be said that the phenomenon excited the interests of classical students of vision—it seems to have drowned in the waters of antiquity. Indeed, it did not resurface until the nineteenth century, when Purkinje's acute vision was directed to it. His first description, in 1820, was overlooked by Wohlgemuth (1911) and so it has been neglected generally. The phenomenon could be seen following observation of various moving objects, although it was the instance of viewing a cavalry parade that was repeated in his second book on subjective visual phenomena (Purkinje, 1825); Wohlgemuth did cite this account. Purkinje provided a clear statement of the direction in which the aftereffect motion was seen—opposite to that of the prior motion. He also suggested an interpretation of it that was to have lasting appeal: during adaptation the eyes move unconsciously in the direction of motion, and they continue this movement when viewing stationary objects; this would result in their motion with respect to the retina as if they were indeed moving in the opposite direction.

The description of the phenomenon that has had the widest currency, and has in fact provided another name for it, was given by Addams; he was a peripatetic lecturer on natural philosophy, and he made his observation during a tour of Scotland (see Wade, 1994b). He fixated on a point behind the descending waters of the lower Fall of Foyers and then averted his eyes to the adjacent rocks, noting that they appeared to ascend. The phenomenon is particularly striking, and Thompson (1880) referred to it as the "waterfall illusion," a term which is still in use. The illustration on page 216 is a view of the lower of the two Falls of Foyers, engraved at about the time Addams made his observation and showing the platform from which he viewed the waterfall; other engravings of the falls can be found in Wade (1994b) and Verstraten (1996). Addams did not cite Purkinje, but the interpretation he provided was also couched in terms of eye movements. At the end of his brief article he mentioned that he had also observed the phenomenon when traveling by ship, and that it could be elicited as a consequence of head movements. Perhaps of greatest significance is the statement that "It is also producible by mechanical means, such as by a rapid unrolling of pieces of calico having some pattern or markings on them" (p. 374); this marks the onset of experimental inquiries into the phenomenon, by isolating it from the natural conditions in which it was initially observed.

Addams's paper was translated into German (Addams, 1835), but it was seldom cited in the later German literature. Müller did discuss his own observations of the phenomenon, and he attempted to explain it in terms of motion afterimages.

Aristotle (ca. 330 B.C.): ... when persons turn away from looking at objects in motion, e.g., rivers, and especially those which flow very rapidly, they find that the visual stimulations still present themselves, for the things really at rest are then seen moving (Ross, 1931, p. 459b)

Lucretius (ca. 56 B.C.): ... when our spirited horse has stuck fast in the middle of a river, and we have looked down upon the swift waters of the stream, while the horse stands there a force seems to carry his body sideways and pushing it violently against the stream, and, wherever we turn our eyes, all seems to be rushing and flowing in the same way as we are. (1975, p. 309)

Purkinje (1820): Another form of eye dizziness can be demonstrated if one observes a passing sequence of spatially distinct objects for a long time, e.g. a long parade of cavalry, overlapping waves, the spokes of a wheel that is not rotating too fast. When the actual movement of the objects stops there is a similar apparent motion in the opposite direction. (pp. 96–97)

Purkinje (1825): One time I observed a cavalry parade for more than an hour, and then when the parade had passed, the houses directly opposite appeared to me to move in the reversed direction to the parade. While the eye did strive to fixate on each individual row of militia during observation, it moved unconsciously in the same direction as the parade; this movement was repeated so often that it became habitual, and continued when the parade had passed. The eye attempted to fixate on the stationary objects in a similar manner to the motion it had become accustomed to fixating, and therefore glided unconsciously over them in the usual direction, so that the objects appeared to slide away in the opposite direction. (pp. 60–61)

Addams (1834): Having steadfastly looked for a few seconds at a particular part of the cascade, admiring the confluence and decussation of the currents forming the liquid drapery of waters, and then suddenly directed my eyes to the left, to observe the vertical face of the sombre age-worn rocks immediately contiguous to the water-fall, I saw the rocky face as if in motion upwards, and with an apparent velocity equal to that of the descending water, which the moment before had prepared my eyes to behold this singular deception. . . . I conceive the effect to be owing to an involuntary and *unconscious* muscular movement of the eyeball, and thus occasioning a displacement of the images on the retina. Supposing the eyes to be intently gazing at any point in a transverse plane passing through a vertically moving body, they will naturally and even irresistibly tend to follow the motion of that body; nor can the muscular apparatus of the eye maintain a stable equilibrium when the sight is fatigued and bewildered with a rapid change of moving forms before the eye. Now in the case of the descending water, the eyes, being directed to a particular part in a horizontal section of it, cannot be prevented moving downwards through a small space: every new form in the moving scene invites the eyes to observe, and for that reason to follow it; but the voluntary powers are engaged to raise the axes of the eyes again to the section. This depression of the axes below the *intentional point of sight* seems to be repeated three or four times per second, whilst looking at the water-fall. Then, when the eyes are suddenly turned upon the rock, the muscles, having been brought into a kind of periodic contraction, will perform at least one of these movements after the exciting cause ceases to act; and thus the axes of the eyes, by moving downwards, will occasion a motion of the image of the rock over the retina in a direction from above downwards, and consequently the object giving that image will *appear* to move the contrary way, that is, upwards, agreeably to observation. (pp. 373–374)

Müller (1838): If, after looking for a long time at the undulations of a stream of water, we suddenly turn our eyes to the ground at its margin, the ground itself appears to move, in the opposite direction to the waves of water. I have frequently remarked this phenomenon, when, after gazing from my window upon the neighbouring river, I have directed my eyes to the pavement of the street. I observed it also when, being on board a steam-packet, and, having looked for some time upon the waves which passed, I suddenly turned my eyes towards the deck of the vessel. If we suppose that in these cases spectra were left in the retina by the impressions of the waves, and that they disappeared in the order in which they arose, their successive disappearance while looking upon the fixed surface of the ground or deck would necessarily cause an apparent motion of this surface in the opposite direction. (1843, p. 1180)

5.5 Vertigo

Purkinje's (1820) description of motion aftereffects was embedded in a long article concerned with vertigo, and his systematic studies did much to clarify the link between visual and vestibular function (see Boring, 1942). Vertigo is a disturbance of balance (giddiness), usually associated with visual motion and eye movements; it can be temporarily induced by a number of actions. One that has been commented on frequently since the time of Aristotle relates to the effects of alcohol: it causes dizziness and apparent motion of stationary objects. Similar effects can be generated by rapidly rotating the whole body and then

Engraving of the lower Fall of Foyers (Beattie, 1838). Addams would have observed the waterfall from the observation point shown.

stopping; Lucretius referred to this as a children's game. Ptolemy likened the visual effects so produced to looking at an object beneath flowing water. Vertigo can be induced visually, as when looking downward from a great height, as Montaigne noted. Aristotle's account was often repeated, almost verbatim in the case of Francis Bacon.

More systematic studies of visual vertigo following body rotation were initiated by Porterfield, although he remarked that the eyes remained stationary during the visual rotation. Robert Darwin suggested that the sequence of light entering the eyes during rotation was visible as moving afterimages when the body came to rest. Both these positions were soundly criticized by Wells, who provided experimental evidence to refute them. Robert Darwin's article of 1786, in which afterimages and vertigo were discussed, was reprinted as an appendix to the first volume of Erasmus Darwin's *Zoonomia* in 1794, and Erasmus added some more comments, essentially supporting his son's theory against the attack of Wells (see Cohen, 1984). There followed two blistering ripostes by Wells. Not only did he demonstrate that visual vertigo occurs with rotation in darkness but he also gave a phenomenological description of the reducing amplitude of postrotational nystagmus, and that it could be suppressed by fixation. Other aspects of this dispute will be found in the discussion of eye torsion (section 5.7). One commentator (T. J.), referring to these "vertiginous philosophers," did add another puzzle—that acrobats can rotate for considerable durations without losing their balance. Purkinje elaborated upon the studies of Wells, noting that the postrotational motion observed was dependent on the head orientation during rotation. He constructed a rotating chair to facilitate studying the visual and oculomotor aftereffects (see Grüsser, 1984).

Aristotle (ca. 330 B.C.): Why is it that to those who are very drunk everything seems to revolve in a circle, and as soon as the wine takes a hold of them they cannot see objects at a distance?.... objects near at hand are not seen in their proper places, but appear to revolve in a circle. (Ross, 1927, p. 892a)

Lucretius (ca. 56 B.C.): The room seems to children to be turning round and the columns revolving when they themselves have ceased to turn, so much so that they can hardly believe all the building is not threatening to fall in upon them. (1975, pp. 307–309)

Ptolemy (ca. 150): A continuous revolution of the visual field results in objects appearing to move. Movement of this kind is produced in visual vertigo. (Lejeune, 1956, p. 73)

Paulus Ægineta (ca. 680): Vertigo is occasioned by a cold and viscid humour seizing upon the brain, whence the patients are ready to fall down from a very slight cause, such as sometimes from looking at any external object which turns round, as a wheel or top, or when they themselves are whirled round, or when their head has been heated, by which means the humours or spirit in it are set in motion. (1844, p. 374)

Ibn al-Haytham (ca. 1040): For if someone turns himself quickly round several times, then stops, he sees all perceptible objects about him turning, though they are stationary. (Sabra, 1989, p. 360)

Montaigne (1580): Put a philosopher into a cage of small thin set bars of iron, and hang him on the top of the high tower of Nôtre Dame at Paris: he will see, by manifest reason, that he cannot possibly fall, and yet he will find (unless he has been used to the plumber's trade) that he cannot help but the sight of the excessive height will fright and astound him: for we have enough to do to assure ourselves in the galleries of our steeples, if they are made with open work, although they are of stone; and some there are that cannot endure so much as to think of it. (1853, p. 278)

Platter (1614): An intense, uniform, and extended movement of the head transfers itself in a similar way to the spiritus. Despite holding the head still afterwards, it appears to continue moving for a while, before it eventually feels still. This is the basis for dizziness, if one rotates the head and body in a circle for a long time. (Koelbing, 1967, p. 89)

Francis Bacon (1627): Drunken men imagine every thing turneth round; they imagine also that things come upon them; they see not well things afar off; those things that they see near hand they see out of their place; and (sometimes) they see things double. The cause of the imagination that things turn round is, for that the spirits themselves turn, being compressed by the vapour of the wine (for any liquid body upon compression turneth, as we see in water); and it is all one to the sight, whether the visual spirits move, or the object moveth, or the medium moveth. And we see that long turning round breedeth the same imagination. (1857a, p. 572)

Hartley (1749): Giddiness, or an apparent irregular Motion in the Objects of Sight, almost always goes before any general Confusion and Privation of Sense and Motion. (p. 200)

Porterfield (1759b): If a Person turns swiftly round, without changing his Place, all Objects about will seem to move in a Circle to the contrary Way, and the Deception continues, not only when the Person himself moves round, but, which is more surprising, it also continues for some time after he stops moving, when the Eye, as well as the Objects, are at absolute Rest. (p. 425)

Robert Darwin (1786): When any one turns round rapidly on one foot, till he becomes dizzy and falls upon the ground, the spectra of the ambient objects continue to present themselves in rotation, or appear to librate, and he seems to behold them for some time still in motion. (p. 315)

Wells (1792): But whoever will make the experiment, will find, that objects about him appear to be equally in motion, when he has become giddy by turning himself round, whether this has been done with his eyes open or shut. (p. 94)

Erasmus Darwin (1794): ... but it is certain, when any person revolves in a light room with his eyes closed, that he nevertheless perceives differences of light both in quantity and colour through his eyelids, as he turns round; and readily gains spectra of those differences. And these spectra are not very different except in vivacity from those, which he acquires, when he revolves with unclosed eyes, since if he then revolves very rapidly the colours and forms of surrounding objects are as it were mixed together in his eyes, as when the prismatic colours are painted on a wheel, they appear white as they revolve. The truth of this is evinced by the staggering or vertigo of men perfectly blind, when they turn round. (pp. 571–572)

Wells (1794a): But what would the event be if we were to turn ourselves in a *dark* room? To this Dr. Darwin says nothing. I can assert, however, from experience, that if any person will turn himself in a dark room till he becomes giddy, having previously remained in it a sufficient time to allow the *spectra* of objects to disappear, he will observe, upon the admission of light, that the surrounding bodies seem to move in the same manner as if the room had been lightened during the whole course of the experiment.... When a person ceases to turn, after he has become giddy, objects at first appear to move through considerable segments of circles. The segments thenceforth gradually become less; and, at length, the objects seem to rest. (p. 795)

Wells (1794b): ... when we have become giddy by turning, if the apparent motions are not considerable, we can stop them altogether by viewing any particular object very stedfastly; but that, if we shortly after withdraw our attention from it, and look carelessly at objects in general, their apparent motions will re-commence. (p. 907)

T. J. (1794): ... I would ask one or both of the vertiginous philosophers, who have lately so much opposed each other in your Magazine on the subject of vision, how the intoxicated man sees double? and, when he has tumbled on the floor, and scrambles lest he should fall lower, what could have occasioned those *moving spectra* to a body at rest? And yet we have seen at Sadler's Wells, and such places, a tumbler spin like a top for a quarter of an hour together, at the same time balancing naked swords and drinking-glasses over his head, and not be affected in the least. (p. 1093)

Purkinje (1820): Visual vertigo is determined by a conflict between unconscious involuntary muscle actions working in the opposite direction to conscious voluntary ones. (p. 95)

Müller (1834): Certain influences acting on the brain give rise, not to rotatory motions, but to sensations of rotation. These are the revolving sensations of vertigo, of which we are principally conscious by the sense of vision. It is a well-known fact that if any person turns round quickly upon his own axis for a short time, he not only begins to lose his recollection, but also, when he ceases to move, seems to see the objects around him still revolving in the same direction. Purkinje has made some very curious observations relative to this phenomenon. It would appear from his experiments that the direction which the revolution of the images shall take can be regulated by the position of the body, and particularly of the head while turning round, and by the position of it afterwards when we have ceased to move round. It is only when the experimenter has kept his head in the ordinary vertical position while turning round, that afterwards, when he stands still with his head upright, the objects appear to revolve in the horizontal direction. If while turning he holds his head with the occiput upwards, and then raises it when he stands still, the apparent motion is like that of a wheel placed vertically revolving around its axis. And thus, according to the difference in the position of the head when turning and when standing still afterwards, the direction of the apparent motion can be altered. If, however, the body lying upon a disk is made to revolve with this disk, an apparent motion of objects in tangents is perceived. It results from a repetition of these experiments that the apparent motion produced, when the rotation of the body has ceased, and the position of the head is changed, always seems to take place in the plane perpendicular to that axis of the head, regarded as a sphere, around which the real motion was performed. (1840, pp. 847*–848*)

5.6 Eye Movements

Vertigo has an oculomotor component to it, but, as was noted in the previous section, this was a matter of dispute for some time. The eye movements that are both easier to observe in others and to produce oneself are voluntary. Both eyes generally move together, and this was a feature commented on by Aristotle, and it will be returned to in section 5.9. The eyes are, of course, moved by muscles, and Galen gave a lucid account of the six extraocular muscles, describing the three axes around which the eye can rotate. The function that eye movements served was also stressed by many writers—namely to direct the eyes to objects of interest so that they can be seen distinctly. Some features of distinct vision will be addressed here, while others can be found in section 7.7.

Descartes (1662) provided a speculative mechanistic account of how the paired muscles work by means of reciprocal action. The illustrations are paired, too, being from *De homine* and *Traité de l'Homme*, but in this case the translation is taken from the Latin; the illustrations are radically different in detail, but the principle that they depict was remarkably prescient, despite reference to Galen's animal spirits. From the late seventeenth century there was greater concern with the detail of eye movements. The involuntary movements that attend fixation were reported by Mariotte, and Robert Darwin provided an expression of this in the formation of afterimages. The rapid inspection of objects by a succession of eye movements was detailed by Scheiner, Condillac, Porterfield, Harris, and Young, who gave the range over which the eye could move. Much of the material concerning motion of the eyes included in Porterfield's *Treatise* of 1759 had been presented over twenty years previously (Porterfield, 1737, 1738). More methodological studies were initiated by Wells, who used an afterimage as a stabilized retinal image. He generated an afterimage and then rotated his body to generate vertigo; after rotation ceased, the afterimage was projected onto a piece of paper, and it was noted to undergo motion (as a consequence of nystagmus)—moving slowly in one direction and returning rapidly. He also observed that the amplitude of afterimage motion gradually decreased. Purkinje and Weber returned to the question of eye movements affording distinct vision. While noting that acuity declines toward the periphery (see section 7.3), Weber suggested that the eye movements over objects are used to assess their dimensions.

Eye movements in the newborn and young were remarked on by Hartley, Buffon, and the anonymous author of the article on eyes in the third edition of the *Encyclopædia Britannica*. Hartley contended that the two eyes moved conjugately from birth, whereas according to Buffon fixation did not develop until six or seven weeks. Interest in the developmental aspects of eye movements was a consequence of the debate between nativists and empiricists, and this will be returned to in sections 6.7, 7.1, and 7.2.

Boring (1942) commenced his discussion of eye movements with Müller's (1826a) book on comparative physiology and eye movements; Carmichael (1926) took issue with this claim, suggesting in its stead Bell's (1823) account of eye movements. The quotations presented in this section indicate that the systematic study of eye movements was a little earlier than either of these worthy scientists. We return to Bell in section 5.8. Heller (1988) provides an account of early studies of eye movements and their later measurement.

Aristotle (ca. 330 b.c.): So when one eye moves, the common source of sight is also set in motion; and when this moves, the other eye moves also. (Ross, 1927, p. 957b)

Galen (ca. 175): . . . since there are four movements of the eyes, one directing them in toward the nose, another out toward the small corner, one raising them up toward the brows, and another drawing them down toward the cheeks, it was reasonable that the same number of muscles should be formed to control the movements. . . . Since it was better that the eye should also rotate, Nature made two other muscles [*obliqui, superior* and *inferior*] that are placed obliquely, one at each eyelid, extending from above and below toward the small corner of the eye, so that by means of them we turn and roll our eyes just as readily in every direction. (May, 1968, p. 483)

Pecham (ca. 1280): Because of the many things that must be quickly perceived, the eye must be round so that it can easily turn and move. (Lindberg, 1970, p. 111)

Scheiner (1619): The movement of the eye is adapted for every variation of position so that the optical axis is directed to a selected part of an object making it most distinctly perceived. (Rohr, 1919, p. 124)

Descartes (1637): ... when our eyes or our head turns in some particular direction, our mind is informed of this by the change which the nerves inserted in the muscles used for these movements cause in our brain. (1965, p. 105)

Descartes (1662): Let us imagine a tube or small nerve, bf, flowing into the muscle D—which I am supposing is one of those which move the eye, and there dividing into several branches, loose in texture, which stretch or shrink according to the quantity of animal spirits which flow in or out of them, and whose ramifications are so arranged that the animal spirits flowing into them cause the whole body of the muscle to swell and shorten, thus pulling the eye to which it is attached towards it; on the other hand when the animal spirits flow out of the muscle it shrinks and lengthens. Besides bf, there is another tube, ef, through which the animal spirits flow into muscle D, and there is yet another tube, dg, by which they flow out of it. Muscle E, which I assume moves the eye in the opposite direction to muscle D, receives the animal spirits from the brain through the tube cg, while those from muscle D flow through dg to E, which returns them to D by ef.... if the animal spirits from the brain tend to flow with more force through bf than cg, they close the valve g and open f— either more or less, in proportion to their strength—through which the spirits in muscle E pass to muscle D by the tube ef, at a speed in proportion to the size of the opening at f, so that the muscle D shortens, because the spirits cannot flow out of it, while E lengthens, and thus the eye is turned towards D. On the other hand if the spirits flow with greater force through cg than through bf they close valvule f and open g, so that the spirits from muscle D go back at once by the tube dg to muscle E, which thus becomes shortened and pulls the eye towards that side. (Fulton, 1966, pp. 263–264)

Mariotte (1683): ... the *Eyes* are always in motion, and very hard to be fixt in one place, though it be desired. (p. 266)

Porterfield (1737): The Motions of the Eye are either external or internal. I call *external*, those Motions performed by its four straight and two oblique Muscles, whereby the whole Globe of the Eye changes its Situation or Direction. And by its *internal* Motions, I understand those Motions which only happen to some of its internal Parts, such as the *crystalline* and *Iris*, or to the

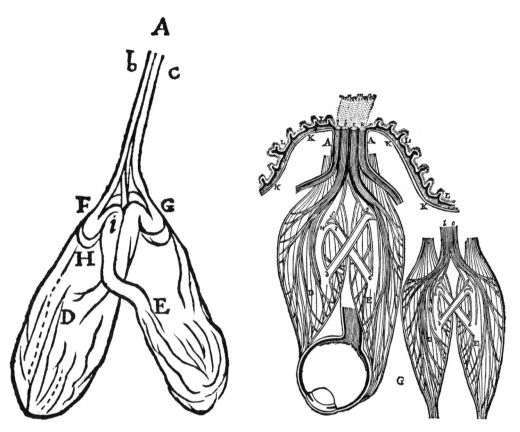

Representations of reciprocal innervation of the eye muscles from Descartes (1662) on the left and Descartes (1664/1909) on the right.

whole Eye, when it changes its spherical Figure, and becomes oblong or flat.... Now, though it is certain that only a very small Part of any Object can at once be clearly and distinctly seen, namely, that whose Image on the *Retina* is in the *Axis* of the Eye; and that the other Parts of the Object, which have their Images painted at some Distance from this same *Axis*, are but faintly and obscurely perceived, and yet we are seldom sensible of this Defect; and, in viewing any large Body, we are ready to imagine that we see at the same Time all its Parts equally distinct and clear: But this is a vulgar Error and we are led into it from the quick and almost continual Motion of the Eye, whereby it is successively directed towards all the Parts of the Object in an Instant of Time; for it is certain, that the Ideas of Objects, which we receive by Sight, do not presently perish, but are of a lasting Nature, as appears from what happens when a Coal of Fire is nimbly moved about in a Circumference of a Circle, which makes the whole Circumference appear like a Coal of Fire, because the Idea of the Coal, excited in the Mind by the Rays of Light, are of a lasting Nature and continue, till the Coal of Fire in going round return to its former Place; and therefore if our Eye takes no longer Time to direct itself successively to all the small Parts of an Object, than what the Coal of Fire takes to go round, the Mind will distinctly perceive all those Parts, without being sensible of any Defect or Insensibility in any Part of the *Retina*, because the Idea of one Part continues, till, by the Motion of the Eye, the Image of the other Parts be successively received upon the same most sensible Part of the *Retina*: And this is the Reason why the Globe of the Eye moves so quickly, and that its Muscles have such a Quantity of Nerves to perform their Motions. (pp. 160 and 185–187)

Hartley (1749): ... new born Children move their Eyes in a congruous Manner; that the Motions are chiefly to the Right and Left, scarce upwards and downwards at all. (p. 216)

Buffon (1749): About a month after birth, the eye seems to have acquired that tension and solidity which are necessary for the proper transmission of the rays of sight; but, even then, infants are incapable of fixing their eyes upon any object: They roll and move them to all sides, without being able to distinguish the objects to which they are directed. In six or seven weeks, however, they begin to fix their attention upon luminous objects. (1780, pp. 2–3)

Condillac (1754): Moreover, it is not enough for the eye to see all of a shape in order to have an idea of it ... It grasps the whole of the simplest shape only when it has analyzed it, that is, when it has observed all its parts in succession. It must make a judgment about each particular part, and another judgment that combines them. It must say to itself: here is one side, here is a second, and here a third; here is a space bounded by these three sides, and from all that, this triangle results. (1982, p. 217)

Reid (1764): ... if this motion of the eyes were got by habit, we should see children, when they are born, turn their eyes different ways, and move one without the other, as they do their hands and legs. I know some have affirmed they are apt to do so. But I have never found it true from my own observation, although I have taken pains to make observations of this kind, and have had good opportunities. I have likewise consulted experienced midwives, mothers, and nurses, and found them agree. (p. 263)

Harris (1775): ... we seldom fix our eyes long to one point, but we nimbly, and insensibly, turn them towards different parts. (p. 116)

Hunter (1786): From all of which we find these three modes of action produced; first, the eye moving from one fixed object to another; then the eye moving along with an object in motion; and last, the eye keeping its axis to an object, although the whole eye, and the head, of which it makes a part, are in motion. (p. 210)

Robert Darwin (1786): When we look long and attentively at any object, the eye cannot always be kept intirely motionless; hence, on inspecting a circular area of red silk placed on white paper, a lucid crescent or edge is seen to librate on one side or the other of the red circle. (p. 341)

Wells (1792): I again produced the spot [afterimage], by looking for some time at the flame of a candle; then turning myself round till I became giddy, I suddenly discontinued this motion, and directed my eyes to the middle of a sheet of paper, fixed upon the wall of my chamber. The spot now appeared upon the paper, but only for a moment; for it immediately after seemed to move to one side and the paper to the other, notwithstanding I conceived the position of my eyes to be in the mean while unchanged. To go on with the experiment, when the paper and spot had proceeded to a certain distance from each other, they suddenly came together again; and this separation and conjunction were alternately repeated a number of times; the limits of the separation gradually becoming less, till, at length, the paper and spot both appeared to be at rest. (pp. 95–96)

Anon (1797): A new-born child shall be observed, perhaps, never to keep its eyes fixed on any one object, but continually changing from one to another, and if you put your hand before them, the child will not wink. Hence some have thought, that new born infants have no sight: but this is a mistake; and the true reason why their eyes are in perpetual motion is, that they have not yet acquired the habit of examining one thing at once with their eyes: their not winking at the approach of the hand, arises from their want of experience how easily their eyes may be hurt; but in a few days they get the habit of winking, so that afterwards their eyes do it spontaneously at the approach of danger. (p. 81)

Young (1801): The motion of the eye has a range of about 55 degrees in every direction; so that the field of perfect vision, in succession, is by this motion extended to 110 degrees. (p. 46)

Saunders (1816): Children thus affected [with congenital cataract] possess various degrees of vision. Some indistinctly see external objects, others can discern only bright colours or vivid lights. If the privation of vision be nearly complete, volition, for want of an external object to attract these organs, is not exercised over the muscles belonging to them, and their actions are not associated, but the eye rolls here and there with rapidity, and trembles as it moves. (pp. 160–161)

Bell (1823): There is a motion of the eye-ball, which, from its rapidity, has escaped observation. At the instant in which the eye-lids are closed, the eye-ball makes a movement which raises the cornea under the upper eye-lid. If we fix one eye upon an object, and close the other eye with the finger in such a manner as to feel the convexity of the cornea through the eye-lid, when we shut the eye that is open, we shall feel that the cornea of the other eye is instantly elevated; and that it thus rises and falls in sympathy with the eye that is closed and opened. (p. 168)

Purkinje (1823a): Firstly, it is to be noted that every movement by means of which the centre of the visual field, where vision is clearest, is directed appropriately back and forth to those contours and lines of external objects. With small objects, e.g. letters, it is only with the greatest difficulty that these movements can be sensed in your own eyes, but it is much easier with larger objects. (pp. 158–159)

Weber (1834): Eyes have the most acute vision in a particular structure: an object that falls on the axis of the retina is distinguished very clearly, but objects located beside it lose clarity the further from the retinal axis their image falls. If we observe very carefully, we notice that the part of the retina where the images are clearly distinguished is very small. So if we wish to perceive any line accurately and take in its length with the eye, we move the eye so that individual parts of the line fall on the retinal axis in turn. We get to know the length of the line by the movement of the eye, of which we are conscious, in the same manner as we discover the length of an object by movement of the finger. (Ross and Murray, 1978, p. 101)

Bell (1836): The remarkable circumstance is, that her eyes are not for a moment at rest, and yet this motion does not disturb her vision. There is a constant tremulous motion in them, which her mother says has continued since her infancy; it is not so much upwards and downwards, as in a transverse direction, but it is irregular in this respect. When she was requested to take her book and read, she read with perfect ease, and yet there was no cessation of the motion in her eye. She threaded her needle without any apparent difficulty, and then shewed how she could sew, which was with the usual nimbleness. (pp. 372–373)

Volkmann (1836): All eye movements are rotations around one point, which is simultaneously the intersection of the optical axis and the visual axis. (p. 35)

5.7   Eye Torsion

Galen had described the oblique muscles that could produce eye rolling or ocular torsion, but evidence of its occurrence was to wait many centuries. Porterfield restated the anatomical position, and toward the end of the eighteenth century John Hunter outlined the function that they could serve: when the head was tilted to one side, the oblique muscles could rotate the eyes in their sockets in the opposite direction. This would now be called ocular countertorsion, and it was treated in greater detail by Hueck, who tried to measure its magnitude. He stated that the eyes remain in the same relation to gravity for head tilts up to 28 degrees; that is, there was perfect compensation for head inclination. Subsequent experiments have not supported this claim: while countertorsion does occur, its magnitude is only a fraction of the head tilt. Hueck also provided a survey of earlier work on ocular torsion.

In addition to the studies of countertorsion, there was the question of post-rotational eye torsion, and this was another battleground between Erasmus Darwin and Wells waged in the pages of the *Gentleman's Magazine*. Darwin conducted a subtle experiment in which he rotated his vertical body with his head tilted backward, looking at a point on the ceiling. When he stopped and looked at the vertical wall he saw rotation in that plane, but said that the eyes did not roll. Wells counterattacked with a stunning experiment. Before rotation he generated an afterimage of a horizontal line; he then rotated his body in the manner described by Darwin; when he was giddy, he stopped and projected the afterimage onto the wall on which a horizontal line had been drawn. He proceeded to see the effects of torsional nystagmus, noting that its duration was shorter than that of lateral nystagmus. Both Erasmus Darwin and Wells were medical men, but the latter had a more secure grasp of the anatomy and function of the extraocular muscles.

Scheiner (1619): ... in an eye movement in which the middle part of the eye remains stationary, it is because it moves by a corresponding head rotation. (Rohr, 1919, p. 126)

Porterfield (1737): The oblique Muscles of the Eye are two in Number.... When it [superior oblique] acts, it rolls the Eye about its *Axis* towards the Nose, and at the same time draws it forwards, and turns its Pupil downwards. The second of these oblique Muscles.... is to roll the Eye about its Axis from the Nose, and at the same time to draw it forwards, and direct its Pupil upwards. (pp. 167–168)

Hunter (1786): Thus when we look at an object, and at the same time move our heads to either shoulder, it is moving in the arch of a circle whose centre is the neck, and of course the eyes would have the same quantity of motion on this axis, if the oblique muscles did not fix them upon the object. When the head is moved towards the right-shoulder, the superior oblique muscle of the right-side acts and keeps the right-eye fixed on the object, and a similar effect is produced upon the left-eye by the action of the inferior oblique muscle; when the head moves in the contrary direction, the other oblique muscles produce the same effect. (p. 212)

Erasmus Darwin (1794): I now come to relate an experiment, in which the rolling of the eyes does not take place at all after revolving, and yet the vertigo is more distressing ... If any one looks steadily at a spot in the ceiling over his head, or indeed at his own finger held high up over his head, and in that situation turns round till he becomes giddy; and then stops, and looks horizontally; he now finds, that the apparent rotation of objects is from above downwards, or from below upwards; that is, the apparent circulation of objects is now vertical instead of horizontal, making part of a circle round the axis of his eye; and this without any rolling of his eyeballs. The reason of there being no rolling of the eyeballs perceived after this experiment, is, because the images of objects are formed in rotation round the axis of the eye, and not from one side to the other of the axis of it; so that, as the eyeball has not the power to turn in its socket round its own axis, it cannot follow the apparent motions of these evanescent spectra, either before or after the body is at rest. (pp. 570–571)

Wells (1794b): As Dr. Darwin gives no proof, from experiment, that the eye does not roll upon its axis during the giddiness which has been produced in the abovementioned situation, I presume he rests his belief of the fact altogether upon the inability of the eye to perform such a motion. But surely the parts which connect the eye-ball to the socket are sufficiently flexible to allow it to move in some degree round its axis; and, whoever bestows the least consideration upon the origin, progress, and termination, of the oblique muscles of the eye must perceive that they have the power of giving it such a motion. That the eye actually does roll upon its axis, is shewn by the following experiment: I placed a long thin rule parallel to the horizon, its edge being towards me, and gave it such a position, in other respects, that it was the only object intervening between my eyes and a bright sky. I afterwards fixed my eyes upon a mark in the middle of its edge, and having obtained in this way a long narrow luminous spectrum [afterimage], I turned myself, having my eyes pointed to a spot over my head, till I became giddy. I then stopped and directed my eyes to the middle of a perpendicular line drawn upon the wall of my chamber. A luminous line, the spectrum of the rule, now appeared upon the wall, crossing the real and perpendicular line at right angles, or nearly so. The two lines, however, did not for a moment preserve the same position with regard to each other, but continually moved round their common point of intersection, in such a manner that the extremities of the one alternately approached and receded from the extremities of the other.... if we have made ourselves giddy while our eyes were directed to a point above us, the apparent motions do not continue nearly so long as if the giddiness had been produced while the head was erect, the body being turned the same number of times in both cases.... when we consider the mechanical resistance to the rolling of the eye upon its axis, and the feebleness of the oblique muscles, which alone can give it this motion, it is natural to expect that, when produced involuntarily, it should continue but for a very short time. (pp. 906 and 907)

Bell (1823): By dissection and experiment it can be proved, that the oblique muscles are antagonists to each other, and that they roll the eye in opposite directions, the superior oblique directing the pupil downwards and outwards, and the inferior oblique directing it upwards and inwards. (p. 174)

Hueck (1838): The observation of actual eye torsion shows that vertical diameter of the eyeball remains vertical if the head is inclined sideways by up to 28°, and therefore during this movement the image maintains its location unchanged. If we incline the head to the side by more than 38°, the eyeball can no longer retain its position with respect to the head rotation. (p. 31)

5.8   Outflow Theory

The visual direction of an object is not determined by visual stimulation alone, as it also involves information about the position of the eyes—otherwise objects would appear to move with every movement of the eyes. Helmholtz (1867) made a distinction between what have become called outflow and inflow theories. The former refers to deriving the eye movement information from efferent (centrally generated) impulses to the eye muscles, whereas the latter reflects use of afferent (sensory) signals from the eye muscles themselves. How this compensation for eye rotation comes about has been a matter of considerable inquiry, the origins of which can be found in the late eighteenth century. Darwin and Wells feature again in this arena, but not disputatiously. Indeed, it is Robert rather than Erasmus who has made a contribution.

One way of examining the association between eye and image movements is to sever the link between them. This can be done by means of a stabilized retinal image, and the afterimage provided the earliest approach to this. Darwin's observation was fortuitous: having formed a peripheral afterimage he sought to examine it, but each eye movement in pursuit of it resulted in its apparent motion in the direction of the eye movement. If the eye remained stationary, then so did the afterimage. No theoretical implications were made regarding this by Darwin, unlike Wells, who appreciated fully the import of this and similar observations. Section 4.1 contained one of Wells's experiments with binocular afterimages: the afterimage followed all voluntary movements of the eyes, but remained single and stationary when one eye was moved passively. The present quotation is closely linked to that one, and it emphasized the importance of voluntary motion in determining visual direction. In short, it provided support for outflow theory (see Ono, 1981). Bell confirmed the distinction between the effects of voluntary and involuntary eye movements on afterimage motion. Although he did not cite Wells's earlier studies, Bell did present an elegant statement of outflow theory (see Wade, 1978b). Brewster, ever an adversary of Bell, raised a pertinent problem relating to passive eye movements: they tend to be translations rather than rotations. Nonetheless, Brewster did not use this fact to discount the apparent motion of real images that follows passive eye movement.

Robert Darwin (1786): If this dark spot [an afterimage] lies above the center of the eye, we turn our eyes that way, expecting to bring it into the center of the eye, that we may view it more distinctly; and in this case the dark spectrum seems to move upwards. If the dark spectrum is found beneath the center of the eye, we pursue it from the same motive, and it seems to move downwards. This has given rise to various conjectures of something floating in the aqueous humours of the eyes; but whoever, in attending to these spots, keeps his eyes unmoved by looking steadily at the corner of a cloud, at the same time that he observes the ocular spectra, will be thoroughly convinced, that they have no motion but what is given to them by the movement of our eyes in pursuit of them. (p. 318)

Wells (1792): Having looked steadily for some time at the flame of a candle, with *one* eye only, I directed afterward, with both eyes open, my attention to the middle of a sheet of paper, a few feet distant; the consequence of which was, that a spot appeared upon it in the same manner, as if I had viewed the flame with both eyes, though somewhat fainter. My attention remaining fixed upon the sheet, I now pushed the eye, by which the spot was seen, successively upward and downward, to the right and to the left, and in every oblique direction; The spot however never altered its position, but kept constantly upon the middle of the appearance of the paper, perceived by the undistorted eye, though the appearance of the paper to the distorted eye, was always separate from the former, and the sheet consequently seen double.... The apparent situation of the spot being ... at the same time affected by the *voluntary* motions of the eye, it must, I think, be necessarily owing to the *action* of the muscles by which these motions are performed. (pp. 68–70)

Charles Bell (1774–1842) after a frontispiece engraving in Bell (1870).

Bell (1823): ... there is an inseparable connection between the exercise of the sense of vision and the exercise of the voluntary muscles of the eye. When an object is seen, we enjoy two senses; there is an impression upon the retina; but we receive also the idea of position or relation which it is not the office of the retina to give. It is by the consciousness of the degree of effort put upon the voluntary muscles, that we know the relative position of an object to ourselves.... If we move the eye by the voluntary muscles, while the impression [of an afterimage] continues on the retina, we shall have the notion of place or relation raised in the mind; but if the motion of the eye-ball be produced by any other cause, by the involuntary muscles, or by pressure from without, we shall have no corresponding change of sensation. (pp. 178 and 179)

Brewster (1826b): When we push up the one eye by the force of the finger, and keep it in its new position, it has not performed an angular movement in its socket, but merely a small vertical ascent, by which, in an upright position of the body and head, it is raised into a more elevated horizontal plane. In this elevated position, the eye can execute horizontal, and even vertical angular movements, and can direct itself to contemplate either one or other of the double images of the objects before it. The consequence of this is that the spectral impression ascends only through a very small space, which it requires nice observation to appreciate, but which increases with the force applied to the eye.... It is a circumstance not a little remarkable, that it never occurred to Dr. Wells, nor, so far as we know, to any person else, to *press upwards the two eyes at the same time.* When this is done, *the two spectral impressions will be seen to move,* and to rise through spaces so obvious to the dullest perceptions, as to put an end for ever to the extraordinary dogma, that they are capable of being put in motion only by the action of the voluntary muscles. (pp. 267 and 268)

5.9 Binocular Eye Movements

Perhaps the most distinctive feature of eye movements is their binocularity—they tend to move together. Such conjoint motion is not always in the same direction, as Aristotle noted in his distinction between version (movements in the same direction) and convergence eye movements. He also commented that there were certain movements that were not possible, namely, divergence beyond the straight ahead, and movements of one eye upward and the other downward. Ptolemy gave a similar description, but he did appreciate the function eye movements served: they resulted in corresponding visual pyramids for the two eyes. As will be seen in sections 6.1 and 6.5, Aristotle believed that there was a single source of control for the movements of both eyes, although he did not make such a clear statement as Ibn al-Haytham regarding what later became called the law of equal innervation (see Howard, 1996). Leonardo related the equal motion of both eyes to the underlying anatomy, suggesting that the source was located in the optic chiasm (see section 2.9). Porterfield, Wells, Bell, and Müller all reinforced the principle of equal innervation which, it will be seen in the next chapter, originated in much earlier speculations about the visual spirits and their control of the eyes. Porterfield considered that this was a functional advantage attained by learning, whereas Müller maintained that it was innate.

Kepler considered that eye movements were involved in distance perception. In this regard he might have been anticipating Descartes's analysis of binocular vision by means of triangulation, rather like a range-finding device. Not all eye movements are symmetrical, and Briggs examined and illustrated the case of viewing an object to one side; asymmetrical convergence occurs in order to stimulate corresponding retinal points. Smith introduced a simple demonstration of corresponding eye movements—the closed eye follows the movements of the open one, as can be ascertained by lightly placing a finger over the closed lid.

Aristotle (ca. 330 B.C.): Why is it that we can turn the gaze of both eyes simultaneously towards the right and the left and in the direction of the nose, and that of one eye to the left or to the right, but cannot direct them simultaneously one to the right and the other to the left? Similarly, we can direct them downwards and upwards; for we can turn them simultaneously in the same direction, but not separately. Is it because the eyes, though two, are connected at one point, and under such conditions, when one extremity moves, the other must follow in the same direction, for one extremity becomes the source of movement to the other extremity. (Ross, 1927, pp. 957b–958a)

Ptolemy (ca. 150): These phenomena [of binocular combination] occur only by virtue of the horizontal separation of the eyes since the height and the depth of the eyes are the same. Both visual axes turn until the bases of the pyramids coincide on the object. It is possible for the eyes to turn in opposite directions left and right, but not up and down. They retain their vertical position, but can converge horizontally. (Lejeune, 1956, p. 34)

Ibn al-Haytham (ca. 1040): When both eyes are observed as they perceive visible objects, and their actions and movements are examined, their respective actions and movements will be found to be always identical. (Sabra, 1989, p. 229)

Leonardo da Vinci (ca. 1500): ... the cause is the intersection of the optic nerves, that is of the equality of movement, if the eyes observe minutely the parts of a circle, and there are nerves which cause them to make a circular movement. (MacCurdy, 1938a, p. 186)

Kepler (1604): The existence of two eyes is to distinguish the distance of objects, so they must have the potential of following each others direction or orientation, which is thus a movement. (Plehn, 1920, p. 13)

Francis Bacon (1627): The eyes do move one and the same way; for when one eye moveth to the nostril, the other moveth from the nostril. The cause is motion of consent, which in the spirits and parts spiritual is strong. (1857a, p. 628)

Briggs (1683): Now let there be plac't an object near the *left Eye* of any person (but not so near that Eye as that the Nose might hinder the rays from falling on the right, because it is to be seen by both), and whilst that person looks on it let a *By-stander* observe the position of both Eyes, and he shall see that the pupil of the right Eye is turn'd in a very oblique manner to the object, whereas the pupil of the left is scarce so at all, whereby there will be three parts to one more in the distance of the pupil of the right Eye from the *external Canthus* (as may be judg'd from the proportion of the White that appears) than there will be in the other; for the position of the right Eye in respect of the left is as in Fig. 5. a. The Object. b. The left Eye. c. The right. d.d. The Pupils. e.e. Two internal-lateral Fibres. f.f. Two external-lateral. g.g. The Optic Nerves. Hereby it appears that if the Object be so plac't that it is seen with both Eyes, the right accommodates it self to the position of the left, that the rays strike correspondent Fibres, and the percussion or Vibration being towards the bottom or Papilla of the Eye (or near its Axis) where I before observ'd Vision to be chiefly perform'd, a small turning of one Eye to another will make that accommodation. (pp. 177–178)

Smith (1738): The habit of directing the optick axes to the point in view is so strong that it is very difficult to do otherwise; insomuch as when one eye is

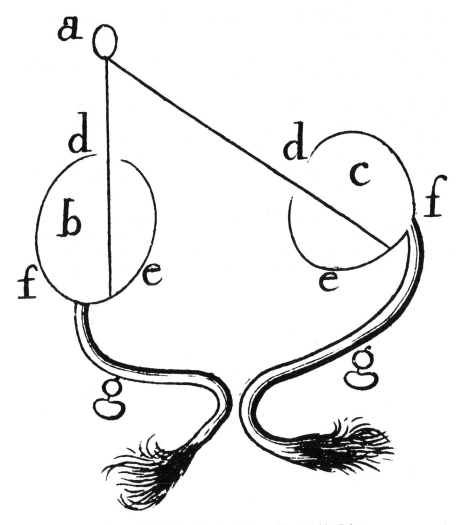

Asymmetrical binocular stimulation as illustrated by Briggs.

shut and the other is in motion one may feel by ones fingers laid upon the eye-lid, that the eye which is shut, always follows the motions of the eye that is open. (p. 46)

Porterfield (1759a): The true Cause of this Uniformity in the Motions of our Eyes to me seems wholly to depend on Custom and Habit: For it is not to be doubted but these Motions are voluntary, and depending upon our Mind, which, being a wise Agent, will them to move uniformly, not from any Necessity in the Thing itself, or for want of Power to move them differently, but because such Motions are most profitable and useful to us. (p. 114)

Wells (1792): Now as all voluntary motions are produced by muscular action, it follows, that every state of action, in the muscles of one eye, has its

corresponding state in those of the other, and that the two are constantly conjoined. (p. 71)

Bell (1803): ... in making the eyes converge or diverge the will is acting upon
both eyes equally. (p. 360)

Brewster (1826b): It is impossible, by the pressure of the finger, to fix the left
eye in a different angular position from the right eye; because, however much
we press upon the former, such is the power of its muscles, and such is the
tendency of both eyes to execute similar movements, that the pressed eye
directs its optic axis as near as possible to the point contemplated by the right
eye, or that which is free. (pp. 267–268)

Müller (1838): The two eyes, whether moved upwards, downwards, or inwards,
must always move together; it is quite impossible to direct one eye upwards
and the other downwards at the same time. This tendency to consensual
motion is evidenced even from the time of birth. (1843, pp. 928–929)

6   Binocularity

Charles Wheatstone (1802–1875) after a frontispiece engraving in *Nature* 13, 1876.

Wheatstone (1838): *... the projection of two obviously dissimilar pictures on the two retinæ when a single object is viewed, while the optic axes converge, must therefore be regarded as a new fact in the theory of vision. It being thus established that the mind perceives an object of three dimensions by means of the two dissimilar pictures projected by it on the two retinæ, the following question occurs: What would be the visual effect of simulta- neously presenting to each eye, instead of the object itself, its projection on a plane surface as it appears to that eye? ... The stereoscope is represented by figs. 8. and 9.; the former being a front view, and the latter a plan of the instrument. (pp. 372–373 and 375)*

In binocularity we find one of the supremely psychological phenomena of vision, hence the constant interest that has been given to it throughout the period under study. Indeed, the period ends with Wheatstone's invention of the stereoscope, the instrument, par excellence, that opened the path to experimental inquiries of vision (see Wade, 1983; Wade and Heller, 1997). Wheatstone was a physicist with interests in acoustics and electricity. Initially he worked in the family business of musical instrument making, and he invented several himself, including the concertina (see Bowers, 1975). His

Fig. 8.

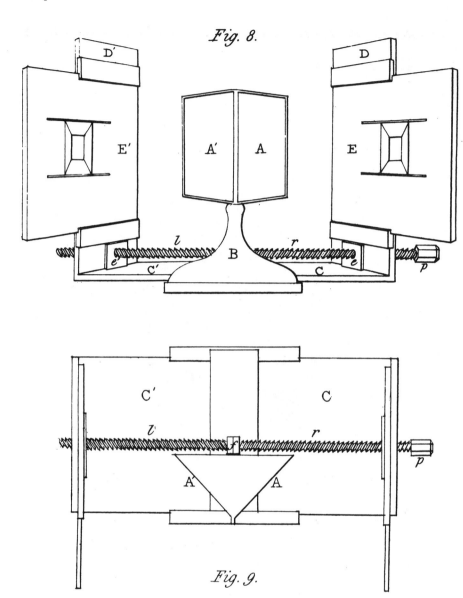

Fig. 9.

Wheatstone's mirror stereoscope.

experiments on visual persistence (section 4.11) fueled his fascination with vision and led to the invention of the stereoscope. The experiments he conducted with it reflected his training: he sought to isolate and manipulate variables in the manner of experiments in physics, thereby disclosing the processes mediating stereoscopic vision. It was his approach that influenced Helmholtz to develop physiological optics as an experimental discipline later in the nineteenth century.

As with many other aspects of visual perception, binocularity has been analyzed in terms of physical optics. Euclid examined binocular vision with the consistency that he had adopted for other aspects of spatial vision—it could be reduced to optical projections. In fact, his discussion of the projections from two eyes was rather cursory, being restricted to three different sizes of sphere with respect to the interocular distance. During the second century A.D. the situation was transformed by Ptolemy in optics, and by Galen in ophthalmology. Of the 180 sections concerned with vision in Books II and III of Ptolemy's *Optics*, 58 are addressed to issues of binocularity. The difference between Euclid and Ptolemy is marked, and the latter can be considered as paving the way for modern approaches to binocular vision (see Crone, 1992; Howard and Wade, 1996). Ptolemy carried out controlled observations of the perceived locations of vertical cylinders; from these he specified the conditions for singleness of vision, the distinction between crossed and uncrossed disparities, and the direction in which objects are seen with two eyes. Remarkably little was added to Ptolemy's analysis until the seventeenth century; it was extended some-what by Ibn al-Haytham, who profited considerably from a thorough knowledge of Ptolemy's work. Ptolemy probably influenced his near contemporary Galen, who pursued similar lines of inquiry, although he did not mention Ptolemy by name (see Siegel, 1970). Chapter 10 of Galen's *De Usu Partium Corporis Humani* (On the Usefulness of Parts of the Body) treats the topics of the eye and vision; his descriptions of binocular vision were particularly astute, since he took Ptolemy's demonstrations into the real world: rather than viewing cylinders arranged on a board, Galen looked at cylinders, or columns, laid out before him.

There are many aspects to binocularity, some of which have been considered already. For example, the pathways from the two eyes to the brain (section 2.9) have obvious implications for the ways in which the signals from the eyes will be combined; the binocular combination of colors (section 3.8) is related to binocular contour rivalry; and binocular eye movements (section 5.9) have been considered above. This chapter examines descriptions that have been made about binocular vision in general, as well as binocular single and double vision. Theoretical issues have directed, as well as having been derived from, observations on binocular rivalry, eye dominance, and strabismus, and comparisons between tasks performed with one or two eyes similarly conflate observation and theory. With such an observational legacy it is truly remarkable that the link between retinal disparity and stereoscopic depth perception came so late.

I have argued elsewhere (Wade, 1987) that the inhibiting factor was the theoretical link between retinal correspondence and binocular single vision, which was only broken by Wheatstone's incontrovertible demonstration of stereopsis seen with two slightly disparate images mounted in the stereoscope.

Surveys of the history of research on binocular vision can be found in Brewster (1856), Nagel (1861), Helmholtz (1867), Gulick and Lawson (1976), Held (1976), Wade (1987), and Howard and Rogers (1995). Only fragments of Wheatstone's first article on binocular vision are presented here; all his papers on vision have been reprinted in Wade (1983).

6.1   Binocular Vision

One of the finest illustrations of the study of binocular vision is to be found as the frontispiece to Book IV of Aguilonius's *Optics*; this and the other five frontispiece engravings were designed by Rubens (see Ziggelaar, 1983). The cosmic observer fixates on the central cross (on the screen), thus producing crossed visible directions of the near object. The putti are pointing to the discs on the screen which mark the locations of the crossed directions. The essence of investigating binocular vision was distilled from the methods adopted for stimulating the two eyes. Rubens's engraving demonstrated the technique of fixating on one object located farther from the eyes than another. This method

Binocular observation of crossed disparities as depicted in the frontispiece to Book IV of Aguilonius (1613).

was introduced by Ptolemy, and elaborated on by Ibn al-Haytham, before its widespread adoption in the seventeenth and eighteenth centuries. Another technique involved placing a septum between the eyes, so that peripheral objects could be seen by one eye but not the other. Galen described this method, and it was pursued by Porta and Aguilonius. Observing distant objects through a small aperture, so positioned that they are aligned each with one eye, was used by Le Clerc and Desaguliers to examine binocular combination and binocular rivalry, respectively. However, by far the most widespread method, and the simplest, was to look with each eye separately and with both eyes.

As was noted in section 5.9, Aristotle considered that both eyes were moved from a single source, and he also stated that vision with one eye was superior to that with two (section 6.6). The quotation that opens this section is a more general statement regarding paired organs which can yield single or double percepts. In comparing vision with touch he related diplopia to the illusion of feeling two objects when one is placed between the crossed fingers. The conclusion he drew from this was exactly the opposite of that adopted centuries later by empiricist philosophers like Berkeley—touch is like vision rather than vice versa. Euclid's analysis was geometrical; he examined three dimensions of sphere that could be observed by two eyes, and simply related them to the amount of the spheres that would be seen. Euclid's use of a sphere was to have unexpected implications: Leonardo examined binocular projections to the eye from a sphere; Wheatstone lamented this, as he felt sure that Leonardo would have appreciated the significance of retinal disparity if he had used any other object (see section 6.9).

Ptolemy appreciated that monocular and binocular visual directions were not necessarily the same. In order to confirm this empirically, he constructed a board on which he could place vertical rods at different distances in the midline (see section 6.2). There followed a description of one of the most commonly used examples of crossed and uncrossed visual directions: with fixation on the far rod, the nearer one appeared double, and to the left with the right eye and to the right with the left eye; the reverse occurred with fixation on the nearer rod. Essentially the same demonstration is now more frequently made with two fingers, rather than rods, held at different distances from the eyes. In the first part of the quotation cited here, Ptolemy stated that singleness of vision with two eyes occurred when the two visual directions corresponded, thus introducing the concept of correspondence into binocular vision. He modified his board to take three rods, in the manner that is shown in the accompanying diagram, and found that objects appeared single to two eyes when they were in the same plane as the fixation point. These facts were interpreted in terms of the visual axes and the common axis.

Ibn al-Haytham made a similar board on which he placed wax cylinders, but of different colors. The text lettering does not correspond exactly with those in the illustration shown because the latter is taken from Alhazen (1572), whereas the text was translated by Sabra from Arabic. The principal differences are in the vertical and horizontal lines: for *HKT* in the text read *kqt*, and for *EZ* read *hz*. Sabra (1966) has also translated a manuscipt by Ibn al-Haytham in which he took issue with Ptolemy's findings using the board and the conclusions he drew from them, demonstrating clearly Ibn al-Haytham's familiarity with Ptolemy's *Optics*.

Galen introduced the method of separating the two eyes by means of a septum, and reported that vision of a peripheral target seen by one eye when both were open was inferior to that with one eye alone. This position was to be repeatedly maintained until the seventeenth century, and it accorded with Galenic theory that the visual spirit, the source of which was at the chiasm, was more concentrated when one eye was open than when both were in use (see section 6.6). In much the same way that Ibn al-Haytham can be seen as developing Ptolemy's optics, Hunain ibn Is-hâq followed Galen's medicine. The quotation given is essentially a repetition of Galen's observation and inter-pretation. In similarly placing the visual faculty at the chiasm, Roger Bacon made the analogy of singleness of vision to a fountain: since a fountain has but one source, so is vision with two eyes single. Galen's method of separating the two eyes was applied with good effect by Porta in the context of eye dominance (section 6.5), and Porta also illustrated Galen's description of the apparent locations of an object seen with each eye separately and with both eyes (section 6.8). Porta stepped out of the Galenic mold by proposing a theory of single vision that was both parsimonious and supported by the phenomenon of binocular rivalry: we see singly with two eyes by using only one at once. Porta's suppression theory could be contrasted with fusion theories of the type advanced by Aguilonius, who introduced the term *horopter* (see section 6.2). Chérubin d'Orléans supported a fusion theory, namely, one which involved the use of both eyes at the same time, because the binocular visual field is so much greater than the monocular. Le Clerc, Haller, and Gall, on the other hand, favored a suppression theory, by arguing that only one eye is used when inspecting objects.

Descartes's analysis of binocular vision was by analogy to a blind man locating an object with two sticks. Essentially the same interpretation (and illustration) is presented in his *Dioptrique* and *Traité de l'Homme*, although a much younger blindfolded person was depicted in *De homine*. His mechanistic analysis of binocular vision has been addressed in the context of the visual pathways (section 2.9) and will be returned to in section 6.2. As was noted in section 2.9, Newton's interests in binocular combination were stimulated by Briggs's mechanical theory, and the quotation used here is taken from his cor-respondence with Briggs. Newton emphasized once more the notion of retinal correspondence in determining binocular single vision, as is amplified in the next section. More empirical demonstrations of binocular phenomena were described by Taylor, Smith, and Le Cat. Smith examined the apparent location of two objects seen separately by each eye (by viewing the extended arms of a compass). He noted that a single object could be seen that had an apparent depth greatly in excess of the physical dimensions of the two components.

Wells investigated experimentally the characteristics of binocular visual direction by means of apertures and also different colored threads extending from each eye to a common point. Ptolemy had constructed a similar arrangement with one of his boards on which colored lines were drawn from the eyes to pass through a common point of intersection. Wells formalized his observations in the three propositions given here. Wells's *Essay upon Single Vision with two Eyes* was one of the first books dedicated to binocular vision, and it is replete with excellent experiments, although stereoscopic vision was not mentioned. Space perception was analyzed in terms of visual direction and visual distance, and most of his empirical concern was with the former. His *Essay* was published in 1792, and republished in 1818 (a year after his death), together with other essays he had written, and an account of his life. Among the other essays was one on dew, in which he gave the first correct theory of its formation, and another on "An account of a white female, part of whose skin resembles that of a negro," in which he anticipated Charles Darwin by some decades in proposing a theory of natural selection (see Shryock, 1944). Darwin, in the fifth edition of his *Origin*, acknowledged that "In this paper he distinctly recognises the principle of natural selection, and this is the first recognition which has been indicated" (p. xxi). Wheatstone was one of the few visual scientists to cite Wells's work on binocular vision, although he did state that "So little does Dr. Wells's theory appear to have been understood, that no subsequent writer has attempted either to confirm or disprove his opinions" (1838, p. 388).

Wheatstone provided a summary of the methods generally employed to combine two similar images in each eye, by under- or overconvergence. In addition, he illustrated binocular combination using two tubes, and a box arrangement. It was because of the difficulties of using such devices that Wheatstone constructed the mirror stereoscope. The box arrangement became called an ocular stereoscope by Brewster (1856), who claimed that James Elliot, a teacher of mathematics in Edinburgh, had made such a device before Wheatstone, and should be accorded priority of inventing the stereoscope. The acrimonious debate between Brewster and Wheatstone over this issue was exposed to the public through the correspondence columns of *The Times*. Elliot had never made such a claim himself and readily retracted, unlike Brewster. This sorry saga can be followed in Wade (1983).

Aristotle (ca. 330 B.C.): Why is it that if the eye be moved sideways a single object does not appear double? Is it because the source of sight is still in the same line? It can only appear double when the line is altered upwards or downwards; and it makes no difference if it is altered sideways, unless it is also at the same time altered upwards or downwards. Why, then, is it possible in sight for a single object to appear double if the eyes are in a certain position in relation to one another, but impossible in the other senses? Is it not possible also in touch that one thing becomes two if the fingers are crossed? But with the other senses this does not happen, because they do not perceive objects which extend to a distance away from them, nor are they duplicated like the eyes. It takes place for the same reason as it does with the fingers; for then the touch is imitating the sight. (Ross, 1927, p. 959a)

Euclid (ca. 300 B.C.): *When a sphere is seen by both eyes, if the diameter of the sphere is equal to the straight line marking the distance of the eyes from each other, the whole hemisphere will be seen.* Let there be a sphere, of which *A* is the center, and on the sphere let the circle *BG* be inscribed about the center *A*, and let *BG* be drawn as its diameter, and at right angles from *B* and *G* let lines be drawn, *BD* and *GE*, and let *DE* be parallel to *BG*, and upon this (*DE*), let *D* and *E* represent the eyes. I say that the complete hemisphere will be seen. Through *A* let *AZ* be drawn parallel to each of the lines, *BD* and *GE*; then *ABZD* is a parallelogram. Now if the inscribed figure is revolved and then restored to the same position whence it started, *AZ* remaining in its place, it will start from *B* and *B* will come over *G*, and the figure formed under *AB* will be a circle through the center of the sphere. So a hemisphere will be seen by the eyes *D* and *E*.

If the distance between the eyes is greater than the diameter of a sphere, more than the hemisphere will be seen. Let there be a sphere, of which *A* is the center, and about the center *A* let the circle *ETDL* be inscribed, and let the eyes be *B* and *G*, and let the distance between the eyes *B* and *G* be greater than the diameter of the sphere, and let *B* and *G* be joined. I say that more than the hemisphere will be seen. Let the rays fall, *BE* and *GD*, and let them be produced beyond the points *E* and *D*; they approach each other because the diameter is less than *BG*. So let them meet at point *Z*. Now, since for some point outside the circle the straight lines *ZE* and *ZD* have touched the circumference, *DTE* is less than a semicircle. But *ELD* is seen by the eyes at *B* and *G*. So, more than half the circle will be seen by the eyes at *B* and *G*.

If the distance between the eyes is less than the diameter of the sphere, less than a hemisphere will be seen. Let there be a sphere, the center of which is the mark *A*, and let the circle *BG* be inscribed around the mark *A*, and let the distance between the eyes, *DE*, be less than the diameter of the sphere, and from the eyes let the same rays be drawn, *DB* and *EG*, touching the sphere. I say that less than a hemisphere will be seen. For let *BD* and *GE* be produced; they will fall upon the section *GLB*, since *DE* is less than the diameter of the sphere. Let them meet at the point *Z*. Now, since from a certain mark, *Z*, the straight lines have fallen, *ZG* and *ZB*, *BLG* is less than a semicircle. But the section of the sphere corresponds to the section *BLG*. So, the rays include less than a hemisphere. (Burton, 1945, pp. 362–363)

Ptolemy (ca. 150): If we join lines *ae, az, zb, eb, ta, tb, bh, ak*, any of *e, d*, and *z* will appear in one location, since *ad* and *bd* are the visual axes, and the visual lines which converge on *e* and *z* are corresponding visual lines because *ae*

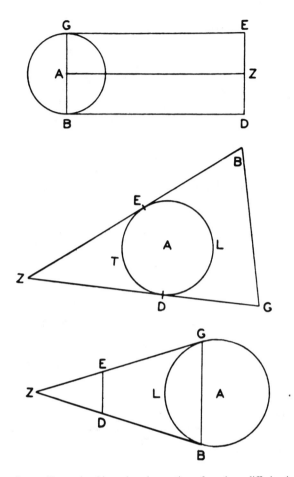

Euclid's three figures illustrating binocular observation of a sphere differing in diameter.

corresponds to *be* and *az* corresponds to *bz*. But *h* and *k*, will appear in one location *t*, since *ah* and *bk* are visual axes. Because *bh* and *ak* are non-corresponding visual lines *h* and *k* will appear at points *l* and *m*: Because visual lines *at* and *bt* are non-corresponding, point *t* will appear in two locations *h, k*. (Lejeune, 1956, pp. 104–105)

Galen (ca. 175): ... if you care to place longitudinally on your nose between your eyes a small piece of wood, your own hand, or anything else that can prevent external objects lying before them from being seen by both eyes, you will see dimly with each eye, but much more clearly if you close one eye, as if the faculty hitherto divided between two were now coming to the other eye. (May, 1968, p. 501)

Hunain ibn Is-hâq (ca. 850): If a man puts his hand lengthwise on his nose in such a manner as to separate both eyes, or if he sets up in the same place another object which prevents the vision of both eyes from falling together on

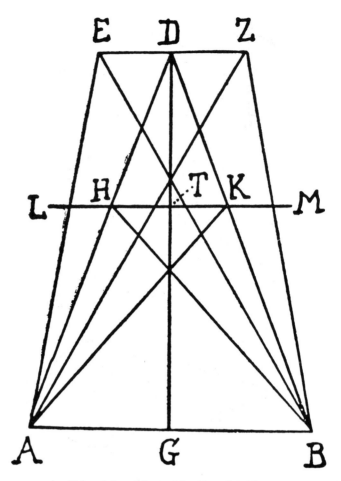

Ptolemy's board for studying binocular vision.

the object on which he directs his gaze, his sight with each eye singly is dimmer and weaker than his sight with both together. If, on the other hand, he shuts one eye, his sight with the other one becomes clearer and sharper. The reason of this is that the whole power which was divided between both of them in two halves now enters into this one eye alone. Therefore, if you look into the pupil of the open eye, when the other is shut, (you will find that) it is considerably enlarged. (Meyerhof, 1928, pp. 24–25)

Ibn al-Haytham (ca. 1040): Take a light-weight wooden board of a pale colour, one cubit in length and four fairly large digits wide. Its surface should be even and smooth and its longitudinal and latitudinal edges should be parallel. Draw two intersecting diameters on it, and from the point of intersection draw a straight line parallel to its length and another straight line perpendicular to the first, middle line. Paint these lines in different bright colours to make them visible, and paint the diameters in one colour. In the middle of the board's shorter edge and at the end of the middle straight line between the diameters, make a round but narrow opening whose wider part at the beginning is large enough for inserting the bridge of the nose so that the board may rest upon it in such a way that the board's corners will be extremely close to the middles of the eyes' surfaces, so close in fact that they almost touch them without actually doing so.... He will see each of the three objects [cylinders of wax painted in different colours] at points H, K and T single, and will also find line HKT to be one. But line EZ, which extends through the length of the board, will appear as two lines intersecting at the middle object. Similarly, when contemplating the diameters while in this situation, he will find them to be four, each of them appearing double. (Sabra, 1989, pp. 237–238 and 239)

Roger Bacon (ca. 1270): ... there must be something sentient besides the eyes, in which vision is completed and of which the eyes are the instruments that give it the visible species. This is the common nerve in the surface of the brain, where the two nerves coming from the two parts of the anterior brain meet, and after meeting are divided and extend to the eyes. Therefore the visual faculty is located here, as is a fountain, and since in this case the fontal faculty is a single one, to which the faculties of the eyes are continued through the medium of the optic nerves, an object can appear as single. (Burke, 1928, pp. 450–451)

Pecham (ca. 1280): The benevolence of the Creator has provided that there should be two eyes so that if an injury befalls one, the other remains. (Lindberg, 1970, p. 117)

Porta (1593): Let A, be the pupil of the right eye, B that of the left, and DC the body to be seen. When we look at the object with both eyes we see DC, while with the left eye we see EF, and with the right eye GH. But if it is seen with one eye, it will be seen otherwise, for when the left eye B is shut, the body CD, on the left side, will be seen in HG; but when the right eye is shut, the body CD will be seen in FE, whereas, when both eyes are opened at the same time, it will be seen in CD. (Brewster, 1856, p. 8)

Aguilonius (1613): When one object is seen with two eyes, the angles at the vertices of the optical pyramids are not always equal, for beside the direct view in which the pyramids ought to be equal, into whatever direction the eyes are turned, they receive pictures of the object under inequal angles, the greatest of which is that which is terminated at the nearer eye, and the lesser that which regards the remoter eye. This, I think, is perfectly evident; but I consider it as worthy of admiration, how it happens that bodies seen by both eyes are not all confused and shapeless, though we view them by the optical axes fixed on the bodies themselves. For greater bodies, seen under greater angles, appear lesser bodies under lesser angles. If, therefore, one and the same body which is in reality greater with one eye, is seen less on account of the inequality of the

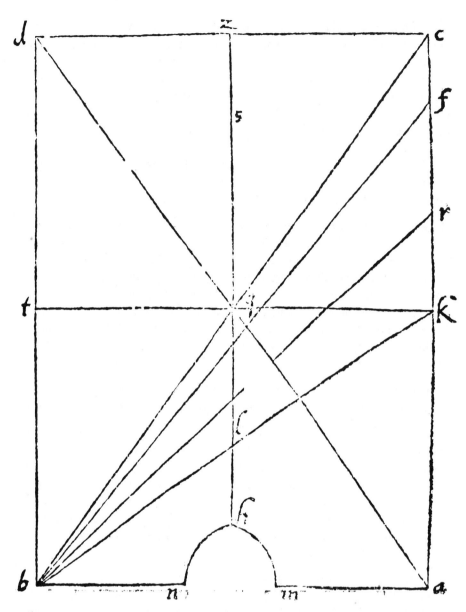

Ibn al-Haytham's board for viewing cylinders of wax by both eyes. The lettering on the figure (which is from Alhazen, 1572) does not correspond exactly with the text: for *HKT* in the text read *kqt* in the figure, and for *EZ*, read *hz*.

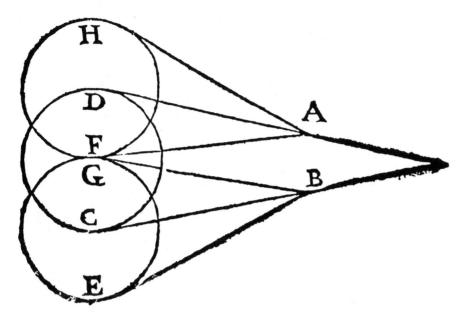

Monocular and binocular views of an object (Porta, 1593).

angles in which the pyramids are terminated, the body itself must assuredly be seen greater or less at the same time, and to the same person that views it; and, therefore, since the images in each eye are dissimilar the representation of the object must appear confused and disturbed to the primary sense. (Brewster, 1856, p. 12)

Descartes (1637): And just as this blind man does not judge that a body is double, although he touches it with both hands, so likewise when both eyes are disposed in the manner which is required in order to carry our attention toward one and the same location, they need only cause us to see a single object there, even though a picture of it is formed in each of our eyes. (1965, p. 105)

Descartes (1664): Notice also that if the two hands *f* and *g* each hold a stick, *i* and *h*, with which they touch object *K*, although the soul is otherwise ignorant of the length of these sticks, nevertheless because it knows the distance between the two points *f* and *g* and the size of angles *fgh* and *gfi*, it will be able to know, as if through a natural geometry, where object *K* is. (Hall, 1972, p. 62)

Chérubin d'Orléans (1677): *The extent of the visual field that one sees on the two sides when we have both eyes open at the same time, is much greater than that which one sees with each eye separately, which is an obvious proof that perfect vision is made by using two eyes together.* (p. 3)

Sébastien Le Clerc (1679): Nature has given us two eyes, but the two eyes have separate functions; and when they only see the same things, both are not necessary: on the contrary, if they act together one mirrors the other . . . because

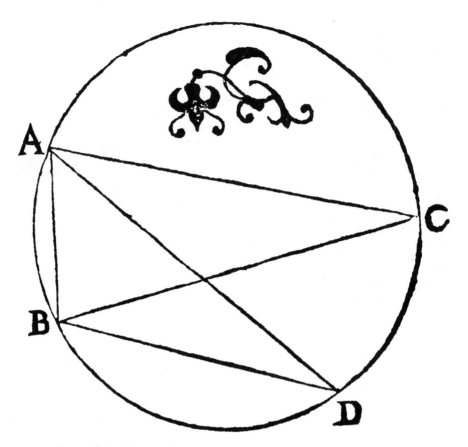

Symmetrical and asymmetrical convergence as represented by Aguilonius (1613).

even though we have two eyes, when it is a question of seeing and examining the objects with which we are presented, we use only one. (pp. 7–8)

Newton (1682): It may be further considered that the cause of an objects appearing one to both eyes is not its appearance of ye same colour form & bigness to both, but in ye same situation or place. Distort one eye & you will see ye coincident images of ye object divide from one another & one of them remove from ye other upwards downwards or sideways to a greater or less distance according as ye distortion is; & when the eyes are let return to their natural posture the two images advance towards one another till they become coincident & by that coincidence appear but one. If we would then know why they appear but one, we must enquire why they appear in one & ye same place & if we would know ye cause of that we must enquire why in other cases they appear in divers places variously situate & distant one from another. (Turnbull, 1960, p. 382)

Taylor (1738): If one places two candles D E, with the same separation as the eyes B C, and if one puts a little plate P Q, in which one had made a small

Binocular vision as an analogy with a blind man holding two sticks (Descartes, 1637/1902).

hole of diameter about half a thumb in A, then one fixes the hole of this plate equally distant from the candles and from the eyes; when one looks straight ahead without paying attention to this plate or to any other thing, one sees two holes and two candles; but if one turns the eyes in the direction of the hole, with the intention of looking at it, then one sees only one hole and only one candle; the reason is that in the first case Gg and Ff was where the optic axes were located, and consequently the impressions which were made on e by the candle E and the hole A, had no communication with the impressions which were made on d, by the candle D and the hole A. The fibres on the side e m do not connect to the fibres on the other side d e, but when the points GF in the middle of each eye are carried to e and d, they receive the impression of the candle and of the hole. (pp. 169–170)

Smith (1738): [If a pair of compass legs are opened] somewhat wider than the interval of your eyes; with your arm extended hold the head or joint in the ball of your hand, with the points outwards and equidistant from your eyes, and somewhat higher than the joint. Then fixing your eyes upon any remote object lying in the line that bisects the interval of the points, you will first perceive two

Tactile analogies of binocular single vision from Descartes (1662) above and Descartes (1664/1909) below.

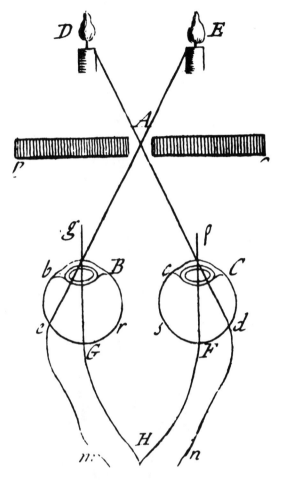

Taylor's method of binocular combination.

pairs of compasses (each leg doubled) with the inner legs crossing each other, not unlike the old shape of the letter W. But by compressing the legs with your hand, the two inner points will come nearer to each other; and when they unite, (having stopt the compression,) the two inner legs will also entirely coincide and bisect the angle under the outer ones; and will appear more vivid, thicker, and longer than they do, so as to reach from your hand to the remotest object in view, even in the horizon itself. (p. 388)

Le Cat (1744): ... look with both Eyes, A, B, Fig. 1. at the Candle C. Beyond this Candle have two Objects fixed, E, F. Look at the Candle with a strong Attention; and see with which of the two Objects, E or F, corresponds. If with Object E, it is with the right Eye you discern this Candle. If it corresponds with the Object F, you see it with the left; or at least your Soul is only attentive to the Image painted in one of your Eyes; and this manner of seeing is the most general. ... The first Time I was convinced that I saw the same Object with both

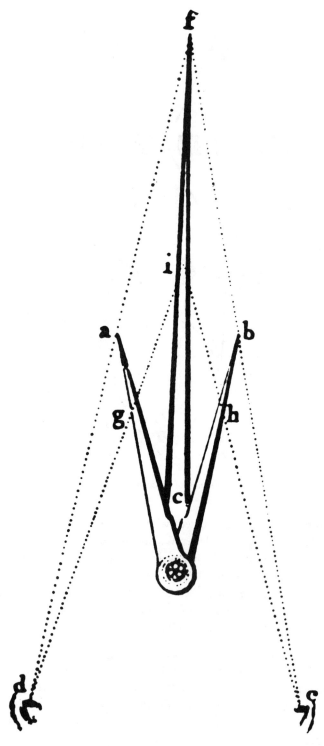

Smith's compass legs demonstration.

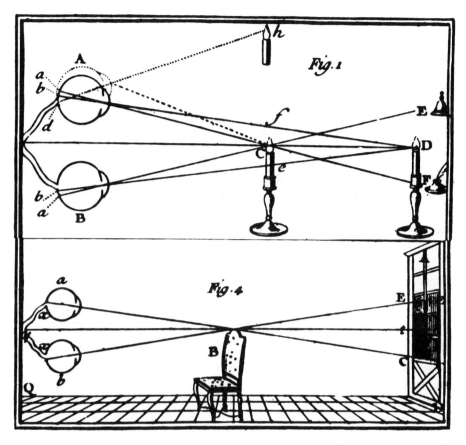

Two diagrams illustrating binocular viewing from Le Cat (1744).

Eyes at once, I lay in Bed on my left Side, both my Eyes determined vertically, as in Fig. 4. had my Body and Feet extended towards Q; over against me was a Window A; and between me and the Window there was the Back of a Chair B. The Back of this Chair hid from me all the lower Part C D of the Window. I looked at the Window and the Chair in a confused manner, that is to say, as one does ordinarily on waking. I saw all the upper Part A C of this Window, but on the lower, C E, I distinguished a Crowd of Vapours e, e, of the Figure of the Back of the Chair.... From this Experiment I conclude, first, that we see Objects with both Eyes at once. Secondly, that one sees better with both Eyes than with one. Because the Portion A E, discerned by both Eyes, always appeared to me clearer and more luminous. Thirdly, that one sees better on looking with Attention, with any Sort of Effort ... Fourthly, that, in case we sometimes see the Object but with one Eye, it is because the Attention is excited rather in this Eye, than in the other, by reason that the Object is on the Side of this Eye, and first strikes it; or else, because we have acquired a particular Habit of putting this Eye on Action rather than the other. (1750, pp. 186 and 192–194)

Buffon (1749): A second error in the vision of infants arises from the double appearance of objects; because a distinct image of the same object is formed on the retina of each eye. It is by the experience of feeling bodies only, that children are enabled to correct this error. (1866, p. 240)

Haller (1767): Whether see we with one eye, or with both? Most frequently with one, and more especially the right eye: but when both are employed together, we see more objects, and more plainly; and we also distinguish more points of the same object, and judge better their distances. (1786, p. 32)

Wells (1792): Objects situated in the Optic Axis, do not appear to be in that Line, but in the Common Axis.... Objects, situated in the Common Axis, do not appear to be in that Line, but in the Axis of the Eye, by which they are not seen.... Objects, situated in any Line drawn through the same Intersection, to a Point in the Visual Base distant half this Base from the similar Extremity of the former Line, towards the left, if the objects be seen by the Right Eye, but towards the right, if seen by the Left Eye. (pp. 40, 46, and 50)

Young (1801): The eyes sympathize perfectly with each other; and the change of focus is almost inseparable from a change of the relative situation of the optic axes. (p. 54)

Brown (1820): When this process, by which our visual sensations become a language significant of things without, has been clearly understood, many supposed mysteries of vision cease to be mysteries. We see an object, for example, as *one*, when we look on it with a single eye, and we see it still as *one*, though we look on it with both eyes. But this single vision, from a double organic affection, has nothing wonderful in it, if we consider single and double affections have been equally *associated*—to use a common phrase—with the tactual and muscular feeling of a single object, when that single object alone was grasped by us, and we gazed on it, usually with both eyes, sometimes perhaps with only one. The visual sensation in this case, suggests what co-existed with it before, and only what co-existed with it before. (p. 149)

Franz Josef Gall (1757–1828) after a frontispiece lithograph in Gall (1835a).

Gall (1822): I have rendered to the sight its just rights, of which the philosophers had deprived it. I have proved that the eye, without the aid of any other sense, and without previous exercise or instruction, can perceive, not only the impressions of light and colors, but likewise those of forms, size, direction, number, and distance of objects. I have established, that the eye is not the organ of the talent of painting, and I have seized the occasion to show the great difference, which exists between the *passive* functions of our organs and their *active* functions. I have also demonstrated that man and animals fix objects, see, and look actively, with one eye only. (1835a, p. 107)

Müller (1826a): It is a sign of poor self observation when Gall and others maintain that we see alternately with only one eye. Who can doubt during

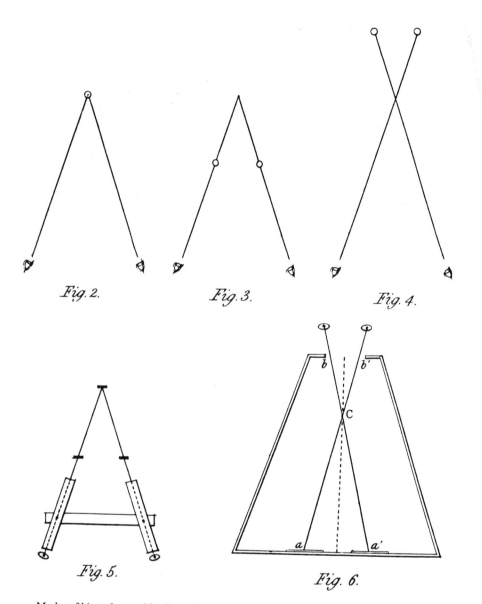

Modes of binocular combination prior to invention of the stereoscope (Wheatstone, 1838).

simultaneous use of both eyes the frequent occurrence of double images of objects. (p. 71)

Wheatstone (1838): Under the ordinary circumstances of vision the object is seen at the concourse of the optic axes, and its images consequently are projected on similar parts of the two retinæ; but it is also evident that two exactly similar objects may be made to fall on similar parts of the two retinæ, if they are placed one in the direction of each optic axis, at equal distances before or beyond their intersection. Fig. 2. represents the usual situation of an object at the intersection of the optic axes. In fig. 3. the similar objects are placed in the direction of the optic axes before their intersection, and in fig. 4. beyond it. In all these three cases the mind perceived but a single object, and it refers it to the place where the optic axes meet.... If the eyes are to converge beyond the objects, this may be afforded by a pair of tubes (fig. 5.) capable of being inclined towards each other at various angles, so as to correspond with the different convergences of the optic axes. If the eyes are to converge at a nearer distance than that at which the objects are placed, a box (fig. 6.) may be conveniently employed; the objects *aa'* are placed distant from each other, on a stand capable of being moved nearer the eyes if required, and the optic axes being directed towards them will cross at *c*, the aperture *bb'* allowing the visual rays from the right hand object to reach the left eye, and those from the left hand object to fall on the right eye; the coincidence of the images may be facilitated by placing the point of a needle at the point of intersection of the optic axes *c*, and fixing the eyes upon it. In both these instruments (figs. 5. and 6.) the lateral images are hidden from view, and much less difficulty occurs in making the images unite than when the naked eyes are employed. (p. 373)

6.2   Binocular Single Vision

Binocular single vision has been discussed at least since the time of Aristotle, and it has been examined experimentally since Ptolemy, who defined lines of visual correspondence for the two eyes. The concept is probably embodied in the mythological Cyclops who forged thunderbolts for Zeus, and in the Homeric *Odyssey*, where Cyclops was a one-eyed giant. The location of the single eye was central in the forehead, and the locus of binocular visual direction is now referred to as the cyclopean eye (see Julesz, 1971). The illustration by Rubens for Book I of Aguilonius's *Optics*, shows the putti performing an operation beyond the scope of modern medicine—the cyclopean eye is being dissected.

Aristotle did discuss binocular single vision, but it tended to be in the context of its breakdown (double vision) either by distorting one eye (section 6.3) or in strabismus (section 6.7). Although the term "correspondence" is used in the translation given, his description of it was general, as opposed to Ptolemy's precise definition of it. Ptolemy opened the way to a systematic analysis of

Dissecting the cyclopean eye, as shown in the frontispiece of Book I by Aguilonius (1613).

single vision, by means of his board. The figure shown here is the first diagram from his *Optics*, and represents a considerably simpler arrangement than that depicted in section 6.1, as does that of Ibn al-Haytham. Nonetheless, with this simple setup he was able to define both corresponding visual lines and crossed and uncrossed visual directions. It was noted in the previous section that Ptolemy effectively defined the plane of binocular singleness as passing through the fixation point and perpendicular to the common axis. This was repeated by Galen, Hunain ibn Is-hâq, Ibn al-Haytham, and by Aguilonius, who named and defined it as the horopter. Galen did add a physiological dimension to the interpretation of singleness, by enlisting the optic chiasm as the source, and this was both amplified and illustrated by Roger Bacon.

The Galenic visual spirits pervaded Descartes's mechanistic analysis of binocular single vision, and he speculated that union of the fibers from the two eyes occurred in the pineal body. As was remarked in the commentary to section 2.9, Descartes considered, on the then available anatomical evidence, that the two optic nerves remained separate at the chiasm; thus, union was achieved in the only unpaired body in the brain. The reference to correspondence in the quotation can be misleading, as it is unlikely that this was intended to imply corresponding points on the two retinas, and "correspondence" is a term that was not present in the original French (see Hall, 1972). The correspondence to which Descartes was referring was probably that between objects in space and the retinal image, following the analogy of eye and camera. In this way, the same text could have been illustrated with either one or two eyes, and the contrast with the description of binocular vision via the analogy to the blind (section 6.1) is not so stark. Nevertheless, the notion of union between two equivalent retinal images was anathema to some, like Sébastien Le Clerc, whose knowledge of perspective enabled him to represent the differences between the images projected to each eye. Rather than taking this to be evidence for stereopsis, it was used to counter Descartes's fusion theory: if the retinal images were different, then there could no fusion of them in the pineal body.

There is no doubting that Newton was using the term "corresponding points" in a physiological sense. He presented the empirical case for partial decussation at the optic chiasm (section 2.9), although he erred in its detail. Correspondence was physical in his analysis, because the corresponding fibers from the optic nerves actually unified at the chiasm. This had implications for his interpretation of binocular rivalry (section 6.4). Newton's approach to binocular single vision was readily adopted by Desaguliers and Smith, who gave it empirical and definitional precision, respectively. Desaguliers also repeated Molyneux's distinction between looking and seeing, which was essentially one between central and peripheral vision. Observations of different stimuli using the aperture technique became increasingly popular, and were employed by Desaguliers, Taylor, and Le Cat.

The situation remained much the same until the beginning of the nineteenth century. Bell (1803) came close to specifying the rule that was made explicit by Prevost in the following year: corresponding points fall on a circle passing through the point of bifixation and the centers of the eyes. That is, the horopter is a circle rather than a plane. This was rediscovered by Vieth and later by Müller (1826a) and has become known as the Vieth-Müller circle. Bell stated that singleness of vision was a result of stimulating points equally displaced from the center of the retina, and his diagram does reflect a circle of correspondence: the point B lies on an arc of a circle passing through A and the eyes. Alternative views were still voiced: Gall and Spurzheim reiterated their belief in suppression theory, and Brewster returned to an analysis in terms of visual directions. For him binocular single vision was in fact double vision of equivalent images. Müller (1838) augmented his geometrical description of the circle of single vision by linking it with identical retinal points. In this way, there were only two possible states of perception—singleness when objects fell on the circumference of the circle and doubleness otherwise, and singleness was served by a fixed organic relation between nerve fibers. Thus, in the year that saw publication of Wheatstone's article on stereoscopic depth perception we find a statement denying its possibility. Wheatstone was well aware of the originality of both his observations and his interpretation of them, hence the meticulousness of his experiments in their support.

Aristotle (ca. 330 B.C.): When, therefore, being so placed that they are in a similar position to one another and midway between an upward and a downward and an oblique movement, the two eyeballs catch the visual ray on corresponding points of themselves.... If the vision of both eyes does not rest on the same point, they must be distorted. (Ross, 1927, p. 958a)

Ptolemy (ca. 150): Let *a* be the left eye and *b* the right eye. Place two rods *g* and *d* on the perpendicular to *ab*. Extend to them from each eye the visual lines *ga, gb, da, db*. Let the eyes be converged on the nearest rod *g*. *ag* and *bg* fall on the visual axes. Of the remaining two visual lines, *ad* is to the left of the visual axis of the left eye and *bd* is to the right of the visual axis of the right eye. Thus *g* is seen in one location, because the two visual axes are corresponding visual lines; but *d* appears double since visual line *ad* is to the left of the visual axis of the left eye, but visual line *bd* is to the right of the visual axis of the right eye. Therefore when we cover the left eye, the left image will not be seen; and when we cover the right eye, the right image will not be seen. If the eyes converge on *d*, it will come about in the opposite way. Because *ad* and *bd* are on the visual axes *d* will be seen as one. *g* will appear double because *ag* is to the right of the visual axis of the left eye, and *bg* is to the left of the visual axis of the right eye. If we cover the left eye the left image will not be seen, and if we cover the right eye the right image will not be seen. (Lejeune, 1956, pp. 29–30)

Galen (ca. 175): ... the axes of the visual cones must be situated in one and the same plane if single objects are not to appear double. These axes of ours

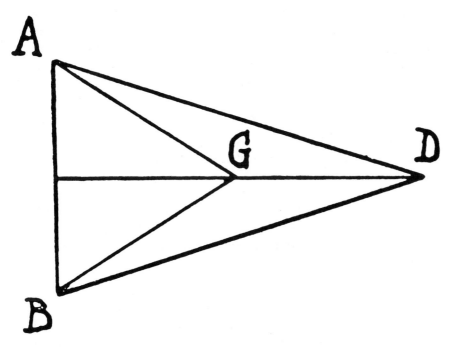

Ptolemy's board for observing crossed and uncrossed visible directions.

have as their beginning the channels from the encephalon.... It is the bringing together of the channels [in the optic chiasm]. For if two straight lines meet at a certain, common point in their apex, they are evidently in one plane, even if they happen to be produced from that point an infinite distance in different directions. (May, 1968, p. 498)

Hunain ibn Is-hâq (ca. 850): Hereby is proved that the following three things must be situated in each eye in one straight line and must follow the same direct course, *viz.* (*a*) the pupil, (*b*) the origin of the whole eye there where the optic nerve begins to appear and to be visible, and (*c*) the place of junction of both nerves from which they begin their course (to the eyes) all lie on the same plane. (Meyerhof, 1928, p. 26)

Ibn al-Haytham (ca. 1040): A beholder, however, perceived visible objects with two eyes. But if vision is brought about through the form that occurs in the eye, and if the beholder perceives the object with two eyes, then the forms of visible objects occur in both eyes, and thus for every object there occur two forms in the eyes. Nevertheless, the beholder perceives each object in most cases as one. The reason for this is that the single object's two forms that occur in the eyes when the object is perceived as one come together when they reach the common nerve and coincide with one another and become one form. And from the form thus united from these two forms the last sentient perceives the form of the object.... if the two axes of the eyes meet at a point on the surface of the object facing the eyes, then that surface will be the common base of the radial cones formed between the centres of the eyes and the object; the point in which both axes meet will have the same position with respect to both eyes because it will be opposite the middles of both eyes, and the axes between it and the two eyes will be perpendicular to the surfaces of those eyes at their middles. As for the remainder of the object's surface, there will exist between every point in it and the centres of both eyes two lines similarly situated in direction with respect to the two axes. (Sabra, 1989, pp. 85 and 229)

Roger Bacon (ca. 1270): We are to understand, moreover, that from the common nerve an imaginary straight line is directed between the two eyes and the object seen, meeting the axes of the eyes in the same part of the object seen, and this line is the common axis, and that point on which these three axes fall is seen with final certitude, as is shown in the figure, and other parts of the object are seen with more or less distinctiveness according to the positions they occupy with respect to this axis. For point *a* is seen most distinctly, because the three axes are concurrent at it, and *b* and *c* are seen more distinctly than *e* and *d*. (Burke, 1928, p. 511)

Aguilonius (1613): If the common radius intercepts the horopter at right angles, the horopter makes the angles equal with the axes. From the two eyes A and B the optic axes project AC and BC and the common radius FC intercepts the

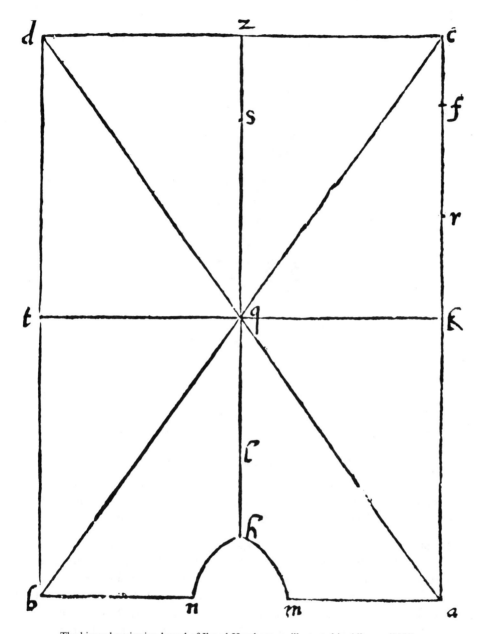

The binocular viewing board of Ibn al-Haytham, as illustrated in Alhazen (1572).

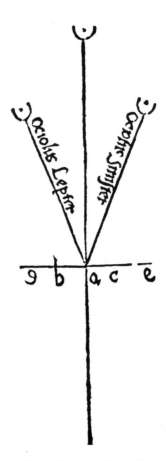

Binocular single vision according to Roger Bacon.

horopter DE at right angles. I say that those angles which the horopter makes with the axes, namely ACD and BCE are equal to one another. (p. 147)

Descartes (1664): . . . the spirits that tend to enter each of the tubules *2*, *4*, *6*, and the like do not come indifferently from all points on the surface of the gland *H* but each from a particular point; those that come from point *a* of this surface, for example, tend to enter tube *2*, those from points *b* and *c* tend to enter tubes *4* and *6*, and so on. As a result, at the same instant that the orifices of these tubes enlarge, the spirits begin to leave the facing surfaces of the gland more freely and rapidly than they otherwise would. And [suppose] that just as [*a*] the different ways in which tubes *2*, *4*, and *6* are opened trace on the internal surface of the brain a figure corresponding to that of object *ABC*, so [*b*] [the different ways] in which the spirits leave the points *a*, *b*, and *c* trace that figure on the surface of this gland. (Hall, 1972, p. 85)

Sébastien Le Clerc
(1637–1714) after an
engraving in Landon
(1805).

Sébastien Le Clerc (1679): Monsieur Descartes having considered that according to his principles external objects should make an impression on both eyes,

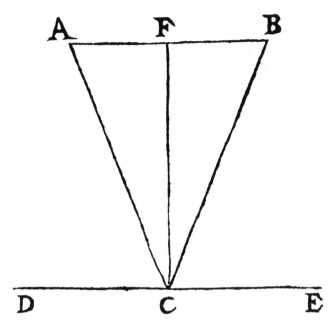

The horopter of Aguilonius.

and that the soul nevertheless had only one perception believed that the images
of the same object found in the two eyes are reunited in the brain; but if this
great genius had reflected a little more on the demonstrations which he gave
in his Treatise on Man, he would have recognised that the images in the two
eyes although produced by the same object, are different, and because of these
differences their reunion is impossible. (pp. 44–46)

Briggs (1682): Why Vision is not *double*, since the organ is so? *Resp*. Because
the *Fibræ concordes* are ... like *unisons* in a *Lute*, and the rayes strike both at
the same time.... Hence also it appears, why upon pressing down one Eye,
an object appears double? viz. Because thereby its rayes fall upon discordant
fibres in the other Eye not prest, so that there are two different sensations.
(p. 172)

Newton (1682): If when we look but wth one eye it be asked why objects
appear thus & thus situated one to another. The answer would be because
they are really so situated among themselves & make their colored pictures
on ye Retina so situated one to another as they are & those pictures transmit
motional pictures into ye sensorium in ye same situation & by the situation of
those motional pictures one to another the soul judges of the situation of things
without. In like manner when we look with two eyes distorted so as to see ye
same object double if it be asked why those objects appear in this situation &
distance one from another the soule judges she sees two things so situate and
distant. And if this be true then the reason why the distortion ceases & ye eyes

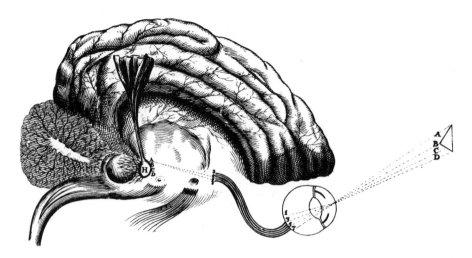

Diagrams of the visual pathways from Descartes (1662) on the left and Descartes (1664/1909) on the right, taken to represent binocular combination.

return to their natural posture the doubled object grows a single one is that the two motional pictures in ye sensorium come together & become coincident. But you will say, how is this coincidence made? I answer, what if I know not? Perhaps in ye sensorium, after some such way as ye Cartesians would have believed or by some other way. Perhaps by ye mixing of ye marrow of ye nerves in their juncture before they enter the brain, the fibres on ye right side of each eye going to ye right side of ye head those on ye left side to ye left. (Turnbull, 1960, pp. 383–384)

Molyneux (1692): For in Vision there is a Difference between *looking* and *seeing*, what ever Object I *look* at with both Eyes appears *single*, and all others more remote or nigher, the I *see*, appear *double*, for upon the Object, I *look* at, the Optick Axes do concur, but not so on those I only *see*. (pp. 288–289)

Desaguliers (1716): Whatever is *seen*, by being *look'd* at with both Eyes, always appears single, by reason of the Communication between the Middle of the *Retina* in one Eye, and the Middle of the *Retina* of the other: there being no such Communication between any other part of the *Retina* in one Eye, and the Correspondent part of the *Retina* in the other, when these correspondent parts are equally distant from the Nose. *There is indeed a Communication between the Nervous Fibres on the Right-side of the* Retina *of one Eye, and the nervous Fibres on the Right-side of the* Retina *of the other Eye, and so of those on the Left: but no single Object can be so painted in each Eye, as to have its Image on the Right or Left Part of one* Retina *that communicates with the Right or Left part of the other, of the same bigness and at the same time as in the other; because in whatever Position the Object is, it must be nearer to one Eye than to*

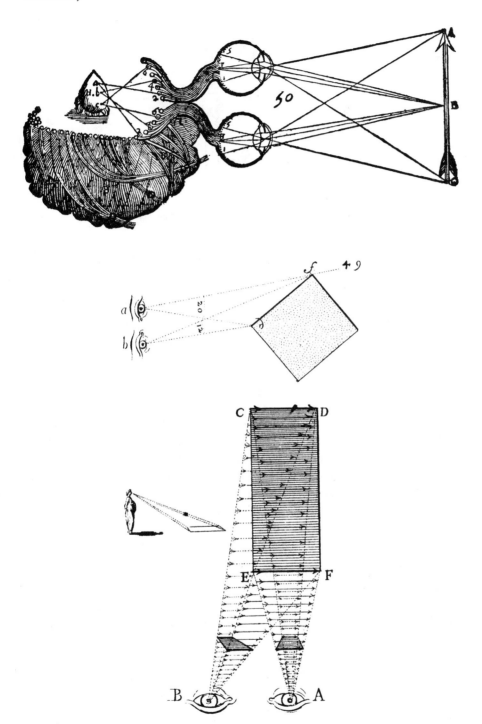

Binocular observation of solid structures, indicating the differences in the projections to each eye as illustrated in Le Clerc (1679).

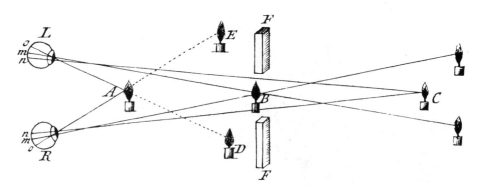

Viewing two candles through an aperture (Desaguliers, 1716).

the other, except it be just in a Line from the Nose betwixt the two Eyes streight forward. Hence it is that if there be two Candles set before one, the First at a Distance of one Foot, and the Second at the Distance of two Feet, from the Eyes; he that looks at the second Candle at *B* will see it single, but see the first Candle or the Candle *A* double; one Appearance being in the Line *ADγ*, the other in *oAE*, because it paints it self upon *oo* in the Retina of each Eye, which Points are not the middle Points, but farther from the Nose than the middles *mm*. So if *B* be the first Candle, and *C* the second, he that looks at *B* will see *C* double, because it is painted in the *Retina* at the Points *nn* nearer the Nose than *mm* . . . If *γ ρ* be two Candles so disposed . . . that by the Interposition of a perforated Board *FF*, *γ* can paint itself only in the Eye *R*, and *ρ* in the Eye *L*. Upon making the Optic Axes meet at *B* and to tend towards *γ*, *ρ* and *γ* will each paint an Image on the Middle of the *Retina* of each Eye, by crossing their Rays at *B*: and thus the two Candles will appear to be but One, or rather to be in one Place, upon the account of the Communication of the Middle of each *Retina.* . . . If the Optic Axes be turn'd directly towards *γ* and *ρ*, as if there was no Board *FF* in the way, there will appear two Holes in the Board, the one having the red hot Iron in it, the other the Candle. (pp. 450–452)

Taylor (1738): Suppose that BC represents the two eyes, A a candle at a distance, say, of two feet, if one looks with the eye B, or with the eye C, and closes them successively, one after the other, or closes both eyes at the same time, the candle will always have the same appearance, and one sees just one, and because the image of this candle is painted on the right eye at E, while it is on the left eye at D; it is necessary that there is a communication from E to D in the *Sensorium commune* (i.e, the soul receives the impression communicated by the sense), otherwise we will see two candles instead of one. (pp. 165–166)

Smith (1738): The axis of the eye is a line drawn through the middle of the pupil and of the crystalline, and consequently falls on the middle of the retina. And the axes of both eyes produced are called the optick axes. When the optick axes are parallel or meet in a point, the two middle points of the retinas, or any two

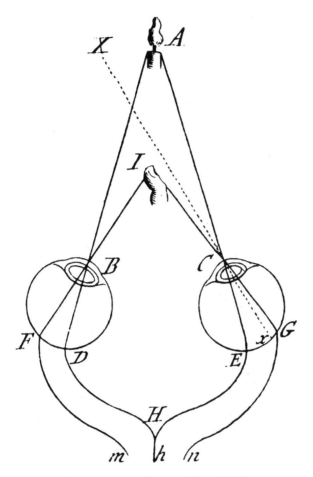

Binocular single vision according to Taylor.

points which are equally distant from them, and lye on the same sides of them either toward the right hand or left hand, or upwards or downwards, or in any oblique direction, are called corresponding points. Now we find by experience that an object or point of an object appears single when its pictures fall upon corresponding points of the retinas, and double when they do not. (p. 46)

Le Cat (1744): Place two Candles, E, F, Fig. 2. at a certain Distace one from the other. You are in C. Look at these Candles thro' a Hole, o, made in a Board, or a Pasteboard A, B, and you will see both Candles, but you will see two Holes, one for each Candle, tho' there be but one Hole for both. The Reason of it is, because when you look at both Candles E, F, the Axes of both Eyes a, G, a, are directed to the Height G, which is the common Point in this Distance.... It is true, that the single Candle which you see on looking sted-fastly on the Hole, is composed of both: and that if you put your Hand before one of the two Candles, you see that which is before your Hand, and see besides

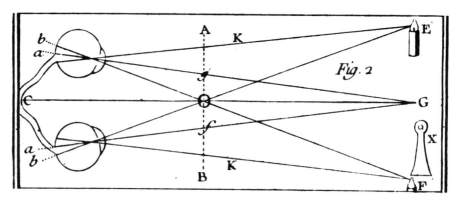

Viewing two candles through an aperture as illustrated in Le Cat (1744).

the Transparency which that, which is behind, produces across your Fingers. (1750, pp. 198–199)

Hartley (1749): When we have attained a voluntary Power over the external Motions of our Eyes, so as to direct them to Objects at pleasure, we always do it in such a manner, as that the same Points of Objects fall upon correspondent Points of the two *Retina's*. And this correspondence between the respective Points of the *Retina's* is permanent and invariable. Thus the central Points, or those where the Optic Axes terminate, always correspond; a certain Point on the right Side of the *Retina* always corresponds (whatever Object we view) to another certain Point on the right Side of the left *Retina*, equally distant from the Centre. (pp. 204–205)

Bell (1803): For example, the object A, in fig. 16., is exactly in the centre of the axis of both eyes, consequently, it is distinctly seen: and it appears single, because the rays from it strike upon points of the retina opposite to the pupils in both eyes. Those points have correspondence; and the object, instead of appearing double, is only strengthened, in the liveliness of the image. Again the object B will be seen fainter, but single, and correct in every respect. It will appear fainter, because there is only one spot in each eye which possesses the degree of sensibility necessary to perfect vision: and it will appear single, the rays proceeding from it having exactly the same relation to the centre of the retina in both eyes. (pp. 352–353)

Prevost (1804): It follows from the stated law that, in the plane of the optic axes, the position of those points seen single with the two eyes is a circumference of a circle which passes through the two centers of the eyes and the intersection of their axes. (Shipley and Rawlings, 1970, p. 1229)

Young (1807): When the images of the same object fall on certain corresponding points of the retina in each eye, they appear to the sense as only one; but if they fall on points not corresponding, the object appears double; and in

Pierre Prevost (1751–1839) after an illustration in Türler, Attinger, and Godet (1929).

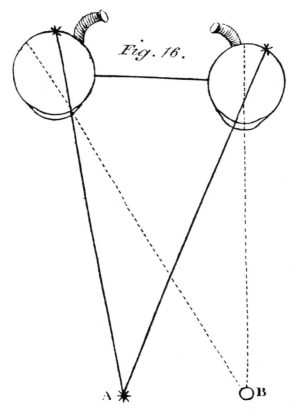

Binocular observation according to Bell. With fixation on *A* the location of corresponding points stimulated by *B* is not on a frontoparallel plane, suggesting an appreciation of a circle of correspondence.

general, all objects at the same distance, in any position of the eyes, appear alike either double or single. The optical axes, or the directions of rays falling on the points of most perfect vision, naturally meet at a great distance; that is, they are nearly parallel to each other, and in looking at a nearer object we make them converge towards it, wherever it may be situated, by means of the external muscles of the eye; while in perfect eyes the refractive powers are altered, at the same time, by an involuntary sympathy, so as to form a distinct image of an object at the given distance. This correspondence of the situation of the axes with the focal length is in most cases unalterable; but some have perhaps a power of deranging it in a slight degree, and in others the adjustment is imperfect: but the eyes seem to be in most persons inseparably connected together with respect to the changes that their refractive powers undergo, although it sometimes happens that those powers are originally very different in the opposite eyes. (p. 453)

Gall and Spurzheim (1810): All these explanations [of binocular single vision] are so different, so contradictory, and also equally insufficient, which indicates

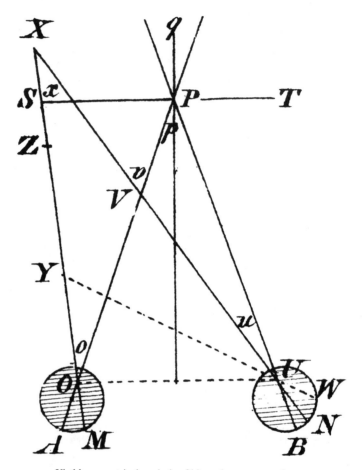

Vieth's geometrical analysis of binocular correspondence.

that much more will be required before the researchers have found the true cause of single vision with two eyes, and we can predict without too much temerity, that one cannot reach this goal as long as touch and the soul enter into the explanation of vision. The soul does not combine this action with any other sense; it perceives only impressions; and necessarily its judgment is tuned to the impression that the senses have communicated to it; thus, to discover successively the cause of the phenomenon of one sense, it is necessary to fix all attention on the organ of this sense, on the type of action, and on the report of the object with which it has a relation. (p. 133)

Vieth (1818): The images *M* and *N* lie in *that* case equally far from *A* and *B*, if the angles *p* and *x* are equal, in which case this applies; if *X* lies *on the circumference of a circle*, which passes through *O* and *U* and *P*, then every angle on this circumference will have the same subtence with respect to *O U*. Therefore if by the expression *corresponding points* one understands such points which lie in the same direction in both eyes and are *equidistant* from *A* and *B*, which appears to me to be the correct meaning, and one asserts one sees things singly whose images fall on such corresponding points, then, according to this rule, one sees that thing singly which is situated within the boundary of a sphere which passes through *O*, *U*, *P*, and does not lie in the plane *ST*, which one names the *horopter*. (p. 238)

Müller (1826a): All differences in the images in both eyes, which deviate from the identical middle point opposite the periphery of the visual field, are not perceived, because they are too small or because, if they are larger, they are in the periphery of the visual field and are not seen clearly.... The objects at all distances which subtend the same visual angles will appear in the same position in the visual field, so it follows that we wish to visualize the location of double images solely through their projection in the plane of convergence. (pp. 171 and 173)

Gerhard Ulrich Anton Vieth (1763–1836) after an engraving kindly supplied by the Bibliothèque nationale de France.

Brewster (1830b): The phenomenon of single vision from two images admits of an explanation equally satisfactory. Let the rays from an object MN, Fig. 20. fall upon the two eyes X and Z, and form two images *m n*, *m′ n′*, on the retina of each. Then, since the point *m* of the image is referred by each eye to the same point M, and the point *n* to the point N, only one object can be seen at MN. If we press the eye X to one side, the image of MN which it forms, will separate from the image MN formed by the other eye, and the object will appear double. In like manner, if the axes of each eye are directed to a point nearer or farther from the eye than MN, the object will in both cases appear double, as each eye does not refer the same points of the image to the same point in absolute space. When an object, therefore, is seen single with two eyes, it is in reality double, but the two images coincide so accurately, that they appear only as one, and having twice the brightness of the image formed by either eye alone. (p. 615)

Johannes Müller (1801–1858) after a lithograph in Stirling (1902).

Müller (1838): The horopter is therefore always a circle, of which the chord is formed by the distance between the eyes, or, more correctly, between the points of decussation of the rays of light in the eyes, and of which the size is determined by three points,—namely, by the two eyes, and the point to which their axes converge. If *a*, *b*, fig. 108, be the distance of the eyes from each other, the circle *f* is the horopter for the object *c*, the circle *g* that for the object *d*, and the circle *h* that for the object *e*. The cause of the impressions on identical points of the two retinæ giving rise to but one sensation, and the perception of a single image, must lie in the organization of the deeper or cerebral portion of the visual apparatus; it must at all events depend on some structural provision; for it is the property of the corresponding nerves of the two sides of the body in no other case to refer their sensations as one to one spot. It is exceedingly improbable that the identical action of the corresponding parts of the two retinæ is the result of a certain habituation, or of the influence of the mind. The cooperation of the two retinæ in one field of vision, whatever is its cause, must

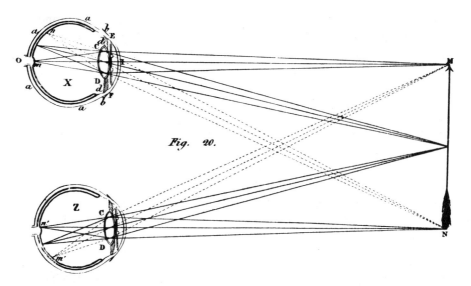

Fig. 20.

Binocular combination of equivalent projections to each eye according to Brewster.

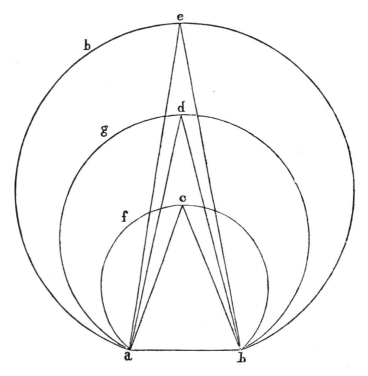

Müller's circular horopters for different fixation distances.

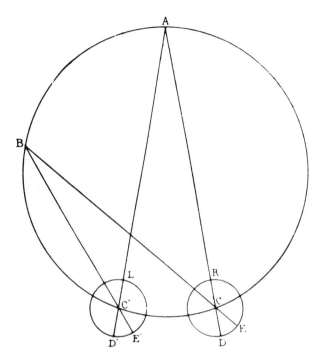

The binocular circle represented by Wheatstone.

rather be the source of all the ideas to which single or double vision may give rise. (1843. p. 1196)

Wheatstone (1838): In the eye itself, the centre of visible direction, or the point at which the principal rays cross each other, is, according to Dr. Young and other eminent optical writers, at the same time the centre of the spherical surface of the retina, and that of the lesser spherical surface of the cornea; in the diagram (fig. 26.), to simplify the consideration of the problem, R and L represent only the circle of curvature of the bottom of the retina, but the reasoning is equally true in both cases. The same reasons, founded on the experiments in this memoir, which disprove the theory of Aguilonius, induce me to reject the law of corresponding points as an accurate expression of the phenomena of single vision. According to the former, objects can appear single only in the plane of the horopter; according to the latter, only when they are in the circle of single vision; both positions are inconsistent with the binocular vision of objects in relief, the points of which they consist appearing single though they are at different distances before the eyes. I have already proved that the assumption made by all the maintainers of the theory of corresponding points, namely that the two pictures projected by any object in the retinæ are exactly similar, is quite contrary to fact in every case except that in which the optic axes are parallel. (p. 390)

6.3   Binocular Double Vision

Many of the statements about binocular double vision are reflections of
the breakdown of binocular single vision. This was the case for Aristotle's
description of one of the most common ways of inducing double vision—by
gently pushing one eye with the finger. The involvement of eye movements
was stressed by Ptolemy, who suggested a functional advantage of double
vision, namely, to bring the two eyes into register with regard to the objects
under inspection. Scheiner noted that double vision often accompanied drunk-
enness, but that it was restricted to horizontal rather than vertical misalign-
ments, assuming an upright head.

The other aspect of binocular vision with which double vision is obviously
connected is strabismus (section 6.7). Aristotle raised it in this context, as did
Smith and Hartley. Porterfield provided a ruler demonstration for double vision
and Smith illustrated the locations in which the two images would appear.
Hartley considered that the recovery of single vision following a traumatic
onset of strabismus was a consequence of learning a new set of retinal corre-
spondences. The issue of aquisition was alluded to by Wardrop, but in the
reverse direction; he had seen cases in which constant use of a telescope had
produced diplopia. However, Müller restated his belief that any object not
falling on the horopter would appear double.

Aristotle (ca. 330 B.C.): . . . if a finger be inserted beneath the eyeball without
being observed, one object will not only present two visual images, but will
create an opinion of its being two objects. (Ross, 1931, pp. 461b–462a).

Ptolemy (ca. 150): It seems too that nature sets up double vision so that we will
look more and so that our viewing will be ordered and brought to a definite
position. It is natural for us to turn our gaze toward diverse locations, our gaze
shifts without our conscious effort with a marvelous and diligent turning
motion, until both visual axes intersect on the centre of the object we wish
to see, and other pairs of corresponding visual lines within the two visual
pyramids are also brought into coincidence. (Lejeune, 1956, p. 27)

Ibn al-Haytham (ca. 1040): . . . an object will be seen double if the rays that meet
on it have the same direction but differ greatly with respect to their distance
from the two axes; that an object will be seen double if it is perceived through
rays of different directions though their distances from the two axes may be
equal. (Sabra, 1989. pp. 241–242)

Pecham (ca. 1280): If the visible object is to the right of one axis and to the
left of the other, the object appears double because of the sensible difference.
(Lindberg, 1970, p. 151)

Aguilonius (1613): This view of the subject [that the images in each eye are
dissimilar and that their visibility must be confused] is certainly consistent

with reason, but, what is truly wonderful is, that it is not correct, for bodies are seen clearly and distinctly with both eyes when the optic axes are converged upon them. The reason of this, I think, is, that the bodies do not appear to be single, because the apparent images, which are formed from each of them in separate eyes, exactly coalesce, but because the common sense imparts its aid equally to each eye, exerting its own power equally in the same manner as the eyes are converged by means of their optical axes. Whatever body, therefore, each eye sees with the eyes conjoined, the common sense makes a single notion, not composed of the two which belong to each eye, but belonging and accommodated to the imaginative faculty to which it (the common sense) assigns it. Though, therefore, the angles of the optical pyramids which proceed from the same object to the two eyes, viewing it obliquely, are inequal, and though the object appears greater to one eye and less to the other, yet the same difference does not pass into the primary sense if the vision is made only by the axes, as we have said, but if the axes are converged on this side or on the other side of the body, the image of the same body will be seen double. (Brewster, 1856, pp. 12–13)

Scheiner (1619): Why do drunkards, and sober persons if they have unusually enlarged eyes, see single things double, but only in breadth and not in height? (Rohr, 1919, p. 122)

Francis Bacon (1627): The eyes, if the sight meet not in one angle, see things double. The cause is, for that seeing two things, and seeing one thing twice, worketh the same effect. (1857a, p. 628)

Rohault (1671): That if we press either of our Eyes with our Finger, so as to make it receive the Image of the Object on a different Part from what it would do by the common Motion of the Muscles; as it is certain, that the Images which are then impressed on the two Eyes, do not fall upon *the Sympathetick Nerves*, nor reunite in the Brain, so we cannot fail to see the Object double. (1723, p. 256)

Porterfield (1737): . . . when our Eyes are restrained from moving uniformly, all Objects are seen double. Neither is it to be doubted, but, when the same *Phænomenon* occurs in drunk or maniac Persons, it proceeds from the same Cause: The uniform Motion of our Eyes requiring an easy and regular Motion of the Spirits, which frequently is wanting in such Cases. . . . The same Way of Reasoning applied to Objects in all Manner of Situations, will shew that all of them must appear double, when placed out of the Plan of the *Horopter*; all which is exactly agreeable to Experience: And this also is the Reason why a double Appearance will be seen when the End of a long Ruler is placed between the Eye-brows, and extended directly forward with its flat Sides respecting Right and Left; for, by directing the Eyes to a remote Object, the right Side of the Ruler seen by the right Eye, will appear on the left Hand, and the left Side on the right Hand. (pp. 194 and 238)

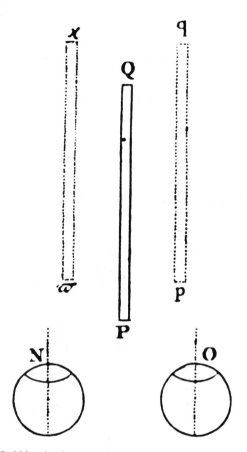

Smith's ruler demonstration of binocular double vision.

Smith (1738): If by squinting or depressing an eye with ones finger, the optick axes are not directed to the same point; in these cases objects appear double: and now it is plain that the pictures are not painted on corresponding places of the retina.... A double appearance will also be seen when the end of a long ruler is placed between the eye brows and extended directly forward with its flat sides respecting the right hand and the left; and by directing the eye to a remote object, the right side of the ruler, seen by the right eye, will appear on the left hand, and the left side on the right hand. (pp. 46–47 and 48)

Le Cat (1744): When you look at the candle C, you turn both eyes towards it so that it can be located at the junction of the axes of the two eyes, and that both images fall on the visual pole, *a, a*, of each eye. In this situation the images of the candle (D) fall on *b, b*, outwith and below the visual poles, and because of this these two images are perceived separately and the candle appears double. (pp. 209–210)

Hartley (1749): After a Person, whose Eye is distorted by a Spasm, has seen double for a certain time, this ceases, and he gains the Power of seeing single

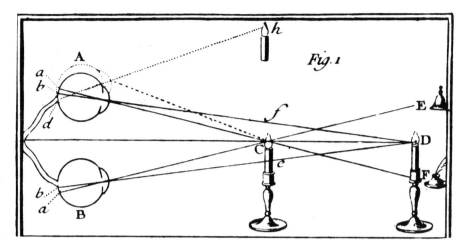

Binocular double vision according to Le Cat (1744).

again, provided the Distortion remains fixed to a certain Degree. For the Association between Points of the two *Retina's*, which corresponded formerly, grows weaker by degrees; a new one also between Points that now correspond, takes place, and grows stronger perpetually. (pp. 205–206)

Buffon (1749): It is equally easy to show that we see all objects double: if, for instance, we look at an object with the right eye, we will find that it corresponds with a certain point of the wall; if we look at the same object with the left, it then corresponds with a different point; and, lastly, when we look at it with both eyes, it appears in the middle of these points. Thus an image of the object is formed on both eyes, one of which appears on the left, and the other on the right; and we perceive it to be single and in a middle situation, because we have learned to correct this error of vision by the sense of touching. (1866, pp. 240–241)

Harris (1775): The double appearance of an object, does perhaps as frequently happen, as we see an object before us that we do not look at, or before we have directed our eyes properly towards it. But we are in no danger of mistaking one object for two, for the sensations we have when we see two objects, and see one double, are very different: In the one case, there are two images ... formed on the retina; in the other, there is but one image in each eye, and the images being on parts of the retina's which do not correspond, may be the cause of that painful sensation we generally have, when we see an object double. The two false objects are always indistinct, have a kind of tremulous motion, approach nearer together, retreat back, or advance forwards, as the case is; and at length, as soon as the eyes have conformed themselves properly for seeing the real object distinctly, the two false objects vanish or coalesce into one, at that determined place where the real object is. (p. 112)

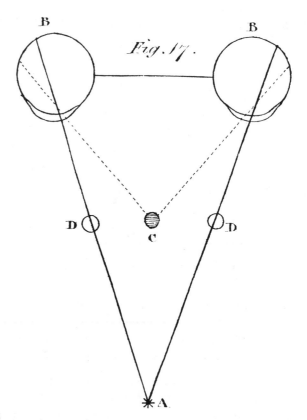

Stimulation of noncorresponding points with resulting double vision.

Bell (1803): But if the eyes are made to fix stedfastly on an object, and if another object should be placed before the eyes within the angle that the axes of the two eyes make with the first object, it will be seen double, because the points on the retina struck by the rays proceeding from the nearer object do not correspond in their relation to the central point of the retina. Thus, the eyes B B, fig. 17., having their axis directed to A, will see the object C double somewhere near the outline D D. Because the line of the direction of the rays from that body C, do not strike the retina in the same relation to the axis A B in both eyes. (p. 353)

Brown (1820): Accordingly we find, that, as often as a double organic affection is produced, different from that which takes place in the retina of each eye in ordinary vision, there is truly a perception of a double object. This is the case, when the common inclination of the axes of the eyes is prevented either by disease or by external force. There is then such an affection of parts of the retina, as could take place, in ordinary circumstances of vision, only when two objects were present; and two objects, therefore, are suggested by the visual sensation. (pp. 149–150)

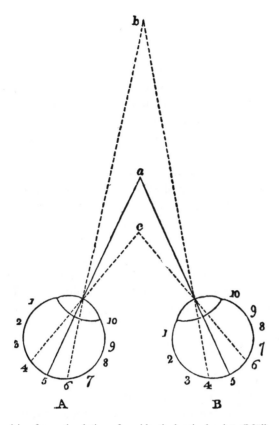

Double vision from stimulation of nonidentical retinal points (Müller, 1838).

Wardrop (1834): Double vision sometimes affects those who make frequent use only of one eye. I have known several cases of officers of the navy who became affected with double vision from fatigue caused by looking through telescopes. (p. 234)

Müller (1838): Whenever an object lies out of the "horopter," its image falls on non-identical points of the retinæ, and it is seen double.... To illustrate this, let *a* be a point towards which the axes of the eyes are directed, and *b* an object more distant from the eyes. An image of *a* will fall upon identical points of the two retinæ,—namely, upon the central points 5, 5; *a* will consequently be seen single. The image of *b* will fall in the left eye at 6, and in the right eye at 4. The points 4 and 6 of the two eyes being non-identical, since the identical point of one eye corresponding to the point 4 of the other is 4, *b* will be seen double.... If *c* be an object nearer the eyes than point *a*, towards which their axes are directed, it is also seen double. (1843, pp. 1201–1202)

6.4   Binocular Contour Rivalry

Binocular single vision occurs when similar images are projected to corresponding parts of each eye. As Brewster noted (section 6.2), it can be difficult to distinguish between seeing two similar things and seeing their combination. No such doubts arise about binocular rivalry, which is one of the reasons that suppression theory has been so ardently supported. When dissimilar images are presented to corresponding areas of each eye they do not combine, but compete. One aspect of this phenomenon (binocular color rivalry) has been surveyed in section 3.8, although there was debate about whether colors fused or engaged in rivalry. Binocular contour rivalry is more clear-cut; the debate has been about whether alternation occurs rather than whether rivalry exists. Ptolemy did arrange stimuli on his board in a manner that would have produced binocular rivalry, but his description of the outcome was ambiguous. Howard and Wade (1996) have suggested that the ambiguity might have been due to the difficulty of maintaining parallel visual directions. The resulting percepts would have followed from converging in the directions *am* and *bl*, and this would have been an appropriate way of presenting dissimilar stimuli to corresponding regions of each eye.

   Porta, applying Galen's partition procedure, placed pages of different books before each eye, and stated that the right eye was dominant. There was, however, the suggestion that alternation could take place, and this was supported by Sébastien Le Clerc's observations. Le Clerc adopted the method of over-convergence to present different figures to each eye, whereas Du Tour held a prism in front of one eye, resulting in the clearest early description of contour rivalry. Either the stimulus presented to one or the other eye would be visible, or some mixture of the two views would present itself. Du Tour's method was superior to that later reported by Müller: a prism merely deviates the path of light entering the eye, but viewing down a microscope would entail a difference in clarity of the images in each eye. Nonetheless, Müller did find that rivalry was occasionally visible when using a microscope.

   All these problems are removed when radically different shapes are viewed with the aid of a stereoscope. Wheatstone surrounded two letters by equivalent circles (to ensure binocular alignment) and noted the rivalry that took place. Not only did he describe the alternation but he examined a stimulus variable (illumination) that could favor one stimulus over the other.

Ptolemy (ca. 150): Position the axes of the eyes parallel to one another and to the common axis, and place a white rod [*l*] on the axis of the left eye [*a*] and a black rod [*m*] in front of the right eye [*b*], so that the distance between the rods is the same as that between the eyes (fig. 3). The two rods will be seen as three. Through the corresponding rays each rod will be seen singly ... But through the non-corresponding rays *am* and *bl* the third, middle image will be

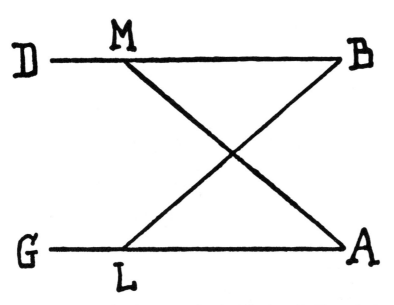

Ptolemy's arrangement for stimulating corresponding visual directions with different stimuli—white and black rods. The report does not describe binocular contour rivalry, probably because of the difficulty of maintaining a parallel gaze.

seen composed of an image of the white rod from the right eye and an image of the black rod from the left eye. (Lejeune, 1956, pp. 31–32)

Porta (1593): Place a partition between the eyes, to divide one from the other, and place a book before the right eye, and read; if another book is placed before the left eye, not only can it not be read, but the pages cannot even be seen, unless the visual virtue is withdrawn from the right eye and changed to the left. (pp. 142–143)

Newton (ca. 1682): ... though one thing may appear in two places by distorting the eyes, yet two things cannot appear in one place. If the picture of one thing fall upon A, and another upon α [its corresponding point in the other eye], they may both proceed to p [in the optic chiasm], but no farther; they cannot both be carried on the same pipes pa into the brain; that which is strongest or most helped by phantasy will there prevail, and blot out the other. (Harris, 1775, p. 110)

Sébastien Le Clerc (1712): If one opens both eyes, the object does not appear in only one of the two places, i.e., in D alone or in C; that is the result of the experiment. From this I conclude again that only one eye sees the object. (p. 17)

Du Tour (1761): If one applies a prism held vertically before one of the eyes, so only refracted rays of light are passed to that eye, and with the other eye open, it is certain that different objects will be projected on corresponding

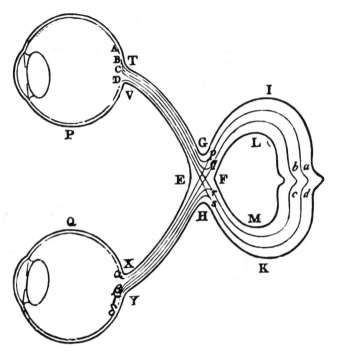

The visual pathways according to Newton, as represented in Brewster (1855). Because the corresponding optic nerve fibers united in the optic tract, either the stimulation of one or the other eye would prevail if different stimuli were present, resulting in binocular rivalry.

portions of the two retinas.... sometimes I would see only objects projected in the bare eye, sometimes only those in the eye covered by the prism, and sometimes the objects projected in one would seem to me to intermingle with the objects projected in the other. (p. 500)

Bell (1803): ... if we look through a glass with one eye, the vision of the other is not attended to. (p. 346)

Müller (1838): That the action of the sensorium can be confined to the sensations of one retina, may be observed also in viewing objects through magnifying glasses; for frequently the eye which is looking into the microscope alone sees or distinguishes its object, while the other eye either distinguishes no object at all, or at least does not see any in that part of the field of vision occupied by the microscopic images perceived by the first eye. Sometimes, on the other hand, the eye not engaged in the microscopic observation resumes its activity, and the image it perceives then seems to float over the image in the microscope and disturbs the observation. (1843, p. 1210)

Wheatstone (1838): If *a* and *b* are each presented at the same time to a different eye, the common border will remain constant, while the letter within it will change alternately from that which would be perceived by the right eye alone

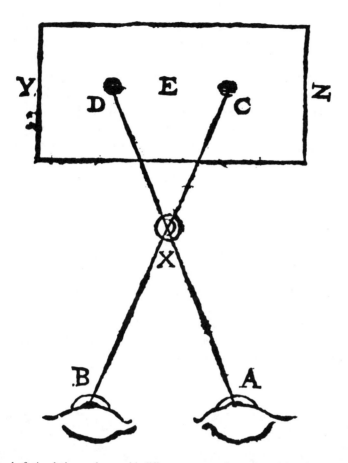

The method of stimulating each eye with different patterns that was used by Sébastien Le Clerc.

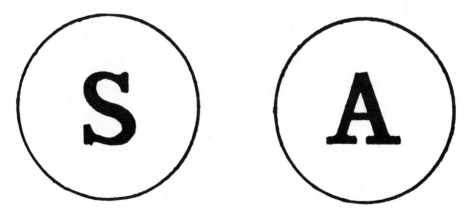

Different letters presented in the stereoscope undergo binocular rivalry (Wheatstone, 1838).

to that which would be perceived by the left eye alone. At the moment of change the letter which has just been seen breaks into fragments, while fragments of the letter which is about to appear mingle with them, and are immediately after replaced by the entire letter. It does not appear to be in the power of the will to determine the appearance of either of the letters, but the duration of the appearance seems to depend on causes which are under our control: thus if the two pictures be equally illuminated, the alternations appear in general of equal duration; but if one picture be more illuminated than the other, that which is less so will be perceived during a shorter time. (p. 386)

6.5   Eye Dominance

It is possible that Rubens represented Porta's sighting test as the frontispiece for Book III of the *Optics* by Aguilonius. The putti are holding a stick that has been sighted by the observer, who then closes one eye to see if the finger and the stick remain in alignment. The eye remaining open is the right, which would accord with Porta's claim that the right eye is preferred in such sighting tests. Porta is usually acknowledged as describing the first test of sighting dominance in 1593 (see Durand and Gould, 1910; Porac and Coren, 1976; Wade, 1998).

Frontispiece of Book III of Aguilonius (1613), possibly depicting Porta's sighting test for eye dominance.

Recorded examples of asymmetries in the functions of members of bilateral pairs, like the hands, extend far into antiquity (see Coren and Porac, 1977), and so eye dominance is generally considered to have entered upon the scene at a much later date. However, eye dominance is not a unitary dimension, and there appear to be at least three different aspects of it: sighting, rivalry, and acuity dominance (Coren and Kaplan, 1973). One measure of sighting dominance is eye closure: it is more difficult to wink with the sighting dominant eye (Crider, 1941). Asymmetries of eye closure were mentioned by Aristotle. However, accurate as this observation might have been, it was not interpreted by Aristotle as reflecting on eye dominance. In fact, he considered that the senses generally did not display the asymmetries associated with motor tasks. The muscles were considered to be active whereas the senses were passive; activities could be modified by practice, unlike the passive senses. Accordingly, right-side superiority was assigned to habit, because it could be changed, as in the case of the ambidextrous.

As noted above, Porta does seem to have been the first to describe binocular rivalry, but the subsequent demonstrations by Le Clerc and Du Tour did not refer to a preference for the stimulus presented to a particular eye. Haller (section 6.1) stated that the right eye is generally preferred in binocular tasks, and that one eye alone is typically employed. Gall was another advocate of suppression theory; while he maintained that there were differences between the eyes, he did not give a general statement regarding which one was typically dominant. Wardrop considered that the right eye was usually dominant, and he suggested a variant of Porta's sighting test to support it.

Acuity dominance is said to be independent of the other forms, and its early history would tend to support this. Although Aristotle stated that both eyes were equally acute, Borelli asserted that the left eye was stronger than the right. He based this claim on the greater distinctness of the view with his left eye, and on the sharper vision of small letters viewed through a tube when using his left eye. He speculated that the asymmetry might be due to the left optic nerve having a better blood supply than the right. The assertion of left eye superiority was challenged by Le Cat, who noted that in addition to those who have a stronger left eye, others had a stronger right, and there were yet others (like himself) who had eyes "perfectly equal." It is not clear what aspects of eye dominance Le Cat was considering, because he seemed to conflate distinctness of vision in an eye with its vigor. Ware made a more subtle link between acuity and dominance, and introduced the possibility that there might be a learned component to it: since most people are right-handed they would be likely to hold an eyeglass with that hand, and place it before the right eye. It is surprising that concern with errors of refraction and their correction (section 2.4) did not disclose differences in the states of the two eyes. Similarly, strabismus provides a potent source of subjects demonstrating oculomotor asymmetries (section 6.7). Interest has understandably focused on diplopia in such individuals, but differences in acuity have also been remarked upon by Jurin and Buffon.

Aristotle (ca. 330 B.C.): Why is it that, though the parts of the body on the right side are more easily moved, the left eye can be closed more easily than the right? (Ross, 1927, p. 959b)

Porta (1593): . . . if one takes a staff and brings it directly opposite some crack in the opposite wall, and notes its location; when he closes his left eye he will not see the staff removed from the opposite crack, the reason being that everyone looks with his right eye as he uses his right hand. (p. 143)

Borelli (1673): . . . the left eye normally sees objects more distinctly than the right. (p. 291)

Sébastien Le Clerc (1679): As we have two hands, it is usually the right one which serves to draw and to write, and that is rendered so by habit. It is the same with the eyes. (p. 7)

Le Cat (1744): I am perswaded, that, in regard of a Number of People, one Eye is ever stronger or more on the watch, than the other, and constantly takes upon itself the greatest Share of the common Task. For instance, Borelli asserts, that the left Eye is stronger, and always discerns more distinctly, than the right. I have verified this Observation by Trials on several Persons: but I have discovered likewise that it is not a general one. There are Eyes perfectly equal, such amongst others are my own. There are, on the contrary, Instances, where the right Eye is the most vigorous. (1750, p. 187)

Bell (1803): If one eye is weaker than the other, the object of the stronger eye, alone is attended to, and the other is entirely neglected. (p. 346)

Gall and Spurzheim (1810): One maintains also, and with reason, that often there exists an inequality between the two eyes and the two ears, and that as a consequence an impression must be more lively and more distinct in one than the other. But we cannot sustain that, in this case, one perceives only the strongest of these impressions. On the contary, experience indicates that one can see better with both eyes, and that one can listen better with both ears; one receives thus the two impressions, and the weaker is not cancelled by the stronger. . . . Nearly always one is weaker, and often in an extreme manner; and nevertheless these people, who see straight ahead or to the side do not perceive the objects as single. Finally, with two eyes completely equal, the object appears double, when one squints. (p. 133)

Ware (1813): I have observed, that most of the near sighted persons, with whom I have had an opportunity of conversing, have had the right eye more near sighted than the left; and I think it not improbable, that this difference between the two eyes has been occasioned by the habit of using a single concave hand-glass; which, being most commonly applied to the right eye, contributes . . . to render this eye more near sighted than the other. (p. 34)

Giovanni Battista della Porta (1535–1615) after an engraving in Mach (1926).

Giovanni Alfonso Borelli (1608–1678) after an illustration in Wolf (1935).

Wardrop (1834): It will generally be found, that not only the right is more perfect than the left eye, but when a person is apparently looking at an object with both eyes, generally only one of the eyes, and that usually the right one, is directed to the object. To demonstrate this, let a spot be covered with the point of a finger when looking at it with both eyes. If the left eye be closed, the point of the finger will continue to appear to cover the spot, and to preserve the same relative situation; but if the right eye be closed and the left opened, then the relative situation of the point of the finger and spot appear altered, the spot being uncovered; proving that, in directing the finger to cover the spot, the right eye had alone been employed. (p. 245)

6.6 Monocular Compared with Binocular Vision

Perceptual experience is typically unified and coherent: objects are seen as having a specific size, shape, color, and location in space. This obtains under conditions of binocular and monocular vision. Indeed, perceptual experience changes little when one eye is closed, but it does change, and the study of the differences has a long history. Contrary to the evidence accumulated since the late seventeenth century, it was long believed that vision with one eye was superior to that with two. Aristotle took this to be self-evident, interpreting it in terms of eye movement control. By speculating that the movement of two eyes is not unified he contradicted his earlier statement in section 5.9. The source of much subsequent comparison was driven by the Galenic theory of visual spirits: it was transmitted to one or two eyes, and thus was more concentrated in monocular viewing. Despite Ptolemy's statement to the contrary, this opinion was repeated over the following centuries, and it was held as late as the seventeenth century, when Francis Bacon attributed the advantages of aiming with one eye to this cause. Galen made some insightful observations on binocular visual direction in those blind in one eye—it did not accord with the visual axis of the functioning eye. Le Clerc recorded an instance of someone who saw well with each eye separately, but poorly with both together; he used this as further evidence that only one eye is employed most of the time.

Many statements were made about tasks that were more difficult to perform with one eye rather than two, or stimuli that were more difficult to see. Leonardo's often quoted comparison between viewing a painting of a scene and the scene itself was an implicit contrast between vision with one eye and two. It is most instructive because the concept of relief or depth is taken to be the distinguishing characteristic of binocular vision. Rohault, Malebranche, Boyle, and Molyneux all directed attention to tasks that could be performed better with two eyes, although they seemed reluctant to speculate on the reasons for this. Porterfield was more forthcoming: apparent size was determined by retinal size and apparent distance (see sections 7.10 and 7.11), and judgments of the latter were degraded with monocular observation. Wheatstone, with his knowledge of binocular depth perception, inquired why more errors are not made in monocular viewing. His answer was that many cues for depth and distance exist, and that alternatives are used when one source is not available. Most particularly, he proposed that motion of the head (motion parallax) was a potent substitute for retinal disparity.

Wheatstone's stereoscope was not the first binocular instrument to be constructed. Chérubin d'Orléans made binocular telescopes and microscopes 150 years earlier, about which more will be said in section 6.8. He argued that binocular vision was superior to monocular vision, as did Briggs, but it was Jurin's experimental investigation of apparent brightness that placed the matter beyond dispute. After the manner of Bouguer's photometer (see section 7.4) Jurin matched the brightness of two candlelit surfaces, seen by one or two eyes, and calculated the amount by which the binocular part was brighter than the monocular.

Aristotle (ca. 330 B.C.): Why can one see more accurately with one eye than with both eyes? Is it because more movements are set up by the two eyes, as certainly happens in those who squint? The movement of the two eyes, therefore, is not one, but that of a single eye is one; therefore one sees less accurately with both eyes. (Ross, 1927, p. 957a)

Ptolemy (ca. 150): Vision with two eyes is better than with one. (Lejeune, 1956, p. 19)

Galen (ca. 175): Persons blind in one eye do not see objects in front of that eye, even if they are brought close to it. (May, 1968, p. 483)

Leonardo da Vinci (ca. 1500): A Painting, though conducted with the greatest Art and finished to the last Perfection, both with regard to its Contours, its Lights, its Shadows and its Colours, can never show a *Relievo* equal to that of Natural Objects, unless these be view'd at a Distance and with a single Eye. (1721, p. 178)

Paré (1579): An argument whereof may be drawn from such as aim at any thing, who shutting one of their eyes see more accurately; because the force of the neighbouring spirits united into one eye, is more strong than when it is dispersed into both. (1649, p. 133)

Leonardo da Vinci (1452–1519) after an engraving in Knight (1835a).

Aguilonius (1613): The hand is placed on the nose in such a way that all the fingers may be extended upright, the thumb alone lying lengthwise upon the nose, and the whole hand cuts off the eyes' vision—like a barrier. Or if you prefer it, imagine in place of the hand, a small stick placed between the two eyes A and B. The object DE whose parts DF, GE are seen by one eye at least, while the central region FG is seen by both, as the lines from the eyes reveal it. (p. 82)

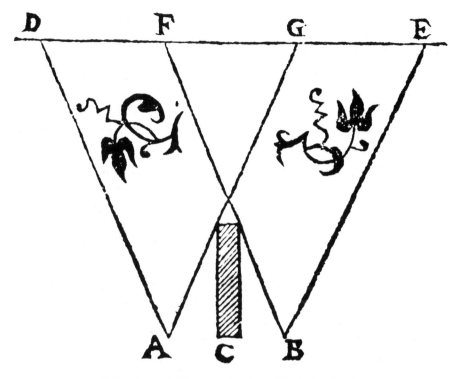

Septum to restrict the view of each eye (Aguilonius, 1613).

Francis Bacon (1627): We see more exquisitely with one eye shut, than with both open. The cause is, for that the spirits visual unite themselves more, and so become the stronger. For you may see by looking in a glass, that when you shut one eye, the pupil of the other eye that is open dilateth. (1857a, p. 628)

Hobbes (1651): . . . they that desire not to miss their mark, though they look about with two eyes, yet they never aim but with one. (1839, p. 250)

Rohault (1671): If we try to touch an Object at three or four Foot distance, with the End of a Stick of about the same Length, we shall find, that if we look at it with one Eye, we shall miss touching it two or three times together; whereas if we look at it with both Eyes, we shall touch it the first Time. (1723, p. 253)

Malebranche (1674): A man would easily be convinc'd of what I say, if he would be at the Trouble of making a very easie Experiment: As, let him hang a Ring at the end of a Thread, so placed that the hoop being turn'd directly towards him, the aperture of it may not appear; or if he please let him drive a Stick in the ground, and take another in his hand, that is curved at the end; let him retreat three or four steps from the Ring or the Stick, and shutting one Eye with one hand, let him try to hit the aperture of the Ring, or with the bent end of the Stick in his hand to touch the other across, at an height that is much upon a level with the Eye; and he will be surpriz'd to find himself incapable of doing it at a hundred tryals, though nothing in the world seems easier. (1700, p. 22)

Chérubin d'Orléans (1677): *The two eyepieces of the binocle are placed in their own tubes in such a way that one can see distinctly through each separately, and they can be adjusted as required for the two visual axes, so that the two eyes, which are looking at the same time, each with its own view, together see just one and the same object.* (p. 73)

Sébastien Le Clerc (1679): I have seen a man who never saw distinctly without closing an eye. When he closed the right eye he saw very well with the left eye, and when he closed the left eye he saw very well with the right eye. But when he opened both eyes he only saw confusedly. It seems to me that this could only happen because of the bad habit he had of looking with both eyes at the same time. (Pastore, 1972, p. 338)

Briggs (1682): Now that Vision is more perfectly performed by *both* Eyes than *one* . . . is obvious to sense, for if any one reads first with *one Eye* and then with *both* (at some distance) he will perceive the character to appear plainly with *both* Eyes, but not so whilst *one is shut*. (p. 173)

Wallis (1687): . . . they who have lost the Sight of one Eye, are at a great disadvantage, as to estimating Distances, from what they could do while they had the use of both. (p. 327)

Boyle (1688): Haveing frequently occasion to pour Distill'd Waters and other Liquors out of one Vial into another, after this Accident [of losing an eye] he often Spilt his Liquors, by pouring quite Beside the necks of the Vials he thought he was pouring them directly Into. (p. 255)

Molyneux (1692): . . . the best [tennis] Player in the World Hoodwinking one Eye shall be beaten by the greatest Bungler that ever handled a Racket; unless he be used to the Trick, and then by Custom get a Habit of using one Eye only. (p. 294)

Porterfield (1737): From this also we may see, why we err so frequently in Judgments we form of the Magnitude of Objects seen only with one Eye: For, since we judge not of Extension or Magnitude from their apparent Magnitude alone, but also from the apparent Distance, it follows that Objects seen with one Eye, must appear smaller or greater, as they are imagined nearer or further off. (pp. 189–190)

Jurin (1738): I set upon a table two candles of equal height, and burning to appearance with equal light, at about a foot distance beyond a slip of white paper lying before me, and about four inches from one another, so as that the distance between the candles was parallel to the slip of paper. Then I set a book

The Oculaire Royale binocular telescope of Chérubin d'Orléans.

upon one end, between the right hand candle and the paper, so as to cast a shade from that candle upon the right hand half of the paper: Thus the left half of the paper was illuminated by the two candles, and the right by one only: consequently, the left half was twice as luminous as the right; and the boundary between those two halves was pretty well defined by the edge of the shade. I then took another book, and applyed it to my left temple, in such a manner as to hide the left and brighter part of the paper from my left eye; so that the left half was seen by my right eye only, while the other half was seen by both eyes: Now I expected, that the right half of the paper, having the light of one candle only thrown upon it, but being seen by both eyes, would appear as luminous as the left half, which had twice as much light cast upon it, but was seen by one eye only. In which I found my self mistaken For the left half appeared much whiter and brighter than the right. Consequently, an object seen with both eyes is nothing near twice as luminous, as when seen with one eye only. Being desirous to know the quantity of this excess of brightness more exactly, I fixed a slip of white paper flat against the wainscot by the help of pins, and at a yard distance I set a candle.... When the second candle is about 11 foot distant from the paper, the right half seen with one eye, and the left half seen with both eyes, must appear of an equal brightness. Consequently, an object seen with both eyes, appears brighter than when seen with one only, by about a thirteenth part.... an object does not appear larger to both eyes than to one. (Smith's *Remarks*, 1738, pp. 108–109)

Le Cat (1744): A one-eyed Person in particular looses a considerable Part of the Objects that present themselves, if he looks on them at never so little a Distance. The Quickness of the Motion of the Eye remedies in a small Degree this Inconvenience, by taking a successive Survey of every Object; but it does not intirely repair it. (1750, p. 160)

Wheatstone (1838): How happens it then, it may be asked, that persons who see with only one eye form correct notions of solid objects, and never mistake them for pictures? and how happens it also, that a person having the perfect use of both eyes, perceives no difference in objects around him when he shuts one of them? To explain these apparent difficulties, it must be kept in mind, that although the simultaneous vision of two dissimilar pictures suggests the relief of objects in the most vivid manner, yet there are other signs which suggest the same ideas to the mind, which, though more ambiguous than the former, become less liable to lead the judgment astray in proportion to the extent of our previous experience.... A person deprived of the sight of one eye sees therefore all external objects, near and remote, as a person with both eyes sees remote objects only, but that vivid effect arising from binocular vision of near objects is not perceived by the former; to supply this deficiency he has recourse unconsciously to other means of acquiring more accurate information. The motion of the head is the principal means he employs. (p. 380)

6.7 Strabismus (Squint)

Distortion of the eyes (strabismus) was recorded in the Ebers papyrus, but its reported association with problems of binocular vision are more recent (see Hirschberg, 1899; Shastid, 1917; Duke-Elder and Wybar, 1973). Aristotle described the ensuing diplopia, and Galen noted that the deviations of the eyes were always nasal or temporal; however, he did also state that persons with strabismus rarely make errors in object recognition. Corrections for the deviation were advocated by Paulus Ægineta: he recommended the use of a mask that has also been attributed to Paré. The most elaborate masks were prescribed by Bartisch, who appreciated that strabismus was more amenable to correction in children than in adults. Jurin and Reid sought to strengthen the weak eye by exercise, and Bell suggested patching the stronger eye.

From the eighteenth century attention was directed to the manner in which persons with strabismus saw objects singly. There were those like Jurin, Buffon, Le Cat, Reid, Bell, and Müller who believed that the nondeviating eye suppressed the signals from the deviating one. This was so despite the opposite being maintained by those with strabismus. The suppression could be a consequence of muscular misalignment (Jurin, Le Cat) or a refractive difference between the eyes (Buffon). Others, like Hartley and Harris, considered that the new pattern of stimulation could be learned (by association) to yield singleness of vision. Accordingly, nature and nurture could vie to determine why the eyes were awry. Hartley stated that the origin of the deviation lay in the asymmetrical cradling of children. Reid devoted many pages of his *Inquiry* to both the theoretical and practical aspects of strabismus. He raised eleven theoretical queries about the condition, and also expressed some success at training persons with strabismus to bifixate on objects. There were also descriptions of cases of alternating dominance (Erasmus Darwin), and the distinction between converging and diverging strabismus was made explicit (Mackenzie).

Studies of strabismus epitomize the emphasis that has been placed on binocular single vision; the many comparisons that were made between strabismic and binocular individuals dwelt on this to the exclusion of other aspects of binocular vision, like stereopsis.

Aristotle (ca. 330 B.C.): Why do objects appear double to those whose eyes are distorted? Is it because the movement does not reach the same point on each of the eyes? (Ross, 1927, p. 958b)

Pliny (ca. 77): Man is the only animal whose eyes are liable to distortion, which is the origin of the family names Squint-eye and Blinky. (1940, p. 527)

Galen (ca. 175): Those who squint, either by an acquired anomaly or from the earliest period of foetal life, never commit any errors in the knowledge of

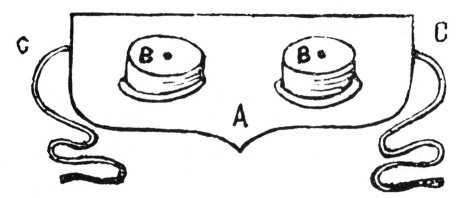

Mask for the correction of squint after Paré (Shastid, 1917).

Ambroise Paré (1510–
1590) after a frontispiece
engraving in Paré
(1649).

objects looked at, for, in such cases, neither of the pupils stands higher than
the other, but the fault consists only in this, that the one eye looks too much
toward the nose or is directed away from it. (Shastid, 1917, p. 9383)

Paulus Ægineta (ca. 680): Congenital squinting is cured by the application of a
mask, so that children are compelled to look straight forwards; for strabismus
is a spasmodic affection of the muscles which move the ball of the eye. (1844,
p. 422)

Paré (1564): The portrayture with the description is in *Ambrose Pare* his
Booke, which without knowledge of the same set downe in the older writers,
hath very wittily and cunningly invented the sayd maske. (Duke-Elder and
Wybar, 1973, p. 225)

Bartisch (1583): Therefore in childhood and youth one can prevent and
modify the usual state [of squinting] more than in adulthood.... If one notices
a child who sees poorly and with a squint, then as soon as this is recognized,
one should make a hood or mask out of thick paper or similar material to cover
the head externally.... If both eyes deviate towards the nose, as is generally
seen, then the two holes for the eyes should be arranged more towards the
ears.... However, if the squint is in one eye alone the hole in the mask should
be made straight ahead for the good eye and to the side for the deviating
eye.... When the eyes are directed outwards towards the ears then a special
mask must be made for this condition. (pp. 14–16)

Francis Bacon (1627): ... for some squint when they will; and the common
tradition is, that if children be set upon a table with a candle behind them, both
eyes will move outwards, as affecting to see the light, and so induce squinting.
(1857a, p. 628)

Briggs (1683): ... the case of cross-eyed persons by birth I have considered ...
and shown why a morbid strabismus, or more violent contortion of those
muscles after great convulsions of the nerves, causes always double vision.
(p. 180)

Georg Bartisch (1535–
1606) after an illus-
tration in Hirschberg
(1908).

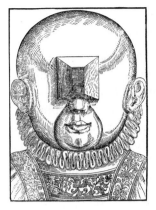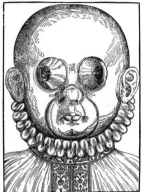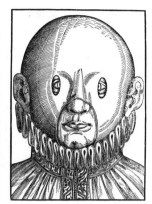

Masks for correcting diverging and converging squints recommended by Bartisch.

Boyle (1688): I have often employed a dextrous artificer, whose right eye (for in his left eye there is nothing more remarkable) is constantly drawn so much aside towards the greater angle of the eye, that the edge of the pupil does almost touch it, and one would think it scarce possible, but that he should see the object double with two eyes, that seem so very differingly turned; and yet he answered me, that he does not see at all, nor that he finds any inconvenience, save the deformity of this unusual situation of his right eye, which hinders him not from reading freely as other men. (Birch, 1966, p. 448)

Robert Boyle (1627–1691) after an engraving in Birch (1743).

Porterfield (1737): *First*, This Disease may proceed from Custom and Habit, while in the Eye itself or in its Muscles, nothing is preternatural or defective ... *Secondly*, The *Strabismus* may proceed from a Fault in the first Conformation, by which the most delicate and sensible Part of the *Retina* is removed from its natural Situation, which is directly opposite to the Pupil, and is placed a little to a side of the *Axis* of the Eye, which obliges them to turn the Eye away from the Object they would view, that its Picture may fall on this most sensible Part of the Organ.... *Thirdly*, This Disease may proceed from an oblique Position of the *Crystalline*.... *Fourthly*, This Disease may arise from an oblique Position of the *Cornea*.... *Fifthly*, This Want of uniformity in the Motions of our Eyes may arise from a Defect, or any great Weakness or Imperfection in the Sight of both, or either of the Eyes.... *Sixthly*, Another Cause from which the *Strabismus* may proceed, lyes in the Muscles that move the Eye. When any of those Muscles are too short or too long, too tense or too lax, or are seized with a *Spasm* or *Paralysis*, their *Equilibrium* will be destroyed, and the Eye will be turned towards, or from that Side where the Muscles are faulty. (pp. 239–240, 241–242, 243, 245, 247, and 249)

Jurin (1738): From these tryals it plainly appears, that the eye is thus distorted, not for the sake of seeing better with it, but rather to avoid seeing at all with that eye, as much as possible.... So that in reality a squinting person sees the object before him distinctly with one eye only; namely that whose axis is pointed directly at the object. (Smith's *Remarks*, 1738, p. 30)

Buffon (1743b): When one of the eyes is much weaker than the other we do not direct it towards the object, but make use of the stronger eye only; for the same reason that we commonly make use of the right arm in preference to the left.... If, for instance, the distance at which a person can read small print be from 8 to 20 inches with the stronger eye, and only 8 to 15 inches with the weaker, distinct vision will be limited to 7 inches, viz. from 8 to 15 inches for both eyes; and as the image of the good eye will be stronger than that in the weak one, the impression upon them will not be so distinct, as if the good eye only had been used. It is no wonder, therefore, that such persons chuse to make use of one eye only, and turn the other aside. (p. 237)

Le Cat (1744): One that is squint-eyed, however, looks at Objects with both Eyes transversely, without seeing them double. It is true. But a squint-eyed Person, without being conscious of it, ever sees but with one Eye, tho' he imagines he looks with both. I lately unfolded this Doctrine to one that squinted very much with his left Eye, who at the same Time firmly believed he saw with both Eyes at once. I assured him that he only saw with his right Eye, and it was thus I convinced him. I made him look with both Eyes, A, B, Fig. 3. at the Object C. I observed his Eyes while he looked at the Object ... I saw then that the sound right Eye, B, of the squinting Person, was actually turned towards the Object, but that the other Eye A, at the same Time was turned towards D. (1750, pp. 204–205)

Hartley (1749): The Circumstances which occasion Squinting in young Children, agree well with the theory proposed here. Thus, if a Child be laid so into his Cradle, as that one Eye shall be covered, the external Influences of Light cannot operate upon it. And if this be often repeated, especially while the Association which confirms the Congruity of the Motions is weak, the Eye which is covered will obey the Influences which descend from the Brain, and

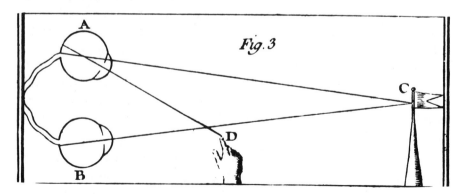

The direction of the eyes in squinting after Le Cat (1744).

turn upwards and inwards for the most part. What turns the Scale in favour of this Position, remains to be inquired.... The persons who squint, preserving the Sight of the squinting Eye, are obliged to move their Eyes in a congruous manner. (pp. 218 and 221)

Morgagni (1761): What? if from the very womb any muscle of either eye be too long, or too short; too strong, or too weak; too detach'd, or too confin'd, in its motions! Will not the children, who have such muscles, be born, more or less, with strabismus, in proportion to the greater or less degree of the unnatural constitution in the muscle? (1769, pp. 302–303)

Reid (1764): We see young people in their frolics learn to squint, making their eyes either converge or diverge, when they will, to a very considerable degree. Why should it be more difficult for a squinting person to learn to look straight when he pleases?... one who squints, and originally saw objects double by reason of that squint, may acquire such habits, that when he looks at an object, he shall see it only with one eye: nay, he may acquire such habits, that when he looks at an object with his best eye, he shall have no distinct vision with the other eye at all.... Having examined about twenty persons that squinted, I found in all of them a defect in the sight of one eye. Four only had so much of distinct vision in the weak eye, as to be able to read with it, when the other was covered. (pp. 351, 356, and 358)

Thomas Reid (1710–1796) after a frontispiece engraving in Stewart (1800).

Harris (1775): Persons that squint have not the axes of both eyes directed to the same object, and yet they see objects single. Squinting may proceed from several causes: one may be, that the correspondent fibres in the two retina's are differently situated, which makes it necessary to turn the eyes different ways, that the images of objects may be received upon the corresponding fibres in both: And this also accounts, why persons having such eyes, see objects single. It is supposed, and that with great probability, that there is a small part of the retina of each eye, which is more sensible than the rest; and that we naturally turn our eyes so, that the impressions of the light may be received upon this most sensible part: Those who have this part awry, or to one side of the optic axes in one or both eyes, necessarily squint; and yet such Persons see objects single like other men, either because they have learnt by experience, that such is the natural direction of their eyes for looking at the same object; or because correspondent fibres of the optic nerves, proceed to those most sensible parts in both eyes. It is manifest, that squinting from either of these two causes, is necessary, and incurable. (p. 113)

Erasmus Darwin (1778): The child was then about five years old, and exceedingly tractable and sensible, which enabled me to make the following observations with great accuracy and frequent repetition. 1. He viewed every object which was presented to him with only one eye at a time. 2. If the object was presented on his right side, he viewed it with his right eye. 3. He turned the pupil of that eye, which was on the same side with the object, in such a direction, that the image might fall on that part of the bottom of the eye where the optic nerve enters it. 4. When an object was held directly before, he turned his head a little to one side, and observed it with but one eye, *viz.* with that most distant from the object, turning away the other in the manner above described; and when he became tired with observing it with that eye, he turned his head the contrary way, and observed it with the other eye alone, with equal facility; but never turned the axes of both eyes on it at the same time. 5. He saw letters, which were written on bits of paper, so as to name them with equal ease, and at equal distances, with one eye as with the other. 6. There was no perceptible difference in the diameters of the irises, nor in the contractibility of them, after having covered his eyes from the light. These observations were carefully made by writing single letters on shreds of paper, and laying wagers with the child that he could not read them when they were presented at certain distances and directions.... A paper gnomon was made, and fixed to a cap; and when this artificial nose was placed over his real nose, so as to project an inch between his eyes, the child, rather than turn his head so far to look at oblique objects, immediately began to view them with that eye which was next to them.... by wearing the artificial nose, he had greatly corrected the habit of viewing objects with the eye farthest from them; and had more and more acquired the voluntary power of directing both his eyes to the same object, particularly if the object were not more than four or five feet from him; and will, I believe, by resolute perseverance, entirely correct this unsightly deformity. (pp. 86–87, 88, and 96)

Wells (1792): ... persons affected with squinting from their earliest infancy, see objects in the same directions with the eye they have never been *accustomed* to employ, as they do with the other they have constantly used. (p. 83)

Hosack (1794): The same fact is also observable in those who squint; the pupil in both eyes equally contracts and dilates, but still the vision of one eye is less perfect than the other. (p. 201)

Bell (1803): Most people who squint, have a defect in one eye, and this is the distorted eye, while the other is directed in the true axis of the object. Now the mind does not attend easily to two impressions, the one being weaker than the other: In a short time the weaker impression is entirely neglected and the stronger only is perceived.—So in squinting, the impression on the weak eye in a short time ceases to be attended to, the strong and vivid impression is alone perceived, and single vision is the consequence. It is evident, then, that those who squint must have a degree of imperfection in the strength of the image; for it is necessary to neglect the impression of one eye, to obtain distinct vision with the other.... In regard to the cure of squinting, it seems the most reasonable, in the first place, to endeavour to strengthen the weak eye by use, by tying up the sound one. In this case, the distorted eye becomes properly directed to the object, and the strength of the impression is in some degree restored. (pp. 357 and 361)

Wollaston (1824a): So long as our consideration of the functions of a pair of eyes is confined to the performance of healthy eyes in common vision, when we remark that only one impression is made upon the mind, though two images are formed at the same moment on corresponding parts of our two eyes, we may rest satisfied in ascribing the apparent unity of the impression to habitual sympathy of the parts, without endeavouring to trace farther the origin of that sympathy, or the reason why, in infancy, the eyes ever assume one certain direction of correspondence in preference to squinting. But, when we regard sympathy as arising from structure, and dependent on connection of nervous fibres, we therein see a distinct origin of that habit, and have presented to us a manifest cause why infants first begin to give the corresponding direction to their eyes, and we clearly gain a step in the solution, if not a full explanation, of the long agitated question of single vision with two eyes. (pp. 229–230)

Purkinje (1825): It is true that in almost all squinters one eye is weaker than the other, or more often one is shortsighted and the other longsighted.. the shortsighted eye alone and exclusively turns inwards in order, so to speak, to disregard the object. (p. 165)

Mackenzie (1835): In this disease, although the patient means to look at the same object with both eyes, one of them, moving involuntarily, and independently of the motions of the sound eye, turns away from its natural direction. If the sound eye be now closed, the other generally returns to the proper position, and so long as it is used alone, can be carried by the will of the patient in any direction he pleases. The instant, however, that the sound eye is again opened, the one affected with strabismus revolves inwards or outwards, and there it remains, not harmonizing in the movements of its fellow, or if it does move along with the sound eye, yet never so as to permit the two axes to be pointed at the same object. Hence the patient sees double, especially in the commencement of this disease; but after it has continued for a length of time, the double vision wears off, the impression on the squinting eye going for nothing. The eye is much more frequently distorted inwards than outwards in this disease, the adductor seeming to overpower the abductor, or the obliqui overcoming the recti. The former case is termed *strabismus convergens*, and the latter *divergens*. In some individuals we find the eyes squint alternately, or both together. In one case only have I seen strabismus directly upwards. (p. 303)

Müller (1838): In cases where the eyes have an unnatural direction inwards, objects situated in the horopter of this position of the eyes ought to be seen single, and it is not easy to conceive for what distance the new identity of the retinæ should be adapted, since the eye which is not affected with the strabismus can accommodate itself to all distances. Moreover, observations made upon

persons affected with strabismus do not show that the original relation of the
identical parts of the two retinæ is disturbed; but that the squinting eye in
general does not co-operate in vision.... Strabismus, indeed, often arises from
the indistinctness of the images in one eye; for the eye thus useless in vision
ceases to be properly directed towards the object which it is desired to view
distinctly, and at last is wholly unemployed. (1843, pp. 1197 and 1210)

6.8 Stereoscopic Vision

All people with two equivalent and undeviating eyes will see objects stereo-
scopically—they appear to have solidity. Thus, stereoscopic depth perception is
not a phenomenon confined to the stereoscope. Rather, the stereoscope made
people aware of what is visible naturally by manipulating the two images in
unnatural ways. Two almost equivalent pictures can assume solidity when
combined in the stereoscope, just as a book looks solid due to the slight differences in the images projected to each eye; it is a natural consequence of having
two frontal and laterally separated eyes. This makes it all the more surprising
that the phenomenon of solid sight, stereopsis, received such cursory attention
before the invention of the stereoscope. Wheatstone himself realized that this
was probably because of the many cues to depth that are available in addition
to retinal disparity. It is also likely that the concentration on binocular single
vision retarded the investigation of depth perception from stimulation of non-
corresponding retinal points (Wade, 1987).

There were occasional descriptions of the enhanced solidity seen with two
eyes, and both Ptolemy and Galen described differences in the views of each
eye. Ptolemy associated binocular vision with distance perception, and this
aspect was amplified by Leonardo, Smith, and Harris. Smith noted that the
perception of relief was more evident for near than for far objects, and that it
was related to the differences in the amounts of the objects seen by one eye and
not the other. Part of the quotation from Harris is well known because it was
reproduced in Boring's (1942) brief account of stereoscopy. Harris remarked
that the visual relief or depth was a consequence of binocular parallax, but
this term was used, as Wallis (1687) had earlier applied it, to the convergence
of the eyes.

Smith's comment about the vividness of objects seen with both eyes was
made in the context of a binocular telescope that he constructed. Accordingly,
as was noted in section 6.6, the stereoscope was by no means the first binocular
instrument. Even earlier than Smith, the Père Chérubin d'Orléans, a Capuchin
monk, had constructed binocular telescopes and microscopes, which were
described in two books published in 1671 and 1677. The quotation cited here
refers to his binocular microscope, which is illustrated. With this instrument
there is the suggestion that more can be seen than with a monocular micro-

scope. This reflects Chérubin d'Orléans's theoretical position that "perfect vision" follows from stimulation of both eyes simultaneously. It is highly unlikely that this instrument would have afforded stereoscopic impressions of minute objects simply because of its construction. Wheatstone (1853) noted that the arrangement of the eyepieces was such that any effects would have been pseudoscopic rather than stereoscopic. That is, the disparities were reversed so that near parts of the specimen would have had uncrossed disparity whereas those for far parts would have been crossed. Wheatstone (1852) devised a pseudoscope in his second memoir on binocular vision, and he had designed a truly binocular microscope at about the same time, although it was not manufactured (see Wade, 1981).

The stereoscopic phenomenon described by Blagden became known as the wallpaper illusion, after Brewster's (1844c) rediscovery of it when observing regularly patterned wallpaper. When equivalent but laterally separated patterns are combined binocularly they seem suspended in the plane of convergence. Blagden converged beyond the plane of the chimneypiece, and remarked that there was a change in apparent size as well as apparent distance.

Wheatstone's fundamental achievement was the *description* of stereoscopic depth perception; the invention of the stereoscope followed from this observation. His accidental encounter with stereopsis was under artificial conditions— the reflected images of a candle flame from a polished metal plate appeared in depth with two eyes, but flat with one. This alerted him to the significance of differences between the two retinal images, and he commenced his studies of them: as Wells (1792) had remarked, "In every part of natural philosophy, accidents often lead to discoveries, which reason alone might not easily have reached" (p. 34). Most of Wheatstone's experiments were conducted with free fusion of the paired images. He constructed both mirror and prism stereoscopes so that others, who did not share his ability to dissociate accommodation from convergence, could experience the phenomena. Wheatstone did not mention the prism stereoscope in his first memoir of 1838 (see Wade, 1983). Not only was perceived depth compelling when stereopairs were combined in the stereoscope but retinal disparities could be exaggerated or reversed, with predictable consequences. It was probably the last two characteristics that established the link between disparity and depth among visual scientists.

Ptolemy (ca. 150): It is fitting that nature should equalize the distance between the two visual axes, and gather them in accordance with the position of the thing to be seen. Therefore the visual axes are seen as falling upon the line through the midpoint between them and the point where the axes converge. This line is equidistant from the two visual axes and the two visual axes appear to coincide with it. Objects on the two visual axes are in different directions since the visual axes are inclined to each other. The only way they can be seen as one is if they are both seen as lying on an axis midway between them. And that middle axis should rightfully be called the common axis. In general, we distinguish between those things that are seen correctly, and those seen displaced. The distance between the apparent location of an object and its true position corresponds to the distance between the common axis and visual axis upon which the object lies. This distance is the perpendicular from the common axis to the object. (Lejeune, 1956, pp. 106–107)

Galen (ca. 175): . . . a thing seen by the right eye alone appears somehow to lie more to the left side when it is close by, and more to the right side when it is farther off; that if it is seen by the left eye alone, it will appear to lie more to the right when it is nearer and to the left when it is farther away; and that if it is seen by both eyes, it will appear to lie in the space between. (May, 1968, p. 493)

Ibn al-Haytham (ca. 1040): The axes of the eyes may meet on an object while the eyes perceive another object differently situated in direction with respect to the eyes. This happens when the object lies closer to the eyes than that on which the axes meet while being situated between the axes, or when it is farther off from the object on which the axes meet while also being situated between the axes if imagined to be extended beyond the meeting point, provided that the object on which the axes meet does not obscure the farther object or obscures [only] part of it. (Sabra, 1989, p. 232)

Leonardo da Vinci (ca. 1500): Things seen with both eyes will seem rounder than those seen with one eye. (MacCurdy, 1938a, p. 264)

Chérubin d'Orléans (1677): *It is known how to construct a novel type of microscope, in order to see the smallest object very agreeably and conveniently, represented entirely to the two eyes together, with a size and distinctness which surpasses all that we have seen until now with this type of microscope.* (p. 77)

Robert Smith (1689–1768) after a painting in Trinity College, Cambridge.

Smith (1738): We have one help to distinguish the place of a near object more accurately with both eyes than with one; in as much as we see it more detached from other objects beyond it; and more of its own surface, especially if it be roundish. . . . But when the objects are pretty large and pretty remote in comparison to the interval between the eyes; all the small helps seem ineffectual. . . . It is a common observation that objects seen with both eyes appear more vivid and strong than to a single eye. (pp. 41–42 and 387)

Harris (1775): We have other helps for distinguishing the prominances of small parts, besides those by which we distinguish distances in general; as their degrees of light and shades, and the prospect we have round them: Both these, I think, contribute their share to the visibility of such small objects as the pile upon a cloth, etc. And by the parallax on account of the distance betwixt our eyes, we can distinguish besides the front, parts of the two sides of near objects, not thicker than the said distance; and this gives a visible *relievo* to such objects, which helps greatly to raise or detach them from the plane, on which they lie: Thus, the nose on a face, is the more remarkably raised by seeing each side of it at once. These observations, I say, are of use to us in distinguishing the figures

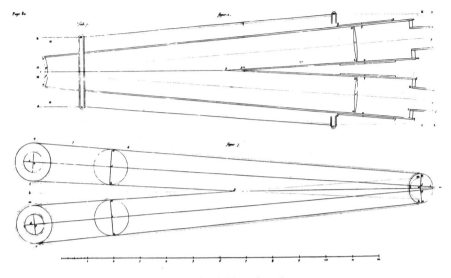

Chérubin d'Orléans's binocular microscope.

of small and near objects; and when the breaks, the prominencies and projections are more considerable, we do not want them. (p. 171)

Blagden (1813): In the house where I then lived was a marble chimney-piece, the upper horizontal block of which was fluted vertically; and the ridge between each convexity of the fluting was about as wide as the cavity itself. When I looked at this range of fluting at the distance of about nine inches, and directed the optic axes to it, I saw of course every ridge and concavity distinctly, and judged rightly of the distance. Adjusting the optic axes as to an object a little further off, I discerned the fluting confusedly and all double, the ridges interfering with the concavities; which was accompanied with the uneasy sensation of squinting. But on widening the direction of the optic axes still more, as to an object about eighteen inches distant; (namely, just so far that the duplication of the images should correspond successively; that is, so that the first ridge and concavity of the fluting, as seen by one eye, should fall in with the second ridge and concavity, as seen by the other;) the fluting appeared as distinctly and single as at first; but it seemed to be about double the distance from the eye that it really was, and to be magnified in proportion; nor had I, in this case, any sensation of squinting. (p. 112)

Wheatstone (1838): When a single candle flame is brought near such a [metal] plate, a line of light appears standing out from it, one half being above, and the other half below the surface; the position and inclination of this line changes with the situation of the light and of the observer, but it always passes through the centre of the plate. On closing the left eye the relief disappears, and the luminous line coincides with one of the diameters of the plate; on closing the

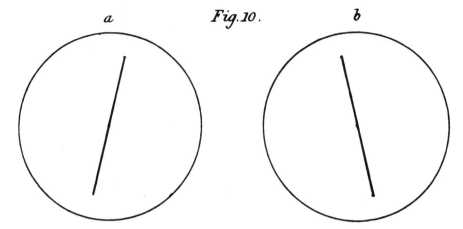

A stereopair of inclined lines which were similar to the reflections of a candle flame from a metal plate seen by each eye (Wheatstone, 1838).

right eye the line appears equally in the plane of the surface, but coincides with another diameter; on opening both eyes it instantly starts into relief. The case here is exactly analogous to the vision of two inclined lines (fig. 10.) when each is presented to a different eye in the stereoscope. It is curious, that an effect like this, which must have been seen thousands of times, should never have attracted sufficient attention to have been made the subject of philosophic observation. It was one of the earliest facts which drew my attention to the subject I am now treating. (p. 379)

6.9   Retinal Disparity

The differences between images at each eye were clearly described by Ptolemy and Galen. Indeed, it is of interest to compare different illustrations based on Galen's text. The one shown here is taken from Calabro's (1550) translation into Latin, and the three views are represented in the same plane. In May's (1968) translation from Greek into English the planes are represented as separated in depth (see Wade, 1987). This provides another example of the way in which prevailing theory can influence representations. As was noted in section 6.1, Euclid gave descriptions of the optical projections to a sphere, and this example was pursued by Leonardo for a sphere with a diameter smaller than the interocular separation. However, he did not restrict his consideration to the sphere alone, but also to the amount of background that would be visible: with one eye part of the background is always obscured from view, but this does not necessarily happen with binocular viewing. The visibility of the whole background when two eyes view a solid object in front of it has become called

Leonardo's paradox. Wheatstone reflected on Leonardo's choice of a sphere, suggesting that his expertise in perspective would have led him to realize projective disparity if a cube had been used; Leonardo might also have appreciated the significance of disparate images. Although Wheatstone declared himself unaware of any subsequent attempt using a cube, one was made by another expert in perspective—Sébastien Le Clerc (see Pastore, 1972). Le Clerc was both an artist and an academician who wrote on perspective. In his two books on vision (published in 1679 and 1712) he described and illustrated the binocular projections from a cube, as well as other objects, and he discussed the differences in the angles subtended by a given side to each eye. He concluded from his elegant inquiries that we see with one eye at once, and that Descartes's theory of binocular union could not be supported (see section 6.2).

Smith and Harris provided illustrations of crossed and uncrossed disparities, with the latter representing the disparate images in the plane of fixation. They were taken to be examples of noncorresponding stimulation and therefore of double vision. It was left to Wheatstone to commision drawings in perspective of cubes from slightly different positions, so that they could be combined binocularly and be seen in three dimensions. The necessity for such detailed drawing was soon obviated. Wheatstone was an acquaintance of Talbot, who provided stereoscopic photographs for him as early as 1839 (see Wheatstone, 1852). Initially the same scene was photographed twice from different locations, but binocular cameras were employed for this purpose later (see Coe, 1978).

Ptolemy (ca. 150): An object appears in one location when seen with only one eye, but when seen with both eyes an object is seen in one location only if it falls on corresponding visual lines, namely those that have corresponding positions with respect to the visual axes. And that comes about when the visual axes converge on the object to be seen, which happens when we see things with a simple gaze and in the way which is natural when we inspect an object. (Lejeune, 1956, pp. 26–27)

Galen (ca. 175): Let the right pupil be at α, the left at β. Let the magnitude seen be $\gamma \delta$ and let visual rays from each pupil fall on γ and on δ, and when they have so fallen, let them be produced. By the right pupil the magnitude $\gamma \delta$ will be seen distinctly over against the magnitude $\varepsilon \zeta$, by the left pupil it will be seen directly over against the magnitude $\eta \theta$, but by both eyes directly over against $\eta \varepsilon$. Hence neither pupil will see the object in the place where the other sees it, and both together will not see it where either sees it separately. (May, 1968, p. 495)

Leonardo da Vinci (ca. 1500): Suppose the two Eyes AB, viewing the Object C, at the Concourse of the two Central Lines, or Visual Rays, AC, B.C.; In this case, I say, that the Lines, or Sides of the visual Angle including these two Central Lines, will see the space GD, beyond, and behind the said Object; and

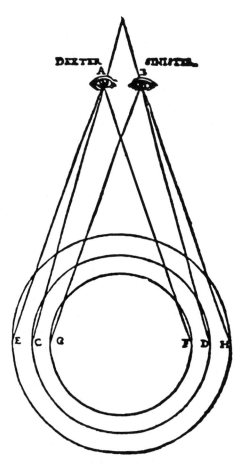

The text, which is a translation from Greek, describes the two views of an object in depth, whereas the illustration is taken from a translation into Latin by Calabro (Galen, 1550).

the Eye A will see the Space FD, and the Eye B, the Space GE; so that the two Eyes will see, behind the Object C the whole Space FE. By which means that Object C, becomes, as it were, transparent, according to the usual Definition of Transparency, which is that, beyond which nothing is hidden. Now this can never happen where the Object is only viewed with a single Eye; and where that Eye is less in Extent than the Object which it views; whence the truth of our Proposition is fairly evinced; a printed Figure intercepting a whole Space behind it; so that the Eye is precluded from the sight of any part of the *Ground* found behind the circumference of that Figure. (1721, pp. 178–179)

Sébastien Le Clerc (1679): It is evident that the images in the two eyes are different from one another; for supposing that the object DEFG is seen by the two eyes A and B, the eye A sees G between the rays AF and AE, the eye B on the other hand sees it between BD, BF and therefore if the eye A sees the point G in the line EF at point H, the eye B sees it in the line FD at point I. (p. 46)

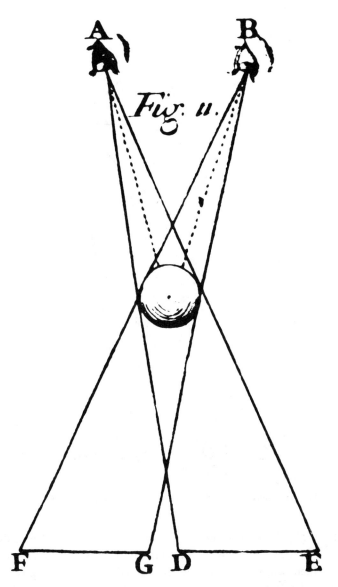

Leonardo's paradox. Viewing a sphere that has a diameter less than the interocular separation can result in the visibility of the whole of the background some distance behind it.

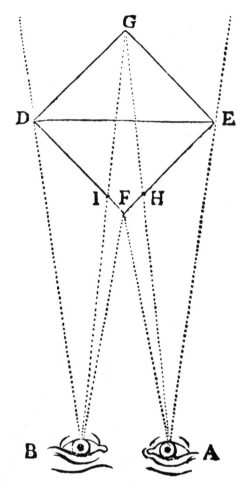

Le Clerc's illustration of binocular projections of a cube to each eye.

Porterfield (1737): Thus, if, while the *Optick Axes*, (See *Fig*. 4) AC, BC, are directed to a Mark C, for viewing it accurately, we attend to an Object *x*, placed any where within the Angle ACB formed of the *Optick Axes*, the Object *x* will appear in two Places; for being seen by the right Eye in the Direction of the visual Line B*x*, it must appear on the left Side of C, and its Distance from C will be measured by the Angle CB*x*; and being seen by the left Eye in the direction of the visual Line A*x*, it must appear on the right Side of C, and its Distance from C will be measured by the Angle CA*x*, and consequently it must appear double, and the Distance between the Places of its Appearance will be measured by the Sum of the Angles CB*x*, CA*x*. (p. 237)

Smith (1738): . . . if while the optick axes NM, OM (Figs 189, 190) are directed to a mark M, we attend to an object or image q, placed anywhere within the angle NMO or its opposite, made by the optick axes produced, the object q will

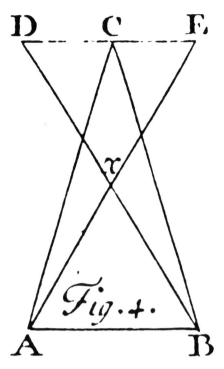

Retinal disparity according to Porterfield.

appear in two places, suppose at a and b, situated in the direction of the visual rays Nq, Oq. For the pictures of the object, q, which lyes between the optick axes, being both inverted with respect to the axes, must fall upon the retinas on contrary sides of the axes, and consequently upon places that do not correspond. (p. 47)

Harris (1775): When the axes of both eyes are directed to some determinate point F, an object G within the optic angle, fig.8, or its opposite, fig. 9, will appear double: that is, instead of being seen at G, it will appear somewhere in the lines rG, sG, as at m and n, either farther or nearer than its true place G, according as G is nearer or further than the horopter PR. (p. 112)

Wheatstone (1838): When an object is viewed at so great a distance that the optic axes of both eyes are sensibly parallel when directed towards it, the perspective projections of it, seen by each eye separately, and the appearance to the two eyes is precisely the same as when the object is seen by one eye only.... But this similarity no longer exists when the object is placed so near the eyes that to view it the optic axes must converge; under these conditions a different perspective projection of it is seen by each eye, and these perspectives are more dissimilar as the convergence of the optic axes becomes greater. This fact may easily be verified by placing any figure of three dimensions, an outline cube for

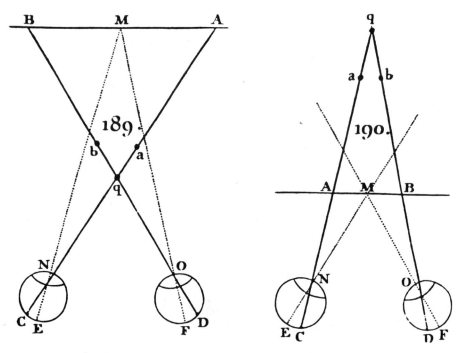

Smith's representation of crossed and uncrossed disparities.

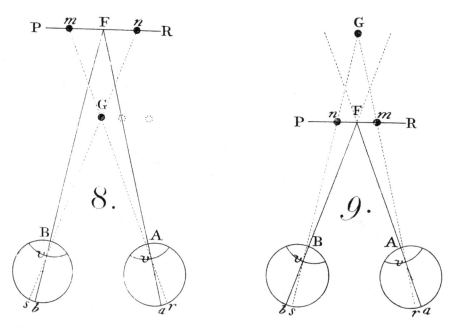

Crossed and uncrossed disparities according to Harris, with the disparate images shown in the plane of fixation.

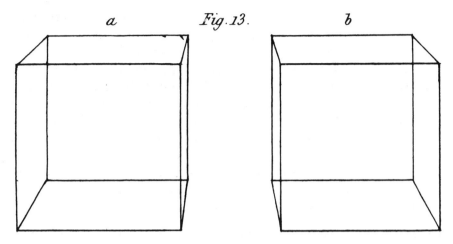

Wheatstone's perspective projections of a cube.

instance, at a moderate distance before the eyes, and while the head is kept
perfectly steady, viewing it with each eye successively while the other is closed.
Fig. 13. represents the two perspective projections of a cube; *b* is that seen by
the right eye, and *a* that presented to the left eye.... Had Leonardo da Vinci
taken, instead of a sphere, a less simple figure for the purpose of his illustration,
a cube for instance, he would not only have observed that the object obscured
from each eye a different part of the more distant field of view, but the fact
would also perhaps have forced itself upon his attention, that this object itself
presented a different appearance to each eye. He failed to do this, and no subsequent writer within my knowledge has supplied the omission. (pp. 371–372)

6.10 Panum's Limiting Case

The influence of retinal disparity on binocular depth perception was studied
in considerable detail in the mid-nineteenth century, particularly by Panum
(1858). Wheatstone (1838) appreciated that there were limits to differences
that could be combined by the two eyes: "No doubt, some law or rule of
vision may be discovered which shall include all the circumstances under
which single vision by noncorresponding points occurs and is limited" (p. 393).
Panum conducted investigations of the range over which fusion occurs.
One stimulus that he presented consisted of a single vertical line to one
eye and two lines to the other. This is like binocular observation of an object
extended along the visual axis of one eye, and it has become referred to as
Panum's limiting case. A variant of it was described by Le Cat, and Müller
illustrated it.

  Of the twelve stereograms Wheatstone presented in his memoir, eleven were
used to demonstrate that stimulation of the two eyes with slightly different

pictures can lead to depth perception. The odd one, which is shown here, was taken to show "that similar pictures falling on corresponding points of the two retinæ may appear double and in different places" (p. 384). This stimulus was the source of considerable controversy, and it is a complex one since it represents Panum's limiting case (see Ono and Wade, 1985). Examining the outcome of viewing this stimulus was later called the Wheatstone experiment by Hering (1862), and was widely investigated by sensory physiologists in Germany.

Wheatstone's examination of Panum's limiting case has been given belated recognition by Westheimer (1976), who has appended Wheatstone's name to Panum's. However, Wheatstone's monumental achievements were the description of stereoscopic depth perception, its isolation by means of the stereoscope, and its systematic examination in a manner equivalent to physical phenomena.

Claude-Nicolas Le Cat (1700–1768) after an engraving in Landon (1805).

Le Cat (1767): . . . when we look with one eye only, we are unable to distinguish distances, and cannot place the end of the finger directly upon an object indicated to us, though it be very near, for the finger hides the object, and appears to correspond to it as exactly when it is at the distance of a foot as if it were only a line removed from it. But, if our other eye be open, it will see the finger and the object from the side, and will therefore discover a considerable interval between them if they are a foot distant from each other, but only a very

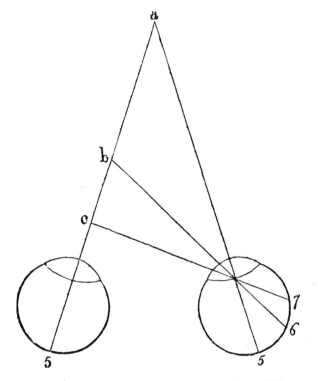

The conditions for Panum's limiting case according to Müller.

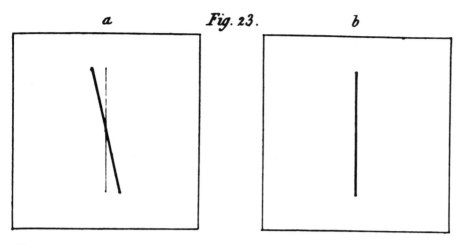

Wheatstone's stereopair showing a variant of Panum's limiting case, which was later called the Wheatstone experiment.

small interval if they are very near; and thus we are enabled to place our finger with certainty upon the desired object. (p. 475)

Müller (1838): Thus, if the axes are directed towards *a*, all other points lying in the line *a, b, c* appear double, because their images fall in the one eye upon the central point of the retina 5, but in the other eye at different points, 6, 7, 8, 9, &c. (1843, pp. 1203–1204)

Wheatstone (1838): Present, in the stereoscope, to the right eye a vertical line, and to the left a line inclined some degrees from the perpendicular (fig. 23.) . . . Draw on the left hand figure a faint vertical line exactly corresponding in position and length to that presented to the right eye, and let the two lines of this left hand figure intersect each other at their centres. Looking now at these two drawings in the stereoscope, the two strong lines, each seen by a different eye, will coincide, and the resultant perspective line will appear to occupy the same place as before; but the faint line which now falls on a line of the left retina, which corresponds to the line of the right retina on which one of the coinciding strong lines, viz. the vertical one, falls, appears in a different place. The place this faint line apparently occupies is the intersection of that plane of visual direction of the left eye in which it is situated, with the plane of visual direction of the right eye, which contains the strong vertical line. (pp. 384–385)

7   Space

René Descartes (1596–1650) after an engraving in Charles Perrault (1696).

Descartes (1637): *Now although this picture, in being so transmitted into our head, always retains the same resemblance to the objects from which it proceeds, nevertheless . . . we must not hold that it is by means of this resemblance that the picture causes us to perceive the objects, as if there were yet other eyes in our brain with which we could apprehend it; but rather, that it is the movements of which the picture is composed which, acting immediately on our mind inasmuch as it is united to our body, are so established by nature as to make it have such perceptions . . . All the qualities that we apprehend in the objects of sight can be reduced to six principal ones, which are: light, color, location, distance, size, and shape. (1965, p. 101)*

Space was not considered as a category of perception in antiquity, but aspects of it have always been at the forefront of inquiry. It impinges on all the headings that have been examined so far, some more so than others. For example, binocular vision has not been examined in isolation, but placed in the context of seeing single objects in particular positions. Ptolemy analyzed the perception of position, size, and form in his *Optics*, and Ibn al-Haytham broadened the range to include distance, position, solidity, shape, size, and separation.

Descartes echoed this classical heritage by listing location, distance, size, and shape as the proper concerns of vision. He had absorbed Keplerian dioptrics of the retinal image and related them to the perception of space. Descartes realized that the two were not equated, and that there was no central eye looking at the retinal image. Accordingly, the mind was involved in interpreting the retinal image. He examined this interaction for many phenomena, including visible direction, size and shape constancies, and cues for distance. His observations and speculations pervade perception's past. His mechanistic analysis of bodily processes did much to hasten the adoption of more precise methods for studying them, but his embrace of a nativist theory of mind polarized philosophers. Locke's empiricism represented an attempt to account for the aquisition of knowledge without recourse to innate ideas. Much of the battle between nativists and empiricists was waged with the weapons of words. Occasionally the disputes rose above the level of polemics and addressed aspects of perception itself. It is these forays into the facets of vision that are considered here. For example, an understanding of retinal image formation posed the problem of upright vision. Related to this were questions of whether newborn babies saw the world upright, and whether those whose sight was restored could recognize objects by vision alone. Both monocular and binocular visual direction were subjected to similar dichotomies, as were questions concerning cues to distance. Boring (1942) and Hochberg (1963) have summarized the main issues dividing nativists and empiricists in the context of perception. The aspects of space perception that were less controversial involved determining its limits, either in terms of visual acuity or distinct and peripheral vision. A final topic that will be touched upon is the representation of space in art; more specifically, some comments on perspective by students of vision will be given.

Space perception represents an arena in which optics and observation were often in conflict. Euclid provided geometrical analyses of almost all the phenomena included in this chapter. Thus, for example, size perception was equated with visual angles, so that the same object would have a different apparent size according to its distance from the eye. Ptolemy's interpretation of the phenomena were more subtle, since he realized that visual angles alone did not accord with the characteristics of observation. By adding distance and orientation to visual angles, he was able to give accounts of size and shape constancies. Ibn al-Haytham pursued this Ptolemaic line, which was given further elaboration by Descartes.

7.1 Space Perception

Euclid's geometrical analysis of size perception did have the virtue of precision, and so it is not unexpected that it should reappear centuries later in the context of artistic representation. Alberti was a mathematician who wrote

a treatise *On Painting* which described the then new technique of linear perspective. In many ways, perspective was a formalization of Euclid's optics, as it is concerned with capturing visual angles of objects at different distances (see section 7.12). We have already encountered statements by Leonardo and Wheatstone concerning the differences in perceiving pictures and scenes; Euclid provided an excellent theory of picture production, but not of space perception.

On a broader note, both Ptolemy and Ibn al-Haytham listed the properties available to vision. In his *Optics* Ibn al-Haytham enumerated eight conditions for perceiving objects accurately, but there were other properties, too. In a later manuscript he took issue with Ptolemy, suggesting that there were twenty-two "things perceived by sight" (Sabra, 1966, p. 147). The additional ones were concerned with features such as texture (roughness and smoothness), light and shade, similarity, and beauty.

Descartes described a situation in which external objects are focused on the retina. The illustrations representing the objects are radically different in *De homine* and *Traité de l'Homme*, with the former showing a very elaborate retinal image, whereas the latter resorts to the traditional arrow in the eye. A number of critical assumptions are made in the text; position and shape are given in terms of retinal correspondence, while size is derived from information concerning distance (provided by accommodation). Descartes enlisted a wider range of cues to distance than accommodation (see section 7.11).

Position was taken by Descartes to be innately coded by retinal location, whereas Locke argued that it was derived by learning; more particularly, by associating vision with touch. Berkeley (1709) strengthened the link between touch and vision, and this empiricist stance found ardent supporters in Buffon, Diderot, and Condillac. Kant sought to integrate the nativist and empiricist theories by proposing that space and time were innate, but other aspects of perception could be learned. Wells adopted a similar interpretation to the perception of location, which was made up of two components—direction and distance. His experiments led him to suppose that visual direction was "given by nature," but visual distance was "learned from feeling." Müller reached a similar conclusion.

Hippocrates (ca. 400 B.C.): To see the sun, moon, heavens and stars clear and bright, each in the proper order, is good, as it indicates physical health in all its signs . . . But if there be a contrast between the dream and reality, it indicates a physical illness. (1923b, p. 427)

Euclid (ca. 323–283 B.C.) after an engraving in *The Historic Gallery of Portraits and Paintings*, Vol. 7. London: Vernor, Hood, and Sharpe, 1811.

Euclid (ca. 300 B.C.): *Objects of equal size unequally distant appear unequal and the one lying nearer to the eye always appears larger.* Let there be two objects of equal size, *AB* and *GD*, and let the eye be indicated by *E*, from which let the objects be unequally distant, and let *AB* be nearer. I say that *AB* will appear larger. Let the rays fall, *EA, EB, EG,* and *ED*. Now, since things seen within greater angles appear larger, and the angle *AEB* is greater than the angle *GED*, *AB* will appear to be larger than *GD*. (Burton, 1945, p. 358)

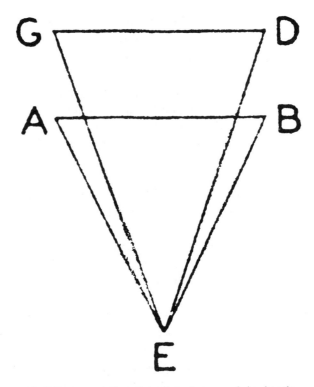

Euclid's representation of visual size in terms of visual angle.

Ptolemy (ca. 150): We say that vision is aware of body, size, colour, form, position, motion, and rest. (Lejeune, 1956, p. 12)

Ibn al-Haytham (ca. 1040): The conditions for perceiving an object as it is are eight: distance, being in a certain position, luminosity, being a sizable magnitude, opacity, transparency of the air, time, and soundness of the eye. (Sabra, 1989, p. 251)

Alberti (1435): Here is a rule: as the angle within the eye becomes more acute, so the quantity appears smaller. (1966, p. 47)

Descartes (1664): Similarly, if eye *D* is turned toward object *E*, the soul will be able to know the *position* of this object, inasmuch as [in the brain] the nerves from this eye are differently arranged than if it were turned toward some other object. And [the soul] will be able to know the *shape* [of *E*], inasmuch as rays from point *I* assembling on the nerve termed optic [the retina] at point 2—and those from point 3 at point 4, and so forth—will trace there a shape corresponding exactly to the shape of *E*. Note also that the soul will be able to know the *distance* of point *I*, for, as has just been mentioned, in order to make all the rays coming from point *I* assemble precisely at point 2 at the centre of the back of the eye, the crystalline humor will be of a different shape than if the object were nearer or farther away.... And [realize] finally that the soul will be able to know the *size* and all similar qualities of visible objects simply through its knowledge of the *distance* and *position* of all points thereof; just as, vice versa, it will sometimes judge their distance from the opinion it holds concerning their size. (Hall, 1972, pp. 59–62)

Locke (1690): As in simple Space, we consider the relation of Distance between any two Bodies, or Points; so in our *Idea* of *Place*, we consider the relation of Distance betwixt any thing, and any two or more Points, which are considered, as keeping the same distance one with another, and so considered as at rest.... The *Idea* therefore of *Place*, we have by the same means, that we get the *Idea* of Space (whereof this is but a particular limited Consideration), *viz.* by our Sight and Touch; by either of which we receive into our Minds the *Ideas* of Extension or Distance. (pp. 76 and 78)

John Locke (1632–1704) after an engraving in Birch (1743).

Le Cat (1744): The Mind not only rectifies the Image of Objects which occurs reversed in the Bottom of the Eye; it not only simplifies the double Impression of these Images in a sole and single Sensation; but judges, moreover, of the Distance and Magnitude of Objects it discerns. What means does it make subservient to this third Operation? The first of these means is the Magnitude of the Image itself, transmitted to the Bottom of the Eye; or as we say, the Magnitude of the visual Angle.... [secondly] By the Confusion of the Image itself, contained in the visual Angle, or by the Body of Vapours, which the Distance raises around the Object, and also by the Length of the optic Angle formed by the Concourse of the optic Axes of each Eye.... A third Rule, whereon the Soul founds its Judgments of the Magnitude and Distance of Objects, is Knowledge we have of the natural Magnitude of certain Objects, and of the Diminution accruing to them from Distance. (1750, pp. 209, 239, and 245)

Hartley (1749): ... in all Cases of Magnitude, Distance, Motion, Figure, and Position, the visible Idea is so much more vivid and ready than the tangible one, as to prevail over it, notwithstanding that our Information from Feeling is more precise than that from Sight, and the Test of its Truth. (p. 204)

Buffon (1749): Infants open their eyes the moment they come into the world; but their eyes are fixed and dull: they have not that lustre and brilliancy they afterwards acquire; neither have they those motions which accompany distinct vision. But they seem to feel the impression of light; for the pupil contracts or dilates, in proportion to the quantity of light. A new-born infant cannot distinguish objects; because the organs of vision are still imperfect: the cornea is wrinkled; and perhaps the retina is too soft and lax for receiving the impressions of external bodies, and for producing the sensations peculiar to distinct vision.... The senses are instruments of which we must gradually learn to use. That of vision is the most noble, and the most wonderful; but, at the same time, it is the most uncertain and elusory. The sensations produced by it, if not rectified every moment by the sense of touching, would uniformly lead us into false conclusions. (1866, p. 210)

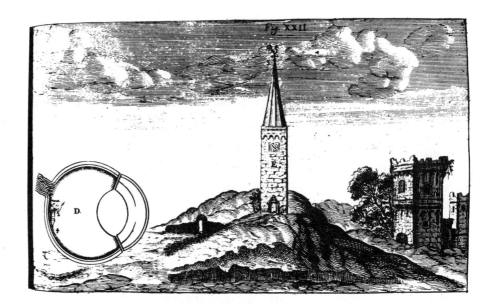

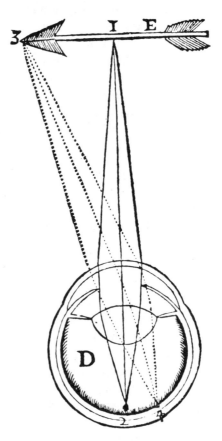

Visible position and distance according to Descartes (1662) above and Descartes (1664/1909) below.

Diderot (1749): It follows ... that we owe to experience the notion of permanent objects; that by touch we acquire that of their distance; that perhaps the eye must learn to see as the tongue to speak; that it would not be astonishing if the aid of one sense were necessary to another; and that touch, which gives us the assurance of external objects when they are before our eyes, is perhaps the sense which is reserved to affirm their distance to us, although I do not go so far as to say that it gives us their shape and other attributes.... The first time one opens the eyes one sees nothing; in the first instants of vision, one is affected only by a multitude of impressions which only become clearer with time and by the practice of reflecting upon what happens inside us. (Morgan, 1977, pp. 51 and 52)

Condillac (1754): It is true that we do not notice the judgments we make in order to grasp the whole of a circle or square.... When someone shows us a very complex picture, we are aware that we study it. We perceive that we count the persons in it, review their poses and features, and make a series of judgments on all these things and that [it] is only after all these operations that we take them all in at a single glance.... But now the quickness with which we habitually run over the sides of a square does not allow us to perceive the sequence of our judgments. It is reasonable to think that, when our eyes were untrained, they had to act in order to see the simplest objects as they now act in order to see the most complex ones. (1982, p. 218)

Kant (1781): Space is not an empirical concept which has been derived from outer experience.... Space is a necessary *a priori* representation, which underlies all our intuitions.... Space is not a discursive or, as we say, general concept of relations of things in general, but a pure intuition.... Space is represented as an infinite *given* magnitude. (1929, pp. 68–69)

Immanuel Kant (1724–1804) after an engraving in Hartenstein (1853).

Haller (1786): But the mind not only receives a representation of the image of the object by the eye, impressed on the retina, and then transferred to the common sensory or seat of the soul; but she learns or adds many things from mere experience, which the eye itself does not really see, and other things the mind confers or interprets to be different from what they appear to her by the eye. (Finger, 1994, p. 78)

Wells (1792): Since visible place contains in it both visible distance and visible direction, it is not necessary that the single appearance of an object, to both eyes, should depend altogether either upon custom, or an original principle of our constitution; for its visible distance to each eye may be learned from feeling, and its visible direction be given by nature; in which case, the unity of its place to the two eyes, will be owing to neither of those causes singly, but to a combination of both. (p. 33)

Müller (1838): The estimation of the form of bodies by sight is the result partly of the mere sensation, and partly of the association of ideas. Since the form of the images perceived by the retina depends wholly on the outline of the part of the retina affected, the sensation alone is adequate to the distinction of mere superficial forms from each other, as of a square from a circle.... The estimation of the different dimensions of bodies from their images on the retina is, on the contrary, a result of experience; since all the ideas obtained by vision refer originally to superficial extent only, and the idea of a solid body, or a body of three dimensions, can only be attained by the action of reason constructing it from the different superficial images seen in different positions of the eye with regard to the object. (1843, pp. 1175–1176)

7.2   Erect Vision

The analogy of the eye with a camera was made long before the dioptrics of the eye were fully understood (see section 2.2). Accordingly, the mismatch between the inverted image produced in a camera and upright perception could be accounted for by appropriate adjustment of the speculative optical properties of the eye: Leonardo represented a double inversion in the eye, so that the image accorded with perception. When Kepler did define the dioptrical properties of the eye, he laid the problem bare. Although he said he was prepared to leave the solution to natural philosophers he did suggest one: the relation of objects to one another was retained in the retinal projection, so if objects were seen in the lines of visual direction, they could then be assigned their appropriate locations. Brewster essentially repeated this interpretation in his analysis of visual direction (section 7.5).

Molyneux displayed a similar reluctance to enter into matters perceptual in a treatise on *Dioptrica Nova*, but he nevertheless gave what was taken to be the correct solution. The terms "up" and "down" are relative, and since it is not the eye itself that sees, the relativities can be resolved in the mind. His solution was repeated by Haller and Harris, and amplified by Bell and Müller. Berkeley considered the problem of the inverted retinal image in the context of a blind person seeing for the first time. He concluded that not only are the terms "up" and "down" relative but they are meaningless without recourse to touch. Despite this seeming consensus, there were those who were more closely bound to the retinal image, like Buffon. He considered that newborn infants saw objects inverted until they had learned otherwise by means of touch.

Leonardo da Vinci (ca. 1500): The pupil always receives the images of objects upside down and the *virtu visiva* sees them as they are. This is due to the rays passing through the center of the crystalline sphere in the middle of the eye. (McMurrich, 1930, p. 222)

Kepler (1604): And so if you are bothered by the inversion of this picture and fear that this would lead to inverted vision, I ask you to consider the following.

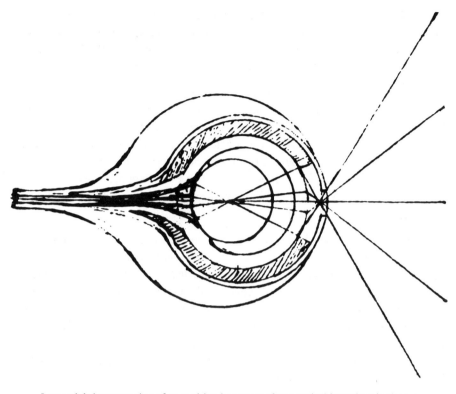

Leonardo's interpretation of erect vision by means of two optical inversions in the eye.

Just as vision is not an action, because illumination is an action but sensation is the contrary of action, so also, in order that the positions may correspond, the sensing things [i.e. the retinal images] must occur opposite to the agents that bring them about. Now the positions are perfectly opposite when all the lines connecting opposite points run through the same centre, which would not have been so if the picture had been erect. And so in the inverted picture, although right and left are interchanged everywhere and with respect to any common line, none the less with respect to themselves the right-hand parts of the object are perfectly opposed to the right-hand parts of the picture, and the upper parts of the object to the upper parts of the picture. (Crombie, 1964, pp. 142–143)

Molyneux (1692): How then comes it to pass that the Eye sees the Object *Erect*? But this Quæry seems to encroach too nigh the enquiry into the manner of the Visive Faculties *Perception*; For 'tis not properly the Eye that *sees*, it is only the Organ or Instrument, 'tis the *Soul* that *sees* by means of the Eye. To enquire then, how it comes to pass, that the Soul perceived the Object *Erect* by means of an *Inverted* Image, is to enquire into the Souls Faculties; which is not the proper subject of this Discourse. But yet that in this Matter we may offer at somthing, I say, *Erect* and *Inverted* are only Terms of *Relation* to *Up* and *Down*, or *Farther from* and *Nigher to* the Centre of the Earth, in parts of the same thing.... The Image of an *Erect* Object being Represented on the Fund of the Eye *Inverted*, and yet the sensitive Faculty judging the Object *Erect*; it follows that when the Image of an *Erect* Object is painted on the Fund of the Eye *Erect*, the sense Judges the Object *Inverted*. (pp. 105–106)

Berkeley (1709): Nor was it possible, by virtue of the Visive Faculty alone, without super-adding any Experience of *Touch*, or altering the Position of the Eye, ever to have known, or so much as suspected, there had been any Relation or Connexion between them: Hence, a Man at first view wou'd not denominate any thing he saw *Earth*, or *Head*, or *Foot*. And consequently, he cou'd not tell by the meer act of *Vision*, whether the Head or Feet were nearest the earth. Nor indeed, wou'd we have thereby any thought of *Earth* or *Man*, *Erect* or *Inverse*, at all. (pp. 119–120)

Porterfield (1737): What hath occasioned some seeming Difficulty in the Business of erect Appearances, is the groundless Supposition, that the Eye, or rather the Soul, by means thereof, sees an inverted Image of the external Object painted on the *Retina*, and that it judges of the Object from what it observes in this Image; But this is a vulgar Error, and I appeal to any one's Experience, whether he ever sees any such thing, and every one is himself best Judge of what he sees; and as the Mind sees not any Image on the *Retina*, so it takes no Notice of the internal Posture of the *Retina*, or the other Parts of the Eye, but useth them as an Instrument only for the Exercise of the Faculty of Seeing. (pp. 213–214)

Buffon (1749): ... the first great error of vision is the inverted representation of objects upon the retina: and, till children learn the real position of bodies by the sense of feeling, they see every object inverted. (1866, p. 240)

Haller (1767): The visible situation of the parts of an object, are judged by the mind to be the same with that which they naturally have in the object, and not the inverted position in which they are painted upon the retina. But it is certainly a faculty innate or born with the eye, to represent objects upright in the mind, whenever they are painted inverted upon the retina: for new-born animals always see things upright. And men who have been born with cataracts, without ever being able to see, are observed, upon couching the cataracts, to see every thing in its natural situation, without the use of any feeling, or previous experiences. (1786, p. 31)

Harris (1775): For although inverted images are, as it were, formed at the bottom of the eye, yet it cannot be supposed that the eye, or the mind sees these images as such: For the eye sees no part of itself, and even external objects must be at a competant distance from it, to become distinctly visible. (p. 102)

Bell (1803): It has been, by some, thought extremely difficult to account for the image appearing to us, as it is, erect, since it is actually figured on the bottom of the eye in an inverted posture; but the terms above and below have no relation to the image on the bottom of the eye, but to the position of our bodies and the surrounding things. (p. 237)

Volkmann (1836): The most natural explanation of upright vision is that it does not require an explanation. (p. 41)

Müller (1838): ... even if we do see objects reversed, the only proof that we can possibly have of it is that afforded by the study of the laws of optics; and that, if every thing is seen reversed, the relative position of the objects of course remains unchanged. It is the same thing as the daily inversion of objects consequent on the revolution of the entire earth, which we know only by observing the position of the stars; and yet it is certain that, within twenty-four hours, that which was below in relation to the stars, comes to be above. Hence it is, also, that no discordance arises between the sensations of inverted vision and those of touch, which perceived every thing in its erect position; for the images of all objects, even of our own limbs, in the retina, are equally inverted, and therefore maintain the same relative position. (1843, p. 1171)

7.3 Visual Acuity

Classical considerations of visual acuity were in terms of distance. When an object recedes from view it eventually ceases to be seen, as Euclid stated. Ptolemy introduced the factor of atmospheric dispersion into this increasing indistinctness. Ibn al-Haytham recognized that, even with near objects, some are too small to be seen, and individuals differ in their ability to distinguish minute detail.

Experimental inquiries into visual acuity were initiated by Hooke. He gave a demonstration to a meeting of Fellows of the Royal Society in which he determined whether they could distinguish a separation of less than one minute of arc; in general they could not. Hooke's study was not reported in the Society's transactions, but it was later noted by Birch in his *History of the Royal Society*. Smith determined the limit of visual acuity to be about two thirds of a minute, a value that was confirmed by Jurin. However, Jurin noted that different values would be obtained for detecting a separation between two pins and detecting a single pin. Such measurements were also dependent on the ambient illumination as well as the nature of the stimulus, as Buffon stated. A wider variety of stimuli were examined by Mayer—dots, gratings, and grids—and the value for visual acuity that he derived was slightly less than those of Smith and Jurin, but it was in close correspondence with that determined by Treviranus.

Hooke's interest was alerted to the problem of acuity because it held such importance for astronomers. The astronomer William Herschel examined his own vision through minute apertures and was able to read letters of text.

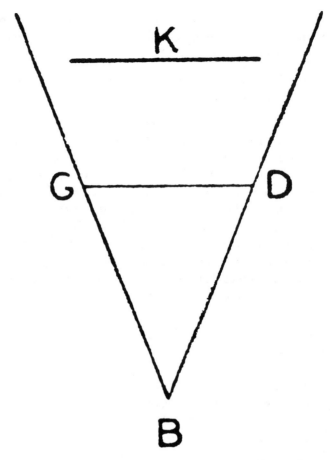

Visual acuity defined as the distance beyond which an object can be seen.

Euclid (ca. 300 B.C.): *Every object seen has a certain limit of distance, and when this is reached it is seen no longer.* For let the eye be *B*, and let the object seen be *GD*. I say that *GD*, placed at a certain distance, will be seen no longer. For let *GD* lie midway in the divergence of the rays, at the limit of which is *K*. And the thing upon which rays do not fall is not seen. (Burton, 1945, p. 357)

Ptolemy (ca. 150): Objects which are near the eyes are seen more clearly ... objects which are far from the eyes are less well perceived because the visual rays carry some of the blackness of the air they cross. For this reason distant objects are seen indistinctly as if behind a veil. (Lejeune, 1956, p. 20)

Ibn al-Haytham (ca. 1040): Further, we find that sight does not perceive any visible object unless the object is of a certain size.... The smallest magnitudes that can be perceived by sight are also related to the strength or weakness of sight. For some small bodies are perceived and sensed by some people but

cannot in any way be seen by many others whose sight is not very strong. . . . .
A more intensely illuminated object may be perceived by sight from a distance
at which objects of equal size are invisible. (Sabra, 1989, pp. 9–10)

Hooke (1674): Mr. Hooke made an experiment with a ruler divided into such
parts, as being placed at a certain distance from the eye, appeared to subtend
a minute of a degree; and being earnestly viewed by all the persons present, it
appeared, that not any one present, being placed at the assigned distance, was
able to distinguish those parts, which appeared of the bigness of a minute, but
that they appeared confused. This experiment he produced, in order to shew,
that we cannot by the naked eye make any astronomical or other observation
to a greater degree of exactness than that of a minute, by reason, that whatever
object appears under a less angle, is not distinguishable by the naked eye; and
therefore he alledged, that whatever curiosity was used to make the divisions of
an instrument more nice, was of no use, unless the eye were assisted by other
helps from optic glasses. (Birch, 1757, p. 120)

Smith (1738): I have been present at making the experiment, when a friend of
mine, who had the best eyes of all the company, could scarce perceive a white
circle on a black ground, or a black circle on a white ground, or against the sky-
light, when it subtended a less angle at the eye than two thirds of a minute; or
which is the same thing, when its distance from the eye exceeded 5156 times its
own diameter: which agrees well enough with Dr. *Hook*'s observation. Hence
I find . . . that the diameter of the picture of that circle upon the retina was
but the 8000th part of an inch at most; and this may be called the sensible
point of the retina. That this point is very small any one may perceive from
hence that the breadth of the finest hair is visible at the length of ones arm.
(p. 31)

Jurin (1738): By trying this experiment with two pins of known diameters, set
in a window against the sky-light, with a space between them equal in breadth
to one of the pins, I find the distance between the pins can hardly be distinguished, when it subtends an angle less than $40''$, though one of the pins alone
can be distinguished under an angle vastly less. (p. 151)

Buffon (1749): The smallest visible angle is about one minute . . . An object,
for example, of a foot square, ceases to be visible at 3436 feet. A man five feet
high is not visible beyond the distance of 17,180 feet, when the sun shines. . . .
In order to determine the utmost distance at which an object can be rendered
visible, three things must be considered: 1. The largeness of the angle formed in
the eye; 2. The degree of light with which the neighbouring and intermediate
objects are illuminated; and 3. The intensity of the light proceeding from the
object itself. Vision is affected by each of these causes; and it is only by estimating and comparing them, that we can determine the distance at which any
particular object can be discerned. (pp. 243–244)

Condillac (1754): An object is only visible insofar as the angle that determines the extent of its image on the retina has a certain value. I suppose that it ought to be at least one minute: but I say this only to anchor our discussion, for the angle must vary according to the eyes. (1982, p. 282)

Mayer (1755): ... there is a certain visual angle below which an object presented to the eye appears either not distinct enough or not even distinct at all, but only confused and as if it had vanished from sight.... We shall call this angle the *limit of vision*, and we shall investigate its magnitude by experiment.... objects seen under such circumstances will not be visible unless they subtend in the eye an angle of more than 34″; those subtending a smaller angle will definitely escape visual acuity. (Scheerer, 1987, pp. 87 and 89)

Johann Tobias Mayer (1723–1762) after an engraving in Augé (1898).

William Herschel (1786): Through a very thin plate of brass I made a minute hole with the fine point of a needle; its magnified diameter, very accurately measured under a double microscope, I found to be ,465 of an inch, while under the same apparatus a line of ,05 in length gave a magnified image of 3,545 inches. Hence I concluded, that the real diameter of the perforation was about the 152d part of an inch. Through this small opening, held close to the eye, I could very distinctly read any printed letters on which I made the trial. Proper allowance must be made for the very inconvenient situation of the eye, which by the unusual closeness to the paper cannot be expected to see with its common facility. Besides, the continual motion of the letters, which is required on account of the smallness of the field of view, must needs take up a considerable time ... In some other pieces of brass I made smaller holes; and among many, that were measured with the same accuracy as in the former experiment, I found one whose magnified diameter was ,29: hence the real diameter could not exceed the 244th part of an inch. Through this opening I could also read the same letters; but the difficulty of managing so as not to intercept all the incident light, as well as the uneasy situation of the eye, were sufficient reason for not carrying the intended experiments any further in this form. Besides, I should hardly have allowed them to be fair, if, on a further contraction of the hole in the brass plate, any indistinctness had come on; as we might well have suspected at least 2 other causes, besides the smallness of the pencils, to contribute to such an imperfection, viz. want of light, and a deflection of it on the contracted edges of the hole. (pp. 501–502)

William Herschel (1738–1822) after an engraving in Knight (1835b).

Treviranus (1828): As a rule, one cannot distinguish two points from one another if they subtend less than 30″ outer visual angle. (p. 49)

Volkmann (1836): It is not determined that 2 retinal images appear as 2 if the space between them is the same size as the diameter of the smallest retinal image.... The limit of the detection of a spider thread–22″; the limit of discrimination between 2 threads–8″. (pp. 201–202)

Fig. 1.

Fig. 2.

Fig. 3.

Fig. 4.

Fig. 5.

Fig. 6.

Fig. 7.

Fig. 8.

Stimuli employed by Mayer to measure visual acuity.

7.4 Weber's Law

Many of Weber's experiments, demonstrating the relativity of discriminations, were conducted with lifted weights. This was so, in part, because weight could be quantified in a manner that was not so readily available for light intensity. He did, however, provide evidence for similar relativity in the context of line length judgments. This allowed him to express the generality of the relationship between difference thresholds and the standards against which comparisons are made. This is now referred to as Weber's law.

   An earlier example of such relativity was in the area of judging brightness. The precise relationship between light intensity and brightness required an adequate method for comparing light sources. This was provided by Bouguer in his *Optical Treatise on the Gradation of Light* (1760); he devised a simple arrangement of screens and apertures, so that candles could be placed at known distances away from them, allowing their brightnesses to be compared. A similar arrangement was alluded to in Rubens's illustration in Aguilonius (1613), but as with all his frontispiece engravings, it is not clear to what extent they were guided by science or art, or whether the ideas conveyed in them originated with Rubens or Aguilonius (see Ziggelaar, 1983). By employing the inverse square law of Kepler (that light intensity declines as the square of the distance from the source) Bouguer was able to determine the differences in

The basic principle of photometry was depicted by Rubens in the frontispiece to Book V of Aguilonius (1613).

intensity that could be discriminated. A difference in brightness could be
detected when one light source was a sixty-fourth more intense than the other.
Bouguer essentially founded experimental photometry; he initially presented his
ideas in his *Essai d'Optique* (1729), but his *Treatise* did not appear until two
years after his death. In the same year that it was published, Lambert (1760)
elaborated the mathematical basis of photometry and placed it on an even surer
footing (see Wilde, 1838; Wolf, 1938). Lambert's photometer was based upon a
comparison of shadows cast by different light sources, and similar principles
were incorporated in Rumford's (1794) photometer (see Walsh, 1958).

Other examples of the relativity of perception were commented on in the
context of color and motion.

Bouguer (1760): We shall put at the head of all our observations those which
have taught us how much intensity a light must have so that its presence may
render the effect of another much more feeble light absolutely invisible....
Having placed a candle at a distance of 1 foot from a very white surface, I
placed beside the candle a ruler of a certain width, and I then placed another
candle of the same size as the first at various distances until I ceased to
distinguish the shadow of the ruler caused by the second candle.... Thus the
difference between the two lights ceased to be visible only when the small part
added was about sixty-four times as weak as the first. (1961, pp. 50–51)

Pierre Bouguer (1698–
1758) after an engraving
in Picard (1924).

Weber (1834): *When noting a difference between things that have been compared,
we do not perceive the difference between the things, but the ratio of the difference
to their magnitude....* My statements concerning weights compared by touch
also hold true for the comparison of lines by vision. For whether you compare
longer or shorter lines, you will find that most subjects cannot perceive the
difference if one line is smaller by a hundredth part. (Ross and Murray, 1978,
p. 131)

7.5 Monocular Visual Direction

Ernst Heinrich Weber
(1795–1878) after a
lithograph in Stirling
(1902).

The body of Euclid's *Optics* is based on visual direction: light was emitted
from the eye to points in space. Both Euclid's visual cone and Ptolemy's visual
pyramid are predicated on this assumption, which was repeated by Galen.
Although the direction of the rays was reversed by Kepler, the same idea is
embodied in his analysis of visual position. Essentially the same definition of
visual direction was given by Smith, Porterfield, Reid, and Brewster, while
Condillac and Brown adopted associationist interpretations. Descartes
considered the consequences of passively moving one eye (after the manner of
Aristotle; section 6.3). Oddly enough, although the text does describe the
projection to the unmoved eye, only one eye was illustrated in *Traité de
l'Homme*. Young espoused a similarly physiological approach to visual
direction, and Müller poured scorn on the directional ray hypothesis.

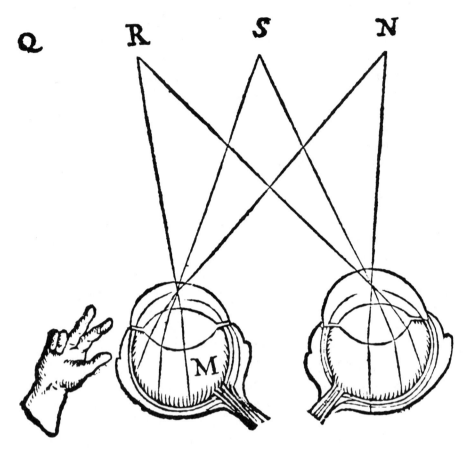

Visual direction according to Descartes (1662) on the left and Descartes (1664/1909) on the right.

Galen (ca. 175): Let there be a circle ... seen by one of the eyes while the other remains closed; from the mid-point of the circle (which is also called its center) think of a straight path to the pupil of the eye that is seeing it. (May, 1968, p. 492)

Kepler (1611): The position of an object is estimated from the direction in which the ray of sight *originally* emerges from the eye, no matter how this direction may be altered by refraction in its path between the eye and the object. For the eye *cannot detect* what happens to the rays outside itself in intervening media, but *assumes* that these proceed in the original direction. (Mach, 1926, p. 44)

Descartes (1664): ... if eye M is turned away by force from object N, and arranged as if looking toward q, the soul will judge that the eye is turned toward R. In this situation rays from object N will enter the eye in the way that those from point S would do if the eye were in fact turned toward R; hence it [the soul] will believe that this object N is at point S and that it is a different object from the one being looked at by the other eye. (Hall, 1972, p. 65)

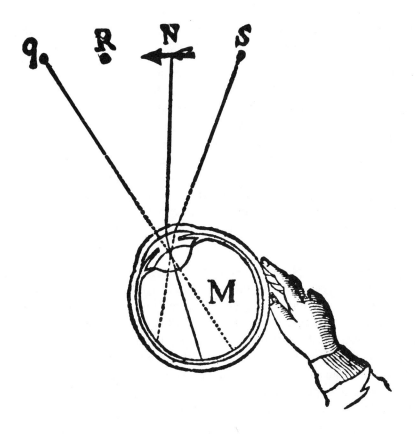

Porterfield (1737): *Every Point of an Object appears and is seen without the Eye, nearly in a straight Line drawn perpendicularly to the* Retina *from the Point of it where its Image forms.* (p. 208)

Smith (1738): Any small object or part of an object, seen by refracted or reflected rays, appears somewhere in the direction of that line, which the visual ray describes after its last refraction or reflection in falling upon the eye. (p. 32)

Condillac (1754): It is the hand that, guiding vision successively over the different parts of a shape, etches all these parts in memory; it is the hand that, so to say, guides the engraving tool when the eyes begin to attribute to the exterior the light and colors that they first experienced as in themselves. They perceive these properties where touch teaches them that they must be: they see as up what touch teaches them to judge as up and they see down what it leads them to judge as down; in a word, they see objects in the same positions as the sense of touch represents them to be. (1982, p. 279)

Reid (1764): ... every point of the object is seen in the direction of a right line passing from the picture of that point on the retina through the centre of the eye. (p. 289)

Young (1801): In the eye, we judge very precisely of the direction of light, from the part of the retina on which it impinges. (pp. 25–26)

Brown (1820): In the same way are we to account for the erect position, in which objects appear to us, when the image on the retina is inverted. It would indeed be truly wonderful, if they were to appear differently; for it is not the image on the retina that appears to us without: it is that which was felt tactually, when similar visual sensations were before excited. The visual sensations, by frequent co-existence, have become representative of what was thus felt; and therefore, if they suggested an inverted object, they would not suggest the object of touch which had before co-existed with such sensations, but one of which the presence had before co-existed with a different sensation, when an object of the same shape, but inverted, was before grasped by us and seen. (p. 150)

Thomas Brown (1778–1820) after a frontispiece engraving in Welsh (1825).

Brewster (1830b): The phenomenon of erect vision, by means of an inverted picture, and of single vision, by means of two images on the retina, have long been a source of difficulty among physiologists. If we conceive that the mind views the inverted picture ... on the retina, in the same manner as an eye behind it would do, the difficulty of accounting for erect vision would be great. Such a hypothesis, however, is certainly without foundation, for we know nothing more than that the mind, residing, as it were, in every point of the retina, refers the impression made upon it at each point to a direction coinciding with the last portion of the ray which conveys the impression. (p. 615)

Müller (1838): The hypothesis, that erect vision is the result of our perceiving, not the image on the retina, but the direction of rays of light which produce it, involves an impossibility, since each point of the image is not formed by rays having one determinate direction, but by an entire cone of rays; and, moreover, vision can consist only in the perception of the state of the retina itself, and not of any thing lying in front of it in the external world. (1843, p. 1171)

7.6   Binocular Visual Direction

Ptolemy examined binocular visual direction experimentally by means of his viewing board (see also sections 6.1 and 6.2). From his observations he concluded that an object on the common axis nearer than the fixation point would appear double, and that objects on the visual axes are seen in three locations: fused on the common axis and laterally separated by twice the distance between the visual axes. These principles of binocular visual direction were redescribed by Ibn al-Haytham and rediscovered by Wells (see section 6.1) and by Hering (1865). Wells was unaware of Ptolemy's experiments but he did cite Alhazen (Ibn al-Haytham), and Hering did not cite Ptolemy, Alhazen, or Wells (see Howard and Wade, 1996; Ono, 1981).

Galen described the apparent movement that occurs when opening and closing the eyes in alternation, and also the different directions in which objects appeared during monocular and binocular viewing (section 6.8). Wells provided a similar demonstration of apparent movement, but with monocular viewing through an aperture. When the distant object was viewed, convergence would have been minimal (even with one eye closed), whereas looking at the aperture would have increased the convergence in the closed eye, with a consequent change in the muscular signals that determine visual direction (see section 5.8). In his discussion of binocular visual direction Descartes returned to the analogy of the paired eyes working like the paired hands. Condillac, on the other hand, had the hand teaching the eyes the direction in which objects appear. The objections that Müller raised with regard to monocular visual direction were repeated in the context of using two eyes: if objects are seen along directed lines, then there is only one point in which the two directions meet.

Ptolemy (ca. 150): Let us speak first about that construction in which the heads of the two visual pyramids [the eyes] are points a and b joined by line ab and divided at the middle at point g [see figure]. Produce from this middle point a perpendicular gd and let the visual axes ad and bd converge on an object at point d. Under these conditions object d is seen as one and in its correct location. If from point d we draw a line edz at right angles to gd, anything positioned on that line, since it is at the head of [in the same frontal plane as] point d, will appear as one and in its correct location. When the line htk is produced parallel to edz, and the eyes are converged on point d, an object at point t will be seen in two locations h and k. Moreover, two objects positioned in h and k will be seen in three locations, t, l, and m. They will both appear superimposed at point t as if they were one thing. In addition, they will appear separately, h at point i and k at point m. Any object on lt and tm will be seen in the same manner on hk. (Lejeune, 1956, pp. 102–103)

Galen (ca. 175): If you look with both eyes at once, it [a pillar] seems to occupy a place midway between the places it appears to occupy to each eye used separately. And if you care to look at one of the stars in the same way or at the moon, especially when it is full and equal on all sides, it will seem to jump suddenly to the right when you open the left eye and close the right, and to the left when you do the opposite. (May, 1968, p. 496)

Ibn al-Haytham (ca. 1040): If, however, the two axes meet on a visible object, while the eyes perceive another object closer to or farther from them.... then: it will be to the right of one of them and to the left of the other; the rays drawn to it from one eye will be on the right of the axis and those drawn to it from the other eye to the left of the axis; therefore, its position relative to the eyes will differ in respect to direction. (Sabra, 1989, p. 237)

Descartes (1637): As to the position, that is to say the direction in which each part of an object lies with respect to our body, we perceive this with our eyes in

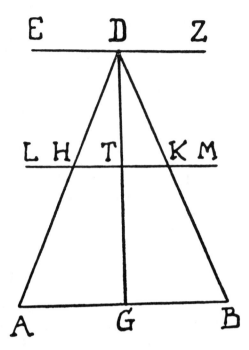

Ptolemy's board used for determining visual directions in binocular vision.

the same way as we would with our hands; and this knowledge does not depend on any image, nor on any action which proceeds from the object, but only on the position of the small points of the brain whence the nerves originate. (1965, p. 104)

Condillac (1754): Each eye fixes on the object that the hand has grasped, each attributes the colors to the same place, at the same distance, and as the reversal of the image does not prevent them from seeing an object in its true location, so the same image, although double, does not prevent them from seeing it as [single].... Thus it is natural for the eyes not to see double. (1982, p. 279)

Wells (1792): Look with one eye, the other being closed, at a remote object through a small hole in a card. If you should afterward suddenly attempt to view the hole itself accurately, with the same eye, you will observe both it and the distant object, particularly the latter, to move from left to right, if the right eye be used; but if the left eye be the one employed, then from right to left. (p. 79)

Müller (1838): All explanations of the direction of vision which are founded on the hypothesis of the projection of the images in decussating lines have one defect in common. The phenomenon of vision with two eyes simultaneously are wholly at variance with them. If visual direction be dependent on an exterior action of the retina in any determinate direction, either in the direction of the

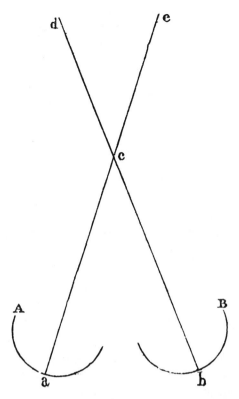

The lines of monocular visual direction depicted by Müller.

point on which the eye revolves, or in a direction perpendicular to the retina, single vision with two eyes becomes inconceivable; for, by the eye A, the image of the object *c* falling on the central part of the retina *a* will be seen in the direction *a c e*; while, by the eye B, the image of the same object, *c*, falling upon the central point of the retina *b*, will be seen in the direction *b c d*. Hence, according to this hypothesis, the image of the same object *c*, will be seen at two different parts of the visual field. (1843, p. 1175)

7.7 Distinct Vision

Euclid related distinct vision to visual acuity, and gave an instance of sharply defined corners appearing rounded when viewed from a distance; Lucretius described a specific example of this. Ptolemy provided a general statement regarding the greater distinctiveness of vision along the axis of the eye as opposed to more lateral rays. He accounted for this by saying that perpendicular rays were stronger than oblique ones. The same principle was enunciated more poetically by Leonardo. An emprical demonstration of distinct vision was given by Ibn al-Haytham. He used the board described previously in section 6.1

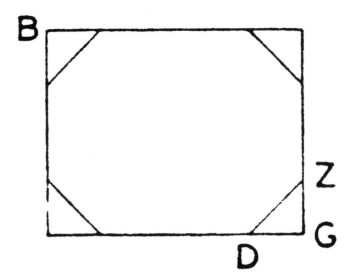

The rounded appearance of distant rectangular objects according to Euclid.

(but note that *H* in the text corresponds to *k* on the diagram shown there), and fixed two equivalently written words in central and peripheral locations; the central word was more easily read than the lateral one. The indistinctness of the lateral word increased with movement into the periphery. The experiment was conducted with monocular and binocular observation with essentially the same results. Thus, Ptolemy's and Ibn al-Haytham's boards were instrumental in rendering spatial as well as binocular vision experimental.

Gassendi drew attention to the large individual differences in vision, by citing an instance of a friend who could see clearly under low levels of illumination. Castelli, Rohault, and Le Cat attributed the distinctness of central vision to the dioptrics of the eye. Harris gave an astute estimate of the size of the field of distinct vision, namely, a visual angle of one to two degrees.

Euclid (ca. 300 B.C.): *Rectangular objects, when seen from a distance, appear round.* For let the rectangle *BG*, standing upright, be seen from a distance. Now, since each thing that is seen has a certain limit of distance beyond which it is no longer seen, the angle *G* is not seen, but only the points *D* and *Z* are visible. In the same way also in the case of the other angles this will happen. So the whole thing will appear round. (Burton, 1945, p. 359)

Lucretius (ca. 56 B.C.): And when afar off we see the foursquare towers of a city, they often appear round, for this reason, because every angle at a distance is seen blunted or rather it is not seen at all. (1975, p. 305)

Ptolemy (ca. 150): . . . that which is perpendicular to the visual axis should be seen more clearly than that which is observed on the sides through the lateral rays. (Lejeune, 1956, pp. 19–20)

Ibn al-Haytham (ca. 1040): Again, let the experimenter take a sheet of paper from which he must cut three small and equal strips; on one of them let him write some arbitrary word in a clear hand, then write the same word with the same size and shape on the other two strips. Let him fix the object at the middle point of the board as before, and also fix one of the two objects at point *H*. He must then attach one of the three strips to the object in the middle of the board and one of the others to the object at point *H*, making sure that [the second strip] has the same position as the first. Let him bring the board close to his eyes, as before, and gaze at the strip on the middle object and contemplate it. He will have a distinct perception of the word written on it, and at the same time perceive the other strip and the word written on it but not as clearly as the corresponding word on the middle strip; the latter will be clearer and more distinct though it has the same shape, configuration and size as the other.... It is evident from this experiment that of the visible objects facing the eyes and binocularly perceived, the clearest is that which lies where the two axes meet, and the closer the object is from this point the clearer it will be. (Sabra, 1989, pp. 244–245)

Leonardo da Vinci (ca. 1500): An object is less distinct if it impresses itself at a greater distance from the center of the pupil, where terminates the median line that goes straight to all the objects of the figure of which one can have a true and certain knowledge. This line is straight, without any intersections, and is the mistress of other lines. (McMurrich, 1930, p. 225)

Gassendi (1624): A friend of mine suffers from exophthalmia; his eyeballs protrude greatly. An object must be brought up very close for him to be able to see it. I believed that he would be far from able to judge soundly about light and color, which are the objects of vision, because of the deformation of his eyes, so different from a normal constitution. But then one day at dusk a letter was brought to me; when I opened it, I, who have fairly good eyesight, could hardly make out whether the paper was blank or covered with lines; but to my stupefaction he read the whole letter through without hesitation, and he could not persuade me that he was not joking until a torch was brought and I proved to myself that it was so and I had also recognized what he read from a book that was brought up. Consequently, I ask you whether his consitution or mine generated the truer appearance of things at that moment. I see more clearly at midday, but he sees more clearly at dusk. Then we are both even, except that he even surpasses me in that he will see the same thing I do at noon if the object is moved up closer, but I cannot see at all at dusk. (Brush, 1972, p. 89)

Castelli (1639): ... when we focus the eye on some object in order to see it, we see it well and distinctly while we see other objects near it with some confusion, a confusion which is increasingly greater as objects are further away from the first one on which we have focussed our sight. This happens because of the pictures inside the eye those which are near the axis of the eye are formed distinctly while other ones which are farther away from the axis show up with increasing confusion. (Ariotti, 1973, p. 9)

Rohault (1671): As to the *Distinctness of Vision*, that evidently depends upon the Refraction of the Rays; and it is then as distinct as possible, when the Refraction is so made, as that all the Rays which come from one and the same Point of the Object, meet together exactly in one and the same Point of the Bottom of the Eye: But this never is precisely so, but in those Rays which come from the Point of the Object which is at the Extremity of the *Optical* Axis; wherefore we cannot at the same time have the most distinct Sensation but in this Place alone, and the rest will be more confused. (1723, p. 249)

Berkeley (1709): The *Visive Faculty* consider'd with reference to it's immediate Objects, may be found to Labour under two Defects. *First*, In respect of the Extent or Number of visible Points that are at once perceivable by it, which is narrow and limited to a certain Degree. It can take in at one view but a certain determinate number of *Minima Visibilia*, beyond which it cannot extend it's Prospect. *Secondly*, Our Sight is defective in that its view is not only narrow, but also, for the most part, confus'd. Of those things that we take in at one Prospect, we can see but a few at once clearly and unconfusedly. And the more we fix our Sight on any one Object, by so much the Darker and more Indistinct shall the rest appear. (pp. 98–99)

Jurin (1738): An object is said to be seen distinctly, when its outlines appear clear and well defined, and the several parts of it, if not too small, are plainly distinguishable, so as that we can easily compare them one with another, in respect to their figure, size and colour. (p. 115)

Le Cat (1744): An Image is distinct, when all the Points of the luminous Cone that form it meet in the same Proportion as they preserve in the Object itself, without Confusion, or Space, without any Mixture of foreign Rays, and without the Organ's being affected by this regular Collection of Rays either in too lively or in too feeble a manner. (1750, p. 250)

Harris (1775): A small object, as a star seen by twilight, is lost by a small motion of the eye from it; and if we look attentively upon a letter in the middle of a word, consisting of capitals of a middling size, the adjacent letters on either side will be somewhat indistinct. These are proofs that the field of distinct vision is very small, and probably does not exceed one degree, or two degrees at most. (p. 118)

7.8 Peripheral Vision

Ptolemy and Ibn al-Haytham discussed distinct and peripheral vision together, and the latter extended his experiments using written words as stimuli to peripheral extents which resulted in the letters being illegible. The magnitudes of these extents were not given, and many centuries were to pass before values were derived. In fact, the measurements of Harris and Young were of the visual field itself. Young estimated the extent of perfect vision as greater than that by Harris (section 7.7). Purkinje referred to central vision as direct and peripheral vision as indirect, and he constructed a perimeter to determine the extent of indirect vision. Most observers had discussed the characteristics of spatial vision that were poorer in the periphery, but both Brewster and John Herschel drew attention to the superiority of the peripheral retina in detecting stimuli of very low intensity. The observations that supported this appreciation were made in the context of astronomy: for Brewster it was occasioned by viewing single stars and Herschel noted it when looking at double stars.

Ptolemy (ca. 150): ... what is seen by the central rays on the visual axis is seen more clearly than objects at the side. (Lejeune, 1956, p. 20)

Ibn al-Haytham (ca. 1040): The experimenter should then gently move the strip [with a word written on it] along the tranverse line in the board, making sure that its orientation remains the same, and, as he does this, direct his gaze at the middle strip while closely contemplating the two strips. He will find that as the moving strip gets farther from the middle, the word that is on it becomes less and less clear.... as the moving strip gets farther from the middle, the word that is on it decreases in clarity until he ceases to comprehend or ascertain its form. Then if he moves it further, he will find that the form of that word becomes more confused and obscure. (Sabra, 1989, pp. 244–245)

Harris (1775): It is not easy to ascertain the precise limits of the field of vision, or the greatest angle that can be seen at one view. When we look attentively at a small distance before us, we have an imperfect glimpse of objects for almost the space of half a sphere, or at least for above 60 degrees each way from the optic axes. But towards the extremity of this space, objects are seen very imperfectly; and they are gradually distincter, the nearer they are to the point we look at. About this point there is some small space, that seems at the same time equally distinct; but to ascertain exactly the quantity of this space is difficult, because the indistinctness around it, is so very gradual. (p. 118)

Young (1801): The visual axis being fixed in any direction, I can at the same time see a luminous object placed laterally at a considerable distance from it; but in various directions the angle is different. Upwards it extends to 50 degrees, inwards to 60, downwards to 70, and outwards to 90 degrees.... but the whole extent of perfect vision is little more than 10 degrees; or, more strictly speaking, the imperfection begins within a degree or two of the visual axis, and at the distance of 5 or 6 degrees becomes nearly stationary, until, at a still greater distance, vision is wholly extinguished. The imperfection is partly owing to the unavoidable aberration of oblique rays, but principally to the insensibility of the retina. (pp. 44–45)

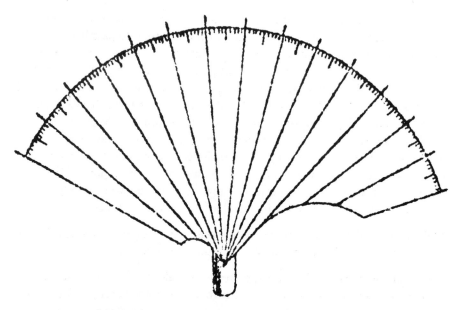

A perimeter for measuring peripheral vision, illustrated in Purkinje (1825).

Purkinje (1823b): In the visual field there is a point in the line of vision where the vision is more distinct than elsewhere. If the visual line terminates in this point we call it direct vision. Indirect vision is the viewing of all other points in the visual field which are outside the visual axis and are seen indistinctly and obliquely.... To observe the extent of direct vision more precisely, compose a segment of three-quarters of a circle divided into degrees, from the center of which the eye, without the movement of the head, can look at a picture which is distinctly seen when moved on the circle's circumference. Let the observer report how far he sees it clearly, that is, until the limitation of the eyeball's movement is reached, when it will be necessary to stop the picture on the periphery and to note the angle in degrees through which the eye moved when the picture was moved to the side.... The extension of indirect vision is given by the space which is less clearly seen when the eye is fixed on one point in the visual field. To determine the range of the extension of indirect vision the apparatus of the preceding experiment can be used, a target being added on which the eye is fixed, while a second, well-illuminated target is moved in or out from the side. (John, 1959, pp. 57–58)

Purkinje (1825): My measurements of the width of indirect vision indicate a temporal angle of 100 degrees (extended to 115 degrees when the pupil is enlarged by Belladonna), 80 degrees downwards, 60 degrees upwards, and the same value for the nasal angle. (p. 6)

Brewster (1832b): The eye has the power of seeing objects with perfect distinctness only when it is directed straight upon them; that is, all objects seen indirectly are seen indistinctly; but it is a curious circumstance, that when we wish to obtain the sight of a very faint star, such as one of the sattelites of Saturn, we can see it most distinctly *by looking away from it*, and when the eye is turned full upon it, it immediately disappears. (p. 15)

John Herschel (1833): It is a very remarkable fact, that the center of the visual area is by far less sensible to feeble impressions of light, than the exterior portions of the retina. Few persons are aware of the extent to which this comparative insensibility extends, previous to trial. To appreciate it, let the reader look alternately full at a star of the fifth magnitude, and beside it; or choose two, equally bright, and about 3° or 4° apart, and look full at one of them, the probability is, he will see *only the other*: such, at least, is my own case. The fact accounts for the multitude of stars with which we are impressed by a general view of the heavens; their paucity when we come to count them. (p. 398)

John Herschel (1792–1871) after an engraving in *The World's Great Men*. London: The London Printing and Publishing Company, 1854.

7.9 Recovery from Cataract Removal

An illustration of a cataract operation of the type performed almost two thousand years ago was reproduced by Thorwald (1962), and written records indicate that such operations had been conducted a thousand years earlier (see Magnus, 1901). Galen used evidence from cataract patients to support the hypothesis that the lens was the seat of vision, because a cloudy lens interfered with vision. Despite this long history, the problem of restoring vision following operations to remove congenital cataracts became of central importance to theories of space perception in the eighteenth century. Empiricist philosophers, like Locke, argued that we learn to perceive visual space by associating it with touch and muscular movement. The catalyst for raising an empirical issue that can be addressed by empiricist philosophy was William Molyneux. He wrote to Locke raising a hypothetical question concerning one who had been born blind, and had sight restored: would the person be able to name a cube and a sphere that had previously been discriminated by touch? This has become known as Molyneux's question, and it has stimulated considerable interest and speculation ever since (see von Senden, 1960; Morgan, 1977).

Locke printed Molyneux's question, and his answer to it, in the second edition of his *Essay Concerning Human Understanding*. His answer, like that of Berkeley a few years later, was negative on theoretical grounds, but some years later postsurgical evidence was brought to bear on the argument. One such case was reported by Grant, but the most celebrated operation was performed by Cheselden. As Voltaire was to note, Cheselden was "one of those famous surgeons, who unite a great extent of knowledge with dexterity in operations" (Morgan, 1977, p. 23). He carried out a number of informal tests on the vision of the boy operated on in order to determine what could be discriminated. The distances, sizes, and shapes of objects could not be differentiated. Moreover, pictures of objects provided particular problems for perception, and it took about two months before they were recognized as representations of other objects.

The outcome of Cheselden's case was commented on briefly by Berkeley in 1732, and more extensively by Jurin and Porterfield, who disagreed with Locke's conclusion. However, it was Voltaire's description of the operation and its theoretical import, in his *Elémens de la Philosophi de Neuton*, that awakened the interests of French philosophers (see Pastore, 1971). Diderot published *An Essay on Blindness* in which he considered the theoretical implications of the Cheselden case in some detail. Both Diderot and Condillac suggested ways in which postoperative testing could be improved to address the theoretical issues raised. Condillac's discussion was in the context of his speculations concerning how a statue could be constructed so that it derived knowledge of the world through its senses.

There was never any question among the philosophers about whether the person with sight restored would be able to see postoperatively, but only whether they could name objects by sight alone. Physicians, on the other hand, were faced with the practicalities of vision in those with sight restored. In the early nineteenth century Ware and Saunders each drew attention to the uniqueness of Cheselden's case, and the difficulties they had in making similar general statements from their own experiences. Ware carried out a number of operations on cataract patients, and applied the technique of depression rather than couching, since children could be so operated on at an earlier age. He gave a detailed account of one case, and on the basis of tests carried out from one day after the operation he reached conclusions opposed to those of Cheselden and the empiricist philosophers. Ware emphasized the residual vision in such patients prior to operation, and Saunders raised the question of the amount of sight initially available postoperatively. It is these aspects that dominated subsequent inquiries (see von Senden, 1960). Wardrop gave an account of a woman who had her sight restored at the age of 45. Her vision from birth was defective, probably due to cataract, but blindness ensued following unsuccessful operations to remove them in her first year. One eye retained gross vision of light and dark, and it was on this eye that Wardrop performed three operations, the results of which he reported. They could be taken to support almost any position. Colors were soon seen, but determining distances demanded much more time.

Galen (ca. 175): ... the crystalline humor itself is the principal instrument of vision, a fact clearly proved by what physicians call cataracts, which lie between the crystalline humor and the cornea and interfere with vision until they are couched. (May, 1968, p. 464)

Molyneux (1694): *Suppose a Man born blind, and now adult, and taught by his touch to distinguish between a Cube and a Sphere of the same metal, and nighly* *the same bigness, so as to tell, when he felt one and other, which is the Cube, which the Sphere. Suppose then the Cube and Sphere placed on a Table, and the Blind Man to be made to see. Quære, Whether by his sight, before he touch'd them, he could now distinguish, and tell, which is the Globe, which the Cube.* To which the acute and judicious Proposer answers: *Not. For though he has obtain'd the experience of, how a Globe, how a Cube, affects his touch; yet he has not yet attained the Experience, that what affects his touch so or so, must affect his sight so or so; Or that a protuberant angle in the Cube, that pressed his hand unequally, shall appear to his eye as it does in the Cube.* (Locke, 1694, p. 67)

William Molyneux (1656–1698) after an engraving kindly supplied by The Wellcome Institute Library, London.

Locke (1694): I agree with this thinking Gent ... and am of opinion, that the Blind Man, at first sight, would not be able with certainty to say, which was the Globe, which the Cube, whilst he only saw them: though he could unerringly name them by his touch, and certainly distinguish them by the difference of the Figures felt. (pp. 67–68)

Berkeley (1709): ... a Man born Blind and made to See, wou'd, at first opening of his Eyes, make a very different Judgment of the Magnitude of Objects intromitted by them, from what others do. He wou'd not consider the Ideas of Sight with reference to, or as having Connexion with, the *Ideas* of Touch. (p. 93)

Grant (1709): The work was performed with great skill and dexterity. When the patient first received the dawn of light, there appeared such an ecstacy in his action, that he seemed ready to swoon away in the surprize of joy and wonder. The surgeon stood before him with his instruments in his hand. The young man observed him from head to foot; after which he survey'd himself as carefully, and seem'd to compare him to himself; and observing both their hands, seem'd to think they were exactly alike, except the instruments, which he took for parts of his hands. When he had continued in his amazement some time, his mother could not longer bear the agitations of so many passions as throng'd upon her, but fell upon his neck, crying out, My Son! My Son! The youth knew her voice, and could speak no more than, Oh Me! Are you my mother? and fainted. (Morgan, 1977, p. 21)

Cheselden (1728): Tho' we say of the Gentleman that he was blind, as we do of all People who have Ripe Cataracts, yet they are never so blind from that Cause, but that they can discern Day from Night; and for the most Part in a strong Light, distinguish Black, White, and Scarlet; but they cannot perceive the Shape of any thing.... When he first saw, he was so far from making any Judgment about Distances, that he thought all Objects whatever touch'd his Eyes (as he express'd it) ... He knew not the Shape of any thing, nor any one thing from another, however different in shape, or Magnitude; but upon being told what Things were, whose Form he before knew from feeling, he would carefully observe, that he might know them again.... We thought he soon

William Cheselden (1688–1752) after an engraving in *The European Magazine and London Review* 46:83, 1804.

knew what Pictures represented, which were shew'd to him, but we found afterwards we were mistaken; for about two Months after he was couch'd, he discovered at once, they represented solid Bodies, where to that Time he consider'd them only as Party-colour'd Planes, or Surfaces diversified with Variety of Paint; but even then he was no less surpriz'd, expecting the Pictures would feel like the Things they represented, and was amaz'd when he found those Parts, which by the Light and Shadow appear'd now round and uneven, felt only flat like the rest; and ask'd which was the lying Sense, Feeling, or Seeing? Being shewn his Father's Picture in a locket at his Mother's Watch, and told what it was, he acknowledged a Likeness, but was vastly surpriz'd; asking, how it could be, that a large Face could be express'd in so little Room, saying, It should have seem'd as impossible to him, as to put a Bushel of any thing into a Pint. (pp. 447–449)

Jurin (1738): Setting aside therefore the authority of Mr. *Locke*, or rather taking it in to my own assistance, I proceed to prove against Mr. Molyneux, that the blind man, now brought to sight, shall be able to distinguish and tell which is the globe, and which the cube, before he touches them.... First. The blind man shall by sight perceive the globe as one thing, distinct from the cube and all other bodies; and shall likewise perceive the cube as one thing distinct from the globe and all other bodies. Secondly. That he shall be told, the two bodies he sees are one a globe and the other a cube; without which information it is to no purpose to ask him which is the globe and which the cube. This being done, I will suppose him to take a careful and repeated view of the two bodies in open sky-light, and by walking round the table on which they are placed, to see and observe them in all their different situations.... Thus, I think the blind man will unerringly distinguish between the two bodies, and by the use of this single principle, that his senses were not given to deceive him; but that the different sensations, which several bodies raise in him, are caused by the differences of those bodies; without which our senses would not only be fallacious, but utterly useless. (Smith's *Remarks*, 1738, pp. 28–29)

Voltaire (François Marie Arouet) (1694–1778) after an engraving in Knight (1833b).

Voltaire (1738): The youth then about fourteen years of age, saw light for the first time. The experiment confirmed all that Locke and Barclay had therein rightly foreseen. For a long time he distinguished neither magnitude, distance, situation, nor even figure. An object of an inch placed before his eyes, that concealed an house from his sight, appeared to him as big as a house. Every thing he saw seemed at first to be upon his eyes, and to touch them, as the objects of the sense of feeling touch the skin. He could not distinguish what he judged round, by the help of his hands, from what he had judged square; nor discern with his eyes, whether what his hands had perceived to be above or below, were really above or below. (Morgan, 1977, pp. 23–24)

Diderot (1749): The question about the man born blind being taken a little more generally than M. Molineux has proposed it, includes two others, which

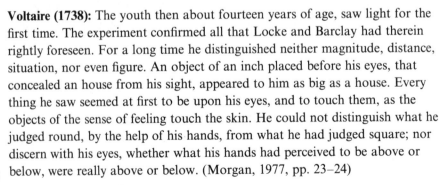

we shall separately consider. It may be asked, 1. Whether he who was born blind will see immediately after his cataracts are couched? 2. Whether, in case he does see, his sight will be such as to distinguish figures; whether, in seeing them, he will be able to give them, with certainty, the same names which he gave them by the touch; and whether he will have any demonstration that these names suit them? . . . I have instead of a sphere put a circle, and a square instead of a cube; because in all appearance, it is only by experiment that we come to judge of distances, and of course, he who uses his eyes for the first time, sees only surfaces, without knowledge of any thing of projecture; the projecture of a body to the sight, consisting in some of its points, appearing more to us than the other. (1750, pp. 79 and 101–102)

Denis Diderot (1713– 1784) after an engraving in Petit de Julleville (1898).

Condillac (1754): One sure way to conduct experiments that can allay all out doubts would be to place the blind man whose cataracts had been removed in a glass chamber. For either he will see objects that are outside of it and judge their shape and size or he will only perceive the space that is bounded by the sides of the chamber and will take all those objects to be only variously colored surfaces that appear to spread out, the closer he brings his hand. In the first case, this would be a proof that the eye judges without having relied on touch; and in the second that it does so only after consulting touch. If, as I presume, this man would not see beyond the boundaries of his chamber, it follows that space that he discovers by eye will be smaller the smaller the size of the chamber. (1980, pp. 294–295)

Etienne Bonnot de Condillac (1714–1780) after an engraving in Petit de Julleville (1898).

Porterfield (1759b): To this Question, both these profound Philosophers [Molyneux and Locke] pronounce in the negative . . . and yet, notwithstanding the great Deference I have for the Opinion of so able Judges, I cannot help thinking that they are mistaken; for, I have already demonstrated, that the Judgments we form of the Situation and Distance of visual Objects depend not on Custom and Experience, but on an original, connate and immutable Law, to which our Minds have been subjected from the Time they were first united to our Bodies; and therefore the blind Person, immediately upon receiving his Sight, must, by virtue of this Law, by his Eyes alone, without any Assistance from his other Senses, immediately judge of the Situation of all Parts of the Globe and Cube. (pp. 414–415)

Wells (1792): For Mr. Cheselden's patient saw objects single, and consequently in the same directions with both eyes, immediately after he had been couched. (p. 83)

Ware (1801): First, When children are born blind, in consequence of having cataracts in their eyes, they are never so totally deprived of sight as not to be able to distinguish colours; and, though they cannot see the figure of an object, nor even its colour, unless it be placed within a very short distance, they nevertheless can tell whether, when within this distance, it be brought nearer to, or carried farther from them. Secondly, In consequence of this power, whilst

James Ware (1756–1815) after an engraving in Pettigrew (1840a).

John Cunningham Saunders (1773–1810) after a frontispiece engraving in Saunders (1816).

James Wardrop (1782–1869) after an engraving in Pettigrew (1838).

in a state of comparative blindness, children who have their cataracts removed, are enabled, immediately on the acquisition of sight, to form some judgment of the distance, and even the outline, of those strongly defined objects with the colour of which they were previously acquainted. Thirdly, When the children have been born with cataracts, the crystalline humour has generally, if not always, been found in a soft, or fluid state. (p. 394)

Saunders (1816): The greatest success attended the operation [of cataract removal] between the ages of eighteen months and four years. . . . The very interesting observations of Cheselden are partly, but not wholly confirmed by the progress in vision of some of these cases. He was in every respect fortunate in his subject for observation—a young gentleman, at an intelligent age, with a favourable cataract, a quick retina, and, it may be concluded from his remarks, a steady eye. The majority of the Author's congential cases were too young for this inquiry, and those at a maturer age had either too much mobility of the eye, if previously quite blind, or an imperfect acquaintance with objects if the circumference of the capsule and lens was transparent, and consequently very few of them were fit subjects for similar experiments. . . . That objects appeared to him [Cheselden's patient] extremely large at first; that he had inadequate conceptions of space and no judgment of distance, in great measure also resulted from inexperience, frequent comparison being indispensably necessary to regulate and even to impart these ideas. But that all objects should seem to touch his eyes, cannot be conceived in the sense in which it is meant, however closely the similitude between sight and touch be drawn. An intelligent girl, on whom the operation had been performed at the age of twelve years, was examined respecting this point, after she had acquired a knowledge of distances; she remarked, that at first she actually could not see objects except they were very near to her eyes, and on this account fell over every thing in her way. These patients are indeed short-sighted for some time after the operation, which affords another reason for their slow acquirement of the knowledge of distance. (pp. 175 and 178–180)

Wardrop (1826): Eighteen days after the last operation had been performed, I attempted to ascertain by a few experiments her precise notions of the colour, size, forms, position, motions, and distances of external objects. As she could only see with one eye, nothing could be ascertained respecting the question of double vision. She evidently saw the difference of colours; that is, she received and was sensible of different impressions from different colours. When pieces of paper one and a half inch square, differently coloured, were presented to her, she not only distinguished them at once from one another, but gave a decided preference to some colours, liking yellow most, and then pale pink. It may be here mentioned, that when desirous of examining an object, she had considerable difficulty in directing her eye to it, and finding out its position, moving her hand as well as her eye in various directions. . . . she not only distinguished small from large objects, but knew what was meant by above and below . . . She could also perceive motions ... She seemed to have the greatest difficulty in finding out the distance of any object; for when an object was held close to her eye, she would search for it by stretching her hand far beyond its position, while on other occasions she groped close to her own face, for a thing far removed from her. (pp. 536–538)

Müller (1826a): If, then, inverted vision is consistent with perception, how can one expect that persons who, having been totally blind since youth, are given their sight by operation, will be conscious of the invertion directly after the operation? There is no accredited, exactly reported case in which this [inversion] was clear to the healed person. (Herrnstein and Boring, 1965, p. 103)

Volkmann (1836): With operated cataract patients one has direct evidence that they are incapable of determining the degree of distance, because the whole visual field lies before them as if painted on a flat table. (p. 23)

7.10 Space Constancy

Space constancy refers to the manner in which the size, shape, and orientation of objects are seen as constant despite the changing projections of each of these dimensions at the eye. A common example, mentioned by Malebranche, is that people do not appear to enlarge as they approach. The possibility of space constancy did not exist in Euclid's geometrical theory of space: visual angles alone defined visual size. Cheselden's patient was said to have seen in this manner, since objects close to the eye subtending equal angles to more distant ones were described as being the same size. Ptolemy realized the limitation of this optical approach to space perception, and added the dimensions of distance and orientation to that of visual angle. He also proposed that the judgment of size was learned, and that it was based on inference. These distinctions were amplified by Ibn al-Haytham, and by the time that they had been absorbed by Pecham it was the established position. Descartes, Rohault, and Malebranche all continued in this line. Indeed, it was taken to be commonplace by Berkeley, who raised the problem of deriving an estimate of one visual dimension (size) from another (distance). Ross and Plug (1998) provide a survey of historical approaches to size and distance perception.

Size perception was raised above the level of polemic by Desaguliers. He compared judgments of the size of stimuli (candles) that were matched for physical distance and visual angle. When two candles of equal size were so perceived (even when one was twice the distance of the other), he substituted a smaller one of equal visual subtense for the far one, with no change in perceived size. Therefore it was apparent distance rather than physical distance that determined apparent size. Note, too, that Desaguliers did not base his conclusions on his own observation but on that of "any unprejudic'd Person," and that he employed a small aperture, presumably to reduce the influence of the background on the judgments.

Orientation constancy was addressed by Hartley and Wells, and both appreciated the importance of a reference orientation, whether postural or visual. Wells suggested an experiment that could distinguish between judgments based on a retinal or a postural reference. Inclining the body by the same amount as an inclined pole would result in its constant retinal orientation, and yet the variations in orientation of the pole could still be perceived.

Ptolemy (ca. 150): Vision knows the true size of objects from the base of the visual pyramid and the distance the object is from us. . . . Suppose there are two extents *ab*, *gd*, which have the same direction and subtend the same angle which is *e*. Accordingly when the distance of *ab* is not equal to that of *gd*, but shorter, *ab* never appears larger than *gd* when judged at its true distance. . . . Similarly, when there are two extents, like *ab*, *gd*, of equal extent, and subtending the same angle which is *e*, but situated in different directions, one with its opposite ends at the same distance [from *e*], the other with its extremities inclined; nevertheless, the one *ab* which is opposite never appears larger than *gd* which is inclined from the one opposite. (Lejeune, 1956, pp. 35 and 40–41)

Ibn al-Haytham (ca. 1040): . . . sight cannot perceive the magnitudes of visible objects by an estimation based on the angles which the objects subtend at the centre of the eye. For the same object does not look different in magnitude when its distance is moderately varied. . . . The magnitude of objects is therefore perceived only by judgement and inference. And the inference through which the object's magnitude is perceived consists in estimating the base of the radial

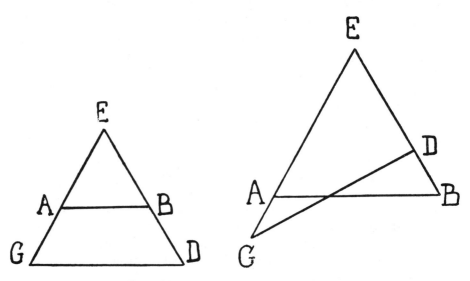

Size and shape constancy according to Ptolemy.

cone, i.e. the object's surface, by the angle of the cone and by its length, namely the distance of the object from the eye. (Sabra, 1989, pp. 174 and 176)

Pecham (ca. 1280): *Perception of the size [of an object] derives from perception of the radiant pyramid and comparison of the base to the length and to the size of the angle....* as experience proves, the faculty that apprehends size considers the magnitude of the distance [from the observer] and not only the angle. For if a one-eyed man looks at a large wall and, after certifying its size, places his hand before his eye, the hand will appear under an angle equal to or larger than that under which the wall is seen; nevertheless, the hand will appear to him smaller than the wall because it is less distant. (Lindberg, 1970, p. 147)

Leonardo da Vinci (ca. 1500): It is possible to bring about that the eye does not see distant things as much diminished as they are in natural perspective. (MacCurdy, 1938b, p. 372)

Descartes (1637): As to the manner in which we see the size and shape of objects.... their size is estimated according to the knowledge, or the opinion, we have of their distance, compared with the size of the images that they imprint on the back of the eye; and not absolutely by the size of these images, as is obvious enough from this: while the images may be, for example, one hundred times larger when the objects are quite close to us than when they are ten times farther away, they do not make us see the objects as one hundred times larger because of this, at least if their distance does not deceive us. And it is also obvious that shape is judged by the knowledge, or opinion, that we have of the position of various parts of the objects, and not by the resemblance of the pictures in the eye; for these pictures usually contain only ovals and diamond shapes, yet they cause us to see circles and squares. (1965, p. 107)

Rohault (1671): When we know the Situation and Distance of an Object, by joining these together, we form a Judgement of the *Bigness* of it. (1723, p. 254)

Malebranche (1674): In beholding a Cube, for Example, it is certain that all the sides we see of it almost cause a Projection, or an Image of an equal dimension in the fund of our Eyes; since the Image of all these sides, when painted in the *Retina*, or the Optick Nerve, nearly resembles a Cube pictur'd in *Perspective*; and consequently the Sensation we have of it, ought to represent the faces of a cube to us as being unequal, since they are in a Cube *unequal*, because they are so in Perspective. This notwithstanding, we see them as equal, nor are we in Error. (1700, p. 18)

Berkeley (1709): It is well known that the same Extension, at a near Distance, shall subtend a greater Angle, and at a farther Distance, a lesser Angle. And by this Principle (we are told) the Mind estimates the Magnitude of an Object, comparing the Angle under which it is seen, with its Distance, and thence inferring the Magnitude thereof. What inclines Men to this Mistake (beside the

Humour of making one see *Geometry*), is that the same Perceptions or *Ideas* which suggest Distance, do also suggest Magnitude. But if we examine it, we shall find they suggest the latter, as immediately as the former. (p. 58)

Malebranche (1712): When I look at a man walking toward me, for example, it is certain that, as he approaches, the image or impression of his height traced in the fundus of my eyes continuously increases and is finally doubled as he moves from ten to five feet away. But because the impression of distance decreases in the same proportion as the other increases, I see him as always having the same size. (1980, p. 34)

Desaguliers (1736a): I took two Candles of equal Height and Bigness AB, CD, and having plac'd AB at the distance of six or eight Feet from the Eye, I placed CD at double that Distance; then causing any unprejudic'd Person to look at the Candles, I ask'd which was biggest? and the spectator said that they were both of a Bigness; and that they appear'd so, because he allow'd for the greater Distance of CD; and this also appear'd to him, when he look'd thro' a small Hole. Then desiring him to shut his Eyes for a Time, I took away the Candle CD, and plac'd the Candle EF close by the Candle AB, and tho' it was as short again as the others, and as little in Diameter, the Spectator, when he open'd his Eyes, thought he saw the same Candles as before. Whence it can be concluded, that when an Object is thought to be twice as far from the Eyes as it was before, we think it to be twice as big, though it subtends but the same Angle. (p. 391)

David Hartley (1705–1757) after a frontispiece engraving in Hartley (1791).

Hartley (1749): The Position of Objects is judged intirely by the Part of the *Retina* on which the Rays fall, if we be in an erect Posture ourselves. If we be not, we allow for our Deviation from it, or make Reference to something judged to be in an erect Posture. If we fail in these, Errors concerning the Position of visible Objects must happen. (pp. 203–204)

Wells (1792): In the estimates we make by sight of the situation of external objects, we have always some secret reference to the position of our own bodies, with respect to the plane of the horizon.... Let a pole be placed upon firm

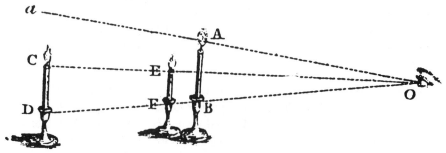

Desaguliers's experiment to demonstrate the influence of familiar size on size constancy.

ground, at right angles to the horizon. If, while we are standing erect, it be inclined upon its lower extremity, successively backward and forward, to the right and to the left, these motions must, without contradiction, be perceived. Suppose now, our bodies to be similarly inclined with the pole, during its different positions, so as to be constantly parallel to it; it is evident, that its motions will be as readily perceived in this case, as they were, when our bodies were erect. (pp. 85 and 89–90)

Müller (1826a): The apparent sizes of objects appear on the true subjective size of the retina, and the sum of the apparent sizes of all objects present in one and the same visual field remains constant through all changes in objects in every visual image. (Herrnstein and Boring, 1965, p. 102)

Müller (1838): Every object seen under the same angle, $a\,x\,b$, has an image of the same size, $a\,b$, upon the retina; the objects d, e, f, g, h, which are of very different size, but are situated at different distances, are seen under the same angle, and therefore produce images of the same size upon the retina: and, nevertheless, these images appear to the mind of very unequal size when the ideas of distance and proximity come into play; for, from the image $a\,b$, the mind forms the conception of a visual space extending to d, e, f, g, or h, and of an object of the size which that represented by the image on the retina appears to have when viewed close to the eye, or under the most usual circumstances. (1843, p. 1166)

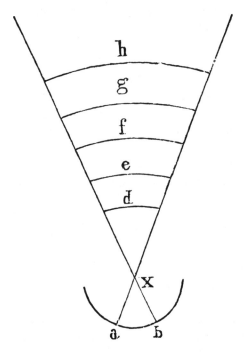

The influence of apparent distance on size judgments for objects of equal visual angle according to Müller.

7.11   Cues to Distance

If size perception was dependent upon the distance away that objects were, then the source of such distance information was of obvious relevance (see Boring, 1942). Variations in distance judgments as a consequence of mist were remarked upon by Aristotle and Ibn al-Haytham, and Leonardo described what has become called aerial perspective. Euclid, in a diagram that forms the essential basis of perspective, related the distance of objects on the ground plane to their height in the visual field. Lucretius noted another perspectival feature, namely the vanishing point, which could be seen in some architectural structures. These were all essentially optical aspects of increasing distance.

Ibn al-Haytham speculated that familiarity with objects assisted in the perception of their distance, and thereafter increasing concern was given to a multiplicity of cues to distance. Descartes specified two others beside familiar size—convergence and accommodation—and he defined the distances over which they operated. Rohault added interposition and motion parallax to these, although the latter was not included in the list provided by Malebranche. La Hire, on the other hand, stated that motion parallax was the most important cue, a view with which Wheatstone concurred (section 6.6).

In stating that perceived distance is inferred, Berkeley argued that (for relatively near objects) the inference was based on convergence, accommodation, and image clarity. Most particularly, the motion of the eyes, in convergence, was of principal importance, and his recourse to motor mechanisms was to influence many eighteenth-century analyses of space perception. Smith, Hartley, Buffon, Condillac, and Reid all espoused the empiricist position in their interpretations of space perception. But it was probably Porterfield who did most both to further acknowledgment of the Berkeleian ideal and to expose its most fundamental paradox. While Berkeley argued that touch teaches vision the dimensions of space, Porterfield countered that neither touch nor vision have priority, and both are "equally incapable of introducing the Idea of any Thing external."

These cues to distance were repeated, with varying emphases, in the early nineteenth century, until Wheatstone added that of retinal disparity.

Aristotle (ca. 330 B.C.): . . . promontories in the sea loom when there is a southeast wind, and everything seems bigger, and in a mist, too, things seem bigger. (Ross, 1931, p. 373b)

Euclid (ca. 300 B.C.): *In the case of flat surfaces lying below the level of the eye, the more remote parts appear higher.* Let the eye be *A*, at a level higher than *BEDG*, and let the rays fall, *AB, AE, AD, AG,* of which rays let *AB* be perpendicular upon the plane below. I say that *GD* appears higher than *DE*, and *DE* higher than *BE*. For somewhere upon the line *BE* let the point *Z* be taken, and let the perpendicular *ZL* be drawn. Since the lines of vision fall upon *ZL*

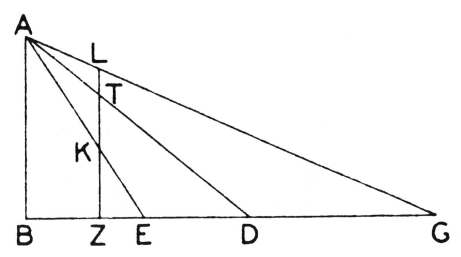

Height in the plane as a cue to distance.

before they reach *ZG*, let the line *AG* meet *ZL* at the point *L*, the line *AD* at the point *T*, and the line *AE* at the point *K*. Now, since *L* is higher than *T* and *T* higher than *K*, but *G* on the same line as *L*, and *D* on the same line as *T*, and *E* on the same line as *K*, and, since between the lines *AG* and *AD*, *DG* is seen, and between the lines *AD* and *AE*, *DE* is seen, *GD* appears higher than *DE*. And, similarly, *DE* will appear higher than *BE*; for things seen by higher rays appear to be higher. (Burton, 1945, p. 359)

Lucretius (ca. 56 B.C.): . . . a colonnade may be of equal line from end to end and supported by columns of equal height throughout, yet, when its whole length is surveyed from one end, it gradually contracts into the point of a narrowing cone, completely joining roof to floor and right to left, until it has gathered all into the vanishing point of the cone. (1975, p. 311)

Ptolemy (ca. 150): We say that an object is nearer when the ray which strikes the middle of it is shorter, and that the distance is greater when the same ray is longer. (Lejeune, 1956, p. 39)

Ibn al-Haytham (ca. 1040): For when sight perceives a familiar object and from a familiar distance, it recognizes both the object and its distance and conjectures the magnitude of that distance. . . . For when sight perceives mountains behind fog or thick air it believes them to be near although they are far (Sabra, 1989, pp. 157 and 340)

Leonardo da Vinci (ca. 1500): You know that in such air the furthest things seen in it—as in the case of mountains, when great quantities of air are found between your eye and the mountains—appear blue. (Kemp, 1989, p. 80)

Descartes (1664): . . . if the two eyes *L* and *M* are turned toward the object *N*, the magnitude of line *LM* and the two angles *LMN* and *MLN* will cause it

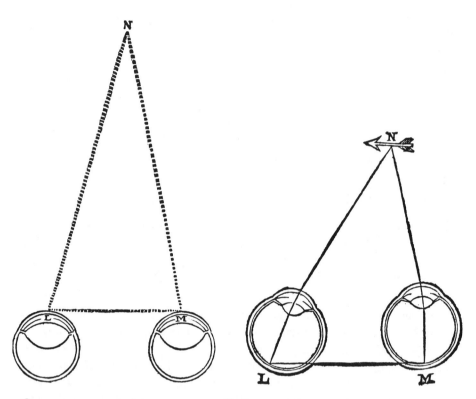

Convergence as a cue to distance as represented in Descartes (1662) on the left and Descartes (1664/1909) on the right.

[the soul] to know where point *N* is. . . . none of the soul's means of knowing the distance of objects will be quite sure, for the following reasons. First, as for the angles *LMN* and *MLN* and so forth, these no longer change appreciably for objects fifteen to twenty feet away or more. Second, as for the shape of the crystalline humor, it changes even less appreciably than the foregoing for objects more than two or three feet from the eye. And finally, what about judging the distances of objects from one's opinion of their size, or from the fact that rays from different points of the object differ in the precision with which they assemble at the back of the eye? (Hall, 1972, pp. 62–63 and 66–68)

Jacques Rohault (1620–1675) after an engraving kindly supplied by The Wellcome Institute Library, London.

Rohault (1671): . . . if we make use of but one Eye, we can know the *Distance* of an Object, provided we move from one Place to another; for we have some kind of Memory of the Position of the *Optical* Axis in the first Station, when we really attend to the Position of it in another Station, so that we imagine two *Optical* Axes, though there be indeed but one, and by that means guess at the Distance where they meet; and to this we refer the Object. . . . Further, the Interposition of a great many other Objects, between us and the Object we look at, makes us think, that the Distance is greater than otherwise we should. . . . It is certain also, that as the Knowledge of the Distance helps us to find out the Bigness, so likewise the knowing of the Bigness helps us to conceive the Distance. Thus, when we know that a Man is about five or six Foot high, when we see him to appear but very little, we conclude him to be at a great Distance. (1723, pp. 252, 254, and 256)

Malebranche (1674): The first, the most universal, and sometimes the safest way we have, whereby to judge of the distance of Objects, is the Angle made by the Rays of our Eyes, whereof the Object is the Vertical Point, that is, the Object is the Point, where the two Rays meet.... The second *Medium* the Soul imploys to judge concerning the Distance of Objects, consists in a Disposition of the Eyes.... The third *Medium* consists in the Greatness of the Image painted on the fund of the Eye, and that makes a Representation of the Objects which we see.... We judge farther the remoteness of an Object by the *Force* wherewith it acts upon our Eyes, because a remote Object acts more languishing and weakly than another; and again by the *Distinctness* and *Clearness* of the Image which is form'd in the Eye.... The sixth then and the Principal *Medium* of all, consists of this, *viz.* that the Eye exhibits not to the Soul a single Object separate from others, but gives her View at once of all those which lye, betwixt us and the Principal Object of our actual Consideration. (1700, pp. 22–24)

La Hire (1694): The parallax of objects is what we use most to recognize distance, but the eyes must change place to recognize which of the two objects is the nearer. For example, if two objects appear near to each other from one position of the eye, when the eye moves to the right the object that thus appears to move away from the other to the right is further away, and the other which remains on the left will be the nearer one. (pp. 236–237)

Berkeley (1709): ... *Distance* of itself, and immediately cannot be seen. For *Distance* being a Line directed end-wise to the Eye, it projects only one Point in the Fund of the Eye, Which Point remains invariably the same, whether the Distance be longer or shorter.... It is certain by Experience, that when we look at a near Object with both Eyes, according as it approaches, or recedes from us, we alter the Disposition of our Eyes, by lessening or widening the Interval between the *Pupils*. This Disposition or Turn of the Eyes is attended with a Sensation which seems to me to be that which in this case brings the *Idea* of greater, or lesser Distance into the Mind. (pp. 2 and 9)

George Berkeley (1685–1753) after an engraving in Fraser (1871).

Smith (1738): The apparent distance of an object, perceived by sight, is an idea of a real distance usually measured by feeling, as by the motion of the body in walking or otherwise. (p. 49)

Hartley (1749): The principal Criterion of Distance is the Magnitude of the Picture, which some known Object makes on the *Retina*. But the five following associated Circumstances seem to have also some Influence on our Judgments concerning Distance, in certain Cases, and under certain Limitation: The

Number of Objects which intervene, the Degree of Distinctness in which minute Parts are seen, the Degree of Bigness, the Inclination of the Optic Axes, and the Conformation of the Eye. (pp. 201–202)

Buffon (1749): The sense of seeing conveys no idea of distances. Without the aid of touching all objects would appear to be within the eye, because it is there alone that the images exist: and an infant, who has had no experience of the sense of touching, must consider all external bodies as existing in itself; they appear larger or smaller only according as they approach or recede from the eye.... A distance ceases to be familiar to us whenever it is too large, or rather when the interval is vertical instead of horizontal.... We have no acquired habit of judging the magnitude of objects which are elevated above, or sunk below us. (1866, p. 241)

Condillac (1754): Then judging size by distance, as it has on other occasions judged distance by size, it sees as larger what it believes to be farther away. Two trees, for example, that provide it with images of the same extension will not appear at all equal nor at the same distance if the image of one is more indistinct than the other: the statue will see as larger and farther the one in which it discerns fewer things.... These principles are known to everyone and painting confirms them. A horse that occupies the same space on the canvas as a sheep will appear larger and in the background provided that it is not painted as distinctly as the sheep. (1982, p. 284)

Porterfield (1759b): But if, by the Touch alone, we can judge thus of the Situation and Distance of external Things, I see not why the same Power should be denied to the Sight.... for the tangible Ideas are as much present with the Mind as the visible Ideas, and, on that Account, must be equally incapable of introducing the Idea of any Thing external. (pp. 301 and 307)

Reid (1764): When a picture is seen with both eyes, and at no great distance, the representation appears not so natural as when it is seen only with one.... In order to remove this perception of distance, the connoisseurs in painting use a method which is very proper. They look at the picture with one eye, through a tube which excludes the view of all other objects. By this method, the principal mean by which we perceive the distance of an object, to wit, the inclination of the optic axes, is entirely excluded. I would humbly propose, as an improvement of viewing pictures, that the aperture of the tube next to the eye should be very small. If it is as small as a pin-hole, so much the better, provided there be light enough to see the picture clearly. The reason of this proposal is, that when we look at an object through a small aperture, it will be seen distinctly, whether the conformation of the eye be adapted to its distance or not. (pp. 442 and 444–445)

Haller (1767): Though the iris has little sensation, and is not endowed with any mechanical irritability; yet in a living man, quadruped, or bird, it is constricted with every greater degree of light, and is dilated on every smaller one; hence it is rendered broader for viewing distant objects, and narrower for viewing such as are near. (1786, p. 11)

Young (1807): We estimate distances much less accurately with one eye than with both, since we are deprived of the assistance usually afforded by the relative situation of the optical axes ... Our idea of distance is usually regulated by a knowledge of the real magnitude of an object, while we observe its angular magnitude; and on the other hand a knowledge of the real or imaginary distance of the object often directs our judgment of its actual magnitude. (p. 453)

Müller (1838): In the sense of vision ... the images of objects are mere fractions of the objects themselves, realized upon the retina, the extent of which remains constantly the same. But the imagination which analyses the sensations of vision, invests the images of objects together with the whole field of vision in the retina, with varying dimensions; the relative size of the images in proportion to the whole field of vision, or of the affected parts of the retina to the whole retina, alone remaining unaltered. (1843, pp. 1167–1168)

Wheatstone (1838): ... the same solid object is represented to the mind by different pairs of monocular pictures, according as they are placed at different distances before the eyes, and the perception of these differences (though we seem to be unconscious of them) may assist in suggesting to the mind the distance of the object. ... The mind associates with the idea of a solid object every different projection of it which experience has hitherto afforded; a single projection may be ambiguous, from its being one of the projections of a picture, or of a different solid object; but when different projections of the same objects are successively presented, they cannot all belong to another object, and the form to which they belong is completely characterized. (pp. 377 and 380)

7.12 Perspective

Science and art meet in perspective. Perhaps it should more accurately be said that the optics of antiquity met the art of the Renaissance in the context of linear perspective. The topic is of interest to this natural history of vision because perspectival theorists produced a theory of image formation long before the dioptrics of the eye were understood. There are many excellent treatments of the history of perspective, but Kemp's (1990) *The Science of Art* is especially instructive because of the way it deals with the emergence of linear perspective in the fifteenth century, and its subsequent development. Greek artists also applied perspective, as Edgerton (1975) argued in his book *The Renaissance Rediscovery of Linear Perspective*. While written records of their use of perspective no longer exist, their graphics and architecture speak to this point. Ptolemy commented on an alluring feature of portraits—that the eyes follow the spectator—that was returned to and illustrated by Wollaston.

Euclid's *Optics* is concerned with describing visual space in terms of visual angles, whereas linear perspective captures those visual angles on a picture plane. The first theorist of perspective in the fifteenth century, Alberti, described a simple method for capturing visual angles: with the eye in a fixed position, the objects in a scene can be traced on a windowpane. This has become called Alberti's window, although Leonardo's name is often given to it, too, following his description of the same procedure. Kircher outlined a simple technique for representing images in a dark room, and Hooke enlisted the camera obscura to draw objects in the environment in precise perspective. Cheselden adopted a similar method for obtaining anatomical drawings of skeletons. The mathematician Brook Taylor (1685–1731) gave an account of perspective which was followed by many, including Kirby.

The differences between viewing a scene and a painting of it were mentioned by Leonardo (section 6.6); the former has a depth that no degree of artifice can add to the latter, for reasons which Wheatstone made abundantly clear. Young raised another difficulty concerning paintings: they do not show the variations in distinctness that would accompany changes of accommodation when viewing a scene. Painting in perspective focuses on the single, static eye with a fixed accommodation in much the same way that the representations of the retinal image (section 2.2) do. Both provided a starting point for the analysis of spatial vision, but neither has proved adequate for understanding its intricacies.

Euclid (ca. 300 B.C.): Let *B* represent the eye and let *GD* and *KL* represent the objects seen; and we must understand that they are equal and parallel, and let *GD* be nearer to the eye; and let the rays of vision fall, *BG*, *BD*, *BK*, and *BL*. For we could not say that the rays falling from the eye upon KL will pass through the points *G* and *D*. For in the triangle *BDLKGB* the line *KL* would be longer than the line *GD*; but they are supposed to be of equal length. (Burton, 1945, p. 357)

Ptolemy (ca. 150): One believes that the eyes of a face painted in a picture turn in the direction of those who look at it. (Lejeune, 1956, p. 79)

Alberti (1435): When they [painters] fill the circumscribed places with colours, they should only seek to present the forms of things seen on this plane as if it were of transparent glass. Thus the visual pyramid could pass through it, placed at a definite distance with definite lights and a definite position of the centre in space and in a definite place in respect to the observer.... I inscribe a quadrangle of right angles, as large as I wish, which is considered to be an open window through which I see what I want to paint.... Then, within this quadrangle, where it seems best to me, I make a point which occupies the place where the central ray strikes. For this it is called the centric point.... I draw

Leon Battista Alberti (1404–1472) after an engraving in Reusner (1589).

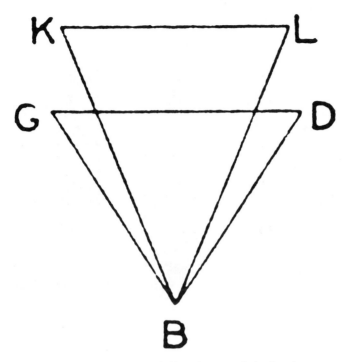

Euclid's representation of objects in terms of visual angles.

straight lines from it to each division placed on the base line of the quadrangle. These drawn lines, [extended] as if to infinity, demonstrate to me how each transverse quantity is altered visually. (1966, pp. 51 and 56)

Leonardo da Vinci (ca. 1500): Linear perspective deals with the action of lines of sight, in proving by measurement how much smaller is a second object than the first, and how much the third is smaller than the second; and so on, by degrees, to the end of things visible. I find by experience that if a second object is as far beyond the first as the first is from the eye, although they are of the same size, the second will seem half the size of the first. (Veltman, 1986, p. 95)

Kircher (1646): [A figure for] Representing things in darkness. (p. 122)

Hooke (1668): Opposite to the place or wall, where the Apparition is to be, let a Hole be made of about a foot in diameter, or bigger; if there be a high Window, that hath a Casement in it, 'twill be so much the better. Without this hole, or Casement open'd at a convenient distance, (that it may not be perceived by the Company in the room) place the Picture or Object, which you will represent, inverted, and by means of Looking-glasses placed behind, if the picture be *transparent*, reflect the rayes of the Sun so, as that they might pass through it towards the place, where it is to be represented; and to the end that no rayes may pass besides it, let the Picture be encompass'd on every side with a board

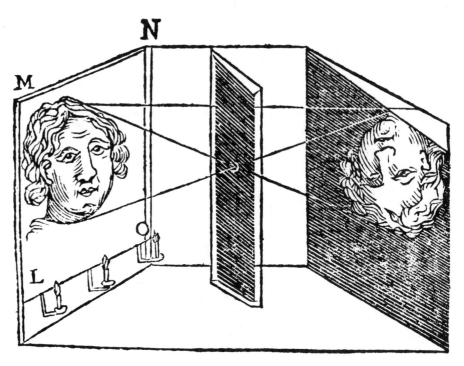

Kircher's technique for representing pictures in a dark room.

or cloath. If the Object be a *Statue*, or some living Creature, then it must be very much enlightn'd by casting the Sunbeams on it by Refraction, Reflexion, or both. Between this Object, and the Place where 'tis to be represented, there is to be placed a broad Convex-glass, ground of such convexity, as that it may represent the Object distinct on the said place; which any one, that hath any insight in the *Opticks*, may easily direct. The nearer it is placed to the object, the more is the Object magnified on the Wall, and the further off, the less; which diversity is effected by Glasses of several spheres. If the Objects cannot be *inverted* (as 'tis pretty difficult to do with Living Animals, Candles, &c.) then there must be *two* large Glasses of convenient Spheres, and they placed at the appropriate distances (which are very easily found by tryals) so as to make the representations *erect* as well as the Object. (pp. 742–743)

Hooke (1694): . . . any person that can but use his pen, and see the profile of what he sees before him, shall be able to give us the true draught of whatever he sees before him, that continues so long a time in the same posture, as while he can nimbly run over, with his pen, the boundaries, or out-lines of the thing to be represented; which being once only truly taken, 'twill not at all be difficult to add the proper shadows and light pertinent thereunto. . . . The instrument I mean for this purpose, is nothing else but a small picture-box, much like that which I long since shewed the *Society* for drawing the picture of a man, or the like. (Gunther, 1930, pp. 755–756)

Kirby (1755):

1. The *Point of Sight*, is that Point where the Spectator's Eye is placed to look at the Picture.... Thus E is the Point of Sight.

2. If from the Point of Sight E, a Line EC is drawn from the Eye perpendicular to the Picture, the the Point C , where the Line cuts the Picture, is called *the Center of the Picture*.

3. *The Distance of the Picture*, is the Length of the Line EC, which is drawn from the Eye perpendicular to the Picture.

4. If from the Point of Sight E, a Line EC be drawn perpendicular to any vanishing Line HL, or JF, then the Point C, where the Line cuts the vanishing Line, is called *the Center of that vanishing Line.*

5. *The Distance of a vanishing Line*, is the Length of the Line EC, which is drawn from the Eye perpendicular to the said Line: and if PO was a vanishing Line, then EJ will be the Distance of that Line.

6. *The Distance of the vanishing Point*, is the Length of a Line drawn from the Eye to that Point: Thus, EC is the Distance of the vanishing Point C, and EJ is the Distance of the vanishing Point J.

7. By *Original Object*, is meant the real Object whose representation is sought: and by *Original Plane*, is meant that Plane upon which the real Object is situated: Thus, the Ground HM is the Original Plane of AB.C.D. (p. 8)

Young (1807): The faculty of judging of the actual distance of objects is an impediment to the deception, which is partly the business of a painter to produce. Some of the effects of objects at different distances may, however, be imitated in painting on a plane surface. Thus, supposing the eye to be accommodated to a given distance, objects at all other distances may be represented with a certain indistinctness of outline, which would accompany the images of the objects themselves on the retina: and this indistinctness is so generally necessary, that its absence has the disagreeable effect called hardness. The apparent magnitude of the subjects of our design, and the relative situations of the intervening objects, may be so imitated by the rules of geometrical perspective as to agree perfectly with nature, and we may still further improve the representation of distance by attending to the art of aerial perspective, which consists in a due observation of the loss of light, and the bluish tinge, occasioned by the interposition of a greater or less depth of air between us and the different parts of the scenery. We cannot indeed so arrange the picture, that either the focal length of the eye, or the position of the optical axes, may be such as would be required by the actual objects: but we may place the picture at such a distance that neither of these criterions can have much power in detecting the fallacy; or, by the interposition of a large lens, we may produce nearly the same effects in the rays of light, as if they proceeded from a picture at any required distance. In the panorama, which has lately been exhibited in many parts of Europe, the effects of natural scenery are very closely imitated: the

Hooke's view box for capturing scenes in perspective.

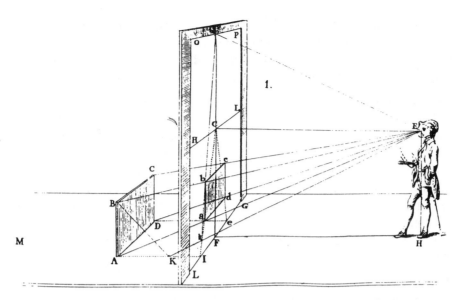

Kirby's description of producing a picture in perspective after the principles of Brook Taylor.

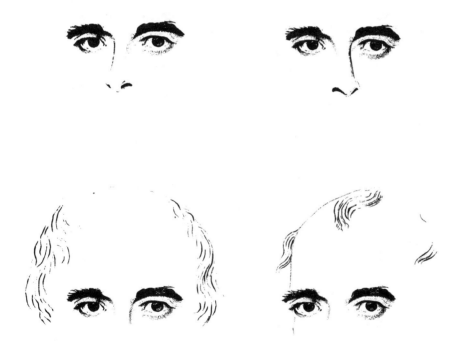

Wollaston's drawings of eyes showing how their apparent direction is changed by the orientation of the head.

deception is favoured by the absence of all other visible objects, and by the faintness of the light, which assists in concealing the defects of the representation, and for which the eye is usually prepared, by being long detained in the dark winding passages, which lead to the place of exhibition. (pp. 454–455)

Wollaston (1824b): When we consider the precision, with which we commonly judge whether the eyes of another person are fixed upon ourselves, and the immediateness of our perception that even a momentary glance is turned upon us, it is very surprising that the grounds of so accurate a judgement are not distinctly known, and that most persons, in attempting to explain the subject, would overlook some of the circumstances by which, it will appear, they are generally guided.... the apparent direction of the eyes to or from the spectator depends upon the balance of two circumstances combined in the same representation, namely, 1st. The general position of the face presented to the spectator; and, 2dly. The turn of the eyes from that position. (pp. 247 and 253–254)

Wheatstone (1838): It will now be obvious why it is impossible for the artist to give a faithful representation of any near solid object, that is, to produce a painting which shall not be distinguished in the mind from the object itself. When the painting and the object are seen with both eyes, in the case of the painting two *similar* pictures are projected on the retinae, in the case of the solid object the pictures are *dissimilar*; there is therefore an essential difference between the impressions on the organs of sensation in the two cases, and consequently between the perceptions formed in the mind; the painting therefore cannot be confounded with the solid object. (p. 272)

8   Illusions

Claudius Ptolemy (ca. 100–170) after an engraving in *The Historic Gallery of Portraits and Paintings*, Vol. 5. London: Vernor, Hood, and Sharpe, 1809.

Ptolemy (ca. 150): *For there are some errors that are caused in all the senses and others that are confined to things seen, of which some are visual and others are in the mind. (Lejeune, 1956, p. 56)*

The inclusion of a chapter headed "illusions" creates certain difficulties. According to Platonic theories, all vision is illusion, and so the category would be all-inclusive. However, Aristotelian rather than Platonic theories have prevailed in the study of vision, and illusions have been, and remain, a source of interest and inquiry. Ptolemy was one of the first writers to provide a detailed account of illusions. Indeed, he devoted over one third of Book II of his *Optics* to the topic; they are classified, and then considered under the headings of color, position, size, shape, and movement. Ibn al-Haytham adopted a similar analysis of the errors of direct vision although he extended the range of phenomena for which they occur—after the manner of the extended categories of vision mentioned in the last chapter.

Many of Ptolemy's descriptions of illusions have appeared already in the contexts of color, motion, binocularity, and space. In fact, he was more strict in his application of the term than was subsequently the case. The changes in perception occasioned by contrasts of color, motion, size, and shape are illusions (mismatches between physical characteristics and their perception), but they are not presently classified as such. The term is now applied mainly to the spatial illusions of size and orientation; the definition of geometrical-optical illusions did not appear until after the end of the period of this survey. Oppel (1855) gave them that title, and they were a central feature of late-nineteenth-century visual science (see Johannsen, 1971; Zusne, 1968).

One illusion stood above these shifting sands of definition—the moon illusion. Astronomy has fueled the fascination with many of the phenomena discussed here, but few so fervently as the variations in the seen sizes of celestial bodies as they coursed over the vault of the heavens. In principle, the illusion is visible with all celestial bodies, but the moon subtends a large enough visual angle and is of low enough luminance to afford direct observation, and so it has been the topic of most such examination. It appears larger when above the horizon than at its zenith. Observation of the sun is not recommended because it leads to prolonged afterimages or damage to the retina (section 4.1). However, both the moon illusion and afterimages have been taken as epitomes of the history of vision.

8.1   Visual Illusions

The modern definition of illusions applies to differences between the perception of figures and their physical characteristics (see Gregory, 1996). Consensus concerning an external reality did not exist in antiquity, and so attention was directed to those instances in which changes in perception occurred. That is, when the same object appeared to have different properties under different conditions. The investigation of aspects of optics derived from precisely these variations: the straight stick appears to bend when partially immersed in water. Ptolemy examined such instances of refraction, and formulated some general properties of it (see section 2.1). The changing perception of objects was the source of Aristotle's interest in phenomena like afterimages, aftereffects, color contrasts, and diplopia. It was precisely such variation in perception that led to the Platonic distrust of the senses. However, the recourse to measurement of physical dimensions itself involves the senses, as both Montaigne and Kepler recognized, and so there is no easy exit from this enigma.

There was one area where the constraints of the physical world did impinge on perception, namely, architecture. Vitruvius noted that parallel columns do not appear so when viewed from the ground, and so their dimensions were modified in order to look uniformly wide (see Johannsen, 1971). This was an example of slight departures from size constancy, whereas Smith described cases where constancy breaks down almost completely—when viewing familiar objects from very great distances.

Ptolemy and Ibn al-Haytham were concerned with more general features of perception, and the latter was more explicit in categorizing the three modes of vision in which illusions can occur. Illusions were to be understood in terms of the breakdown of the process of inference. Nonetheless, the categories Ibn al-Haytham gave for the errors of sight were fewer than the visible properties he listed. Errors of inference were confined to distance, position, illumination, size, opacity, transparency, duration, and condition of the eye.

After the Renaissance, perspective was one of the techniques of visual illusion that could be manipulated, as is evident from the remarks of Francis Bacon and Descartes. The mythical House of Salomon, described by Bacon in his *New Atlantis*, displayed "all delusions and deceipts of sight" as evidence of the advancement of science. Size and distance played a prominent role in such "deceipts." Despite the difficulties in fooling the eye by means of paintings, Young noted (see section 7.12) that the nineteenth-century fashion of panoramas was more successful because the figures were painted on many surfaces at different distances. Perhaps Bacon was referring to such three-dimensional structures with his "perspective-houses." Size, distance, and motion were combined in Malebranche's general statement about illusions. It was noted in section 7.10 that size constancy was considered to be mediated by the perception of the distance objects were away from the observer. Malebranche indicated how errors in distance estimates would also modify motion perception.

The art of perspective was concerned with generating a false but fixed apparent depth on a two-dimensional surface. Nonetheless, ambiguous figures have a long history in art; that is, contours that can be interpreted in more than one way. They moved to center stage in perceptual psychology as a result of the Gestalt movement earlier this century (see Wade, 1982), and Rubin's vase/ face motif was particularly popular. Perceptual psychology came to this point rather late, since much more subtle variations on the same theme had been played since at least the eighteenth century. Pastore (1971) reproduced an engraving entitled *The Mysterious Urn* in which not only the urn defined two (different) profiles but others could be discerned in the foliage of the tree above it. Müller provided an example of a simple geometrical figure that could undergo changes in its organization. Fluctuations in the depth seen in geometrical drawings are discussed in section 8.7.

Vitruvius (ca. 27 B.C.): It is on account of the variation in height that these adjustments are made. For the sight follows gracious contours; and unless we flatter its pleasure, by proportionate alterations of the modules (so that by adjustment there is added the amount to which it suffers illusion), an uncouth and ungracious aspect will be presented to the spectators. As to the swelling which is made in the middle of the columns (this among the Greeks is called *entasis*), an illustrated formula will be furnished at the end of the book to show how the entasis may be done in a graceful and appropriate manner.... All

Vitruvius (ca. 88–26 B.C.) after an engraving kindly supplied by the Bibliothèque nationale de France.

the features which are to be above the capitals of the columns, that is to say, architraves, friezes, cornices, tympana, pediments, acroteria, are to be inclined toward their fronts by a twelfth part of their height; because when we stand against the fronts, if two lines are drawn from the eye, and one touches the lowest part of the work, and the other the highest, that which touches the highest, will be the longer. Thus because the longer line of vision goes to the upper part, it gives the appearance of leaning backwards. When however, as written above, the line is inclined to the front, then the parts will seem vertical and to measure. (1962, pp. 179–181 and 195–197)

Ptolemy (ca. 150): For the errors of vision it is necessary to distinguish not only between this kind [concerning the limits of perception] but also those which result in objects appearing differently. (Lejeune, 1956, p. 58)

Ibn al-Haytham (ca. 1040): . . . there exist three modes of visual perception: pure sensation, recognition, and inference and discernment at the time of perceiving the object. Therefore, an error that occurs in what sight perceives by pure sensation will be an error in the sensation itself. And an error in what sight perceives by recognition will be an error of recognition. And an error in what sight perceives by inference and discernment at the moment of vision will be an error in inference and discernment or in the premisses on which the inference and discernment are based. (Sabra, 1989, pp. 258–259)

Montaigne (1580): To judge of the appearances we receive of subjects, we ought to have a deciding instrument; to verify this instrument we must have a demonstration; to verify this demonstration an instrument; and here we are round again upon the wheel, and no further advanced. Seeing the senses cannot determine our dispute, being full of uncertainty themselves, it must then be reason that must do it; but no reason can be erected upon any other foundation than that of another reason; and so we run back to all infinity. (1853, p. 281)

Kepler (1604): Whilst the diameters of the planets and the amounts of eclipses of the sun are recorded by astronomers as basic values, some deception of vision arises partly from the instruments of observation . . . and partly from vision itself; and this, as long as it is not counteracted, makes considerable trouble for investigators and detracts from scientific judgement. The source of the errors in vision is to be sought in the structure and functioning of the eye itself. (Crombie, 1964, pp. 144–145)

Francis Bacon (1624): We have also perspective-houses, where we make demonstrations of all lights and radiations; and of all colours; and out of things uncoloured and transparent, we can represent unto you all several colours; not in rain-bows, as it is in gems and prisms, but of themselves single. We represent also all multiplications of light, which we carry to great distance, and make so sharp as to discern small points and lines; also all colorations of light:

all delusions and deceipts of sight, in figures, magnitudes, motions, colours: all demonstrations of shadows. We find also divers means, yet unknown to you, of producing light originally from divers bodies. We procure means of seeing objects afar off; as in the heaven and remote places; and represent things near as afar off, and things afar off as near; making feigned distances. (1857b, pp. 161–162)

Francis Bacon (1561–1626) after an engraving in Knight (1837).

Descartes (1664): The example of *tableaux de perspective* shows us amply how easy it is to be deceived. For if [peculiarities of] shape make us overestimate the size of visual objects, or if their colors are somewhat obscure, or their outlines somewhat indefinite, these things make them appear to be more distant, and larger, than they actually are. (Hall, 1972, p. 68)

Malebranche (1674): I come now to give a General Demonstration of all the Errors, into which our Sight leads us, in respect of the Motion of Bodies. Let A be the Eye of the Spectator; C the Object, which I suppose at a convenient distance from A. I say, that though the Object remains fix'd in C, it may be thought to be remov'd as far as D, or to approach as near as B. And though the Object recedes towards D, it may be believ'd immoveable in C, and even to approach towards B; and on the contrary, though it approaches towards B, it may be thought to be immoveable in C, or even to recede towards D. That though the Object be advanc'd from C as far as E or H, or to G or K, it may be thought to have mov'd no farther than from C to F or I. And again on the other hand, that though the Object be mov'd from C unto F or I, a Man may think it mov'd to E or H, or else unto G or K. That whereof the Spectator is the Centre; though that Object be mov'd from C to P, it may be thought to be mov'd only from B to O: and on the contrary, though it be mov'd only from B to O, it may be thought mov'd from C to P. If beyond the Object C there happens to be another Object, suppose M; which is thought at rest, and which notwithstanding is in motion towards N: Though the Object C remains unmoved, or is mov'd with a more gentle motion towards F, than M is mov'd towards N, it will yet seem to be oppositely mov'd towards Y.... It is plain that the proof of all these Propositions except the last, in which there is no difficulty, depends on one Supposition only, namely this, that we cannot with any assurance determine concerning the Distance of Objects. For if it be true that we cannot judge thereof with any certainty, it follows that we cannot be assured whether C is advanc'd on towards D, or has approached towards B, and so of the other Propositions. (1700, pp. 21–22)

Smith (1738): We are frequently deceived in our estimates of distance by any extraordinary magnitudes of objects seen at the end of it: as in travelling towards a large city or a castle or a cathedral church or a mountain larger than ordinary, we think they are much nearer than we find them to be upon tryal.... Animals and all small objects seen in valleys, contiguous to large mountains,

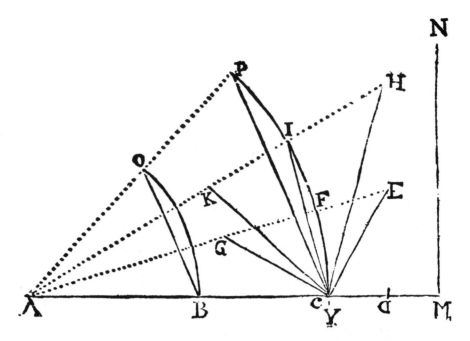

The dependence of motion perception on apparent distance according to Malebranche.

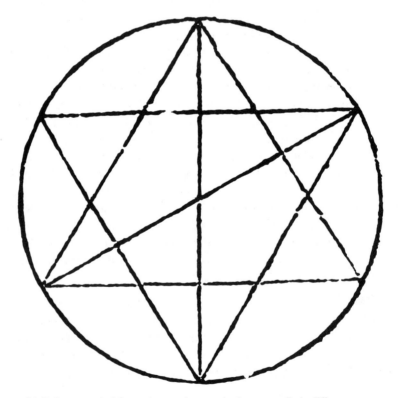

Müller's geometrical figure that can be organized perceptually in different ways.

appear extraordinarily small; because we think the mountain is nearer to us than if it was smaller. (p. 61)

Le Cat (1744): Our Senses are subjected to a thousand Mistakes; and yet we know nothing, but what they apprize us of, or what they give us Grounds to conjecture, by comparing those Hints with what they demonstrate to us. (1750, p. 286)

Müller (1838): Any compound mathematical figure produces a different impression, according as the attention is directed exclusively to one or the other part of it. Thus, in figure 103, we may in succession have a vivid perception of the whole, or of distinct parts only; of the six triangles near the outer circle, of the hexagon in the middle, or of three large triangles. The more numerous and varied the parts of which the figure is composed, the more scope does it afford for the play of attention. (1843, p. 1179)

8.2 Brightness Contrast

Variations in brightness usually followed from changes in light intensity, and this had been remarked upon in antiquity (Magnus, 1901). However, Ptolemy recognized that artists manipulated brightness to suggest depth. Leonardo was an astute enough observer to realize that the background played a role, too, and that such general principles should be utilized in painting. More prosaic demonstrations of the effect were described by Goethe and Müller. Klotz provided an example of contrast with a black-and-white pattern.

Ptolemy (ca. 150): ... the painter who wishes to represent two figures paints the salient one in a vivid colour, and the other in a more veiled and darker colour, so that it appears more distant. (Lejeune, 1956, p. 76)

Leonardo da Vinci (ca. 1500): In order to attain a color of the greatest possible perfection, one has to place it in the neighborhood of the directly contrary color: thus one places black with white, yellow with blue, green with red. (Boring, 1942, p. 165)

Goethe (1810): A grey object on a black ground appears much brighter than the same object on a white ground. If both comparisons are seen together the spectator can hardly persuade himself that the two greys are identical. (1840, p. 15)

Klotz (1816): The circle of contrast as a black and white figure. (Plate 3)

Müller (1838): A grey spot upon a white ground appears darker than the same tint of grey would do, if it alone occupied the whole field of vision. (1843, p. 1187)

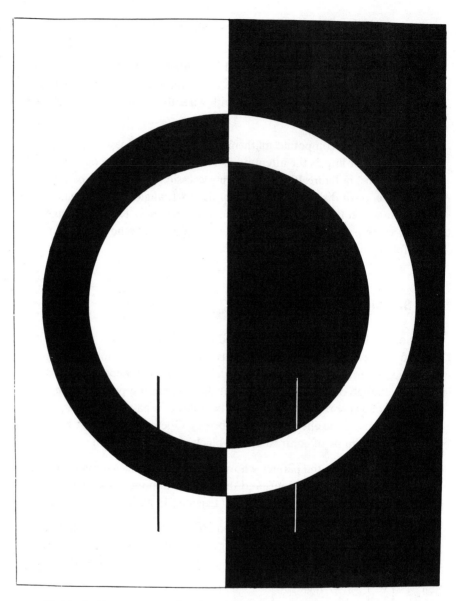

Black and white arcs contrasted against white and black backgrounds (Klotz, 1816).

8.3 Size Illusions

Many early descriptions of size illusions were related to errors in distance perception. This was not the case for Aristotle, who referred to what would now be called the filled-space illusion: dividing an extent into parts modifies the perception of its extent. However, Aristotle described the illusion as being in the opposite direction to the filled-space illusion, in which the divided extent appears greater than the undivided (see Johannsen, 1971). Ptolemy also described this illusion in the same manner as Aristotle, and both related it to the contrast between the parts and the whole extent.

The relationship between apparent size and apparent distance is found in Euclid and Ptolemy. For Euclid an increase in size resulted in a decrease in distance, whereas Ptolemy considered objects of the same size but at different apparent distances. Castelli, Malebranche, and Hartley all described a situation essentially similar to Ptolemy's.

Ptolemy also stated that brighter objects appeared nearer, and would consequently appear larger. Leonardo, Descartes, and Goethe commented on the effects of brightness on apparent size.

Aristotle (ca. 330 B.C.): Why is it that magnitudes always appear less when divided up than when taken as a whole? (Ross, 1927, p. 914b)

Euclid (ca. 300 B.C.): *Objects increased in size will seem to approach the eye.* Let *AB* represent the size of the object seen, and let *G* be the eye, from which let the rays fall, *GA* and *GB*. And let *BA* be increased, and let it be *BD*, and let the ray fall, *GD*. Now, since the angle *BGD* is greater than the angle *BGA*, *BD* appears greater than *BA*. But things thought to be greater than themselves seem to be increased, and things nearer the eye appear greater. (Burton, 1945, p. 372)

Ptolemy (ca. 150): When objects have equal visual angles the more distant one will appear larger. (Lejeune, 1956, p. 75)

Ibn al-Haytham (ca. 1040): The reason why sight perceives an object at an excessively great distance to be smaller than its real magnitude is that the size of an object is perceptible only by estimating the object's size by the angle of the cone that surrounds it together with the magnitude of the object's distance. (Sabra, 1989, pp. 283–284)

Leonardo da Vinci (ca. 1500): I once saw a woman dressed in black with a white shawl over her head. This shawl seemed twice as broad as the darkly clad shoulders. The crenels in the battlements of fortresses are of exactly the same width as the merlons and yet the former appear to be appreciably wider than the latter. (Minnaert, 1940, p. 105)

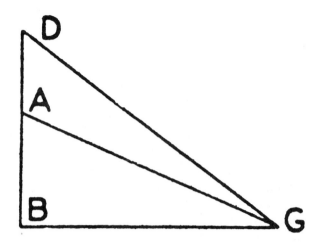

Euclid's analysis of size changes leading to variation in apparent distance.

Descartes (1637): We are also deceived because white or luminous bodies, and generally all those possessing much power to affect the sense of sight, always appear to be a little closer and larger than they would if they had less such power. Thus the reason for their appearing closer is that the movement with which the pupil contracts in order to avoid the force of their light is so joined with the power that disposes the entire eye to see close objects distinctly (by which we judge their distance) that the one can hardly occur but that the other does not also occur to some extent ... And the reason why these white or luminous bodies appear larger consists not only in the fact that the estimate we make of their size depends on that of their distance, but also on the fact that they impress bigger images on the back of the eye. (1965, pp. 111–112)

Castelli (1639): ... if two objects are truly equal and are truly placed at equal distances from our eye but one of them is judged by us to be farther away, it will be appraised larger. (Ariotti, 1973, p. 17)

Malebranche (1674): Thus we see daily on the Earth two things or more, of whose magnitude we can have no exact assurance; because to make a judgment of this Nature, 'tis necessary to know the precise distance of these Bodies, which is very difficult to be known. (1700, p. 17)

Hartley (1749): ... where the Picture on the *Retina* is of a just Size, and also the previous Judgment concerning Distance just, our Estimate of tangible Magnitude by Sight will be just likewise. But if the Picture on the *Retina* be magnified or diminished by Glasses, or our previous Judgment concerning the Distance be erroneous, our Estimate of tangible Magnitude will be erroneous in like manner. (pp. 200–201)

Goethe (1810): A dark object appears smaller than a bright one of the same size. Let a white disc be placed on a black ground, and a black disc on a white ground, both being exactly similar in size; let them be seen together at some distance, and we shall pronounce the last to be about a fifth part smaller than the other. If the black circle be made larger by so much, they will appear equal. (1840, p. 6)

8.4 Moon Illusion

The moon illusion—its larger appearance near the horizon than high in the sky—is, of course, a size illusion, but it has also been interpreted as a distance illusion. Historically, attention has been directed specifically to the illusion associated with the sun, moon, and stars, and it is treated separately here. Nonetheless, it will be noted that the interpretations that have been applied to it generally draw on other phenomena that have been mentioned elsewhere in this book. The moon illusion presented an enigma in the past and it is one that still persists. Modern attempts at explaining it remain problematical (see Hershenson, 1989). It was analyzed initially as a problem of physics, then physiology, and finally psychology. An annotated bibliography of 285 studies on the moon illusion has been compiled by Plug (1989); prior to 1840 there were 77 published accounts of the phenomenon.

The illusion was known long before Aristotle. Plug and Ross (1989) describe a seventh century B.C. cuneiform inscription on a clay tablet from Nineveh that could be interpreted as describing the phenomenon. Aristotle's account was clear, and he related the variations in size to the effects of mist (see section 7.11). This was essentially the explanation given by Ptolemy (1952) in the *Almagest*, despite his appreciation that objects in a rarer medium appear smaller. However the *Almagest* was written before his *Optics* (see Smith, 1996), in Book III of which a psychological account of the phenomenon, in terms of apparent distance, is given. There has been some dispute concerning the relationship between the two passages from Ptolemy (see Ross and Ross, 1976; Sabra, 1987). Ibn al-Haytham absorbed both interpretations and acknowledged that both can apply, but that differences in apparent distance were the principal cause of the illusion. His account is in Book VII of his *Optics*, which was transcribed at a later date than Books I–III. Ibn al-Haytham's explanation was absorbed into the medieval texts on optics, like Pecham's, where it was repeated virtually unchanged.

With the invention of more powerful optical instruments in the seventeenth century, astronomers provided detailed measurements of the dimensions of celestial bodies. Descartes commented that they subtend equal angles throughout their transit, and Molyneux gave the angular dimensions of the full moon. Molyneux took issue with the explanations advanced by both Descartes and Gassendi; the former subscribed to the apparent distance approach, whereas the latter adopted a physiological interpretation. Leonardo had observed that pupil size changed with light intensity and speculated that this was associated with apparent size. Gassendi repeated this claim (see Plug and Ross, 1989). The most popular explanation remained one based on variations in apparent distance. Some attributed the misjudgment of distance to the visibility of interposed objects near the horizon (Castelli, Rohault, Malebranche, Wallis, Le Cat), whereas others assigned it to aerial perspective (Hobbes, Berkeley). Berkeley devoted a substantial section of his *New Theory of Vision* to the moon illusion, particularly to Wallis's account, which he took as representative of the "fruitless and unsatisfactory" received opinion.

Many of the apparent distance protagonists made informal observations of the horizon moon viewed with and without interposed objects. Plug and Ross (1989) ruefully remarked "that almost invariably the effect found by each writer was in line with his explanation of the illusion" (p. 19)! Ibn al-Haytham's suggestion that the vault of the heavens was flat resulted in attempts to represent it in the eighteenth century. Desaguliers and Smith both produced diagrams implying its shallow dome, and the manner in which this could be used to account for the moon illusion, as did Young. Desaguliers and Young drew the spherical dome within the elliptical one, while Smith adopted the opposite strategy in order to show the estimated difference in the magnitude of the illusion. Desaguliers was stimulated to conduct an experiment on changes in apparent size with the distinctness of objects.

Ptolemy implied that viewing objects directly above was unusual and that alone might be a factor involved in the illusion. Gauss pursued this possibility and provided some experimental confirmation of it.

Aristotle (ca. 330 B.C.): ... the sun and the stars seem bigger when rising and setting than on the meridian. (Ross, 1931, p. 373b)

Ptolemy (ca. 142): It is true that their true sizes [of the celestial bodies] appear greater at the horizon; however, this is caused not by their shorter distance, but by the moist atmosphere surrounding the earth, which intervenes between them and our sight. It is just like the apparent enlargement of objects in water, which increases with the depth of immersion. (Ross and Ross, 1976, p. 378)

Ptolemy (ca. 150): For generally, just as the visual ray, when it strikes visible objects in [circumstances] other than what is natural and familiar to it, senses all their differences less, so also its sensation of the distances it perceives [in those circumstances] is less. And this is seen to be the reason why, of the celestial objects that subtend equal angles between the visual rays, those near the point above our head look smaller, whereas those near the horizon are seen in a different manner and in accordance with what is customary. But objects high above are seen as small because of the extraordinary circumstances and the difficulty [involved] in the act [of seeing]. (Sabra, 1987, p. 225)

Ibn al-Haytham (1083): What sight perceives regarding the difference in the size of the stars at different positions in the sky is one of the errors of sight. It is one of the constant and permanent errors because its cause is constant and permanent. The explanation of this is [as follows]: Sight perceives the surface of the heavens that faces the eye as flat, and thus fails to perceive its concavity and the equality of the distances [of points on it] from the eye ... Now sight perceives those parts of the sky near the horizon to be farther away than parts near the middle of the sky; and there is no great discrepancy between the angles subtended at the eye-center by a given star from any region of the sky; and sight perceives the size of an object by comparing the angle subtended by the object at the eye-center to the distance of that object from the eye; therefore, it perceives the size of the star (or interval between two stars) at or near the horizon from comparing its angle to a large distance, and perceives the size of that star (or interval) at or near the middle of the sky from comparing its angle (which is equal or close to the former angle) to a small distance. (Sabra, 1987, p. 241)

Pecham (ca. 1280): *Stars appear larger on the horizon than in any other part of the sky....* Since an object on the horizon is perceived by the eye under the same angle as [that under which it would be perceived] elsewhere in the sky and since it appears to be at a greater distance, the object is judged larger when on the horizon. (Lindberg, 1970, p. 153)

Leonardo da Vinci (ca. 1500): Every object we see will appear larger at midnight than at midday, and larger in the morning than at midday. This happens because the pupil of the eye is much smaller at midday than at any other time. (Plug and Ross, 1989, p. 13)

Descartes (1637): ... for ordinarily when these heavenly bodies are very high in the sky toward midday, they seem smaller than when, as they rise or set, various objects are between them and our eyes, which causes us to take better notice of their distance. And, by measuring them with their instruments, the astronomers definitely prove that the fact that they appear greater at one time than at another does not come from their being seen under a larger angle, but from our judging them to be farther away. (1965, p. 111)

Castelli (1639): I was in a carriage with Monsignor Cesarini and others of his noble party and we were along the Tiber, the full Moon was rising and appeared to peep out over the Aventine hill on the other side of the river. Almost everyone in a single voice said of the Moon—how large it is, how beautiful. I, taking advantage of the opportunity, asked how large it appeared. I was answered that it seemed as of four or five yards in diameter. At this point, by interposing the brim of my hat between the eye of the Monsignor and the Moon, I completely covered the view of the Aventine hill but in such a way that the moon could still be seen over the edge of the hat brim and again asked how much the diameter of the Moon appeared to be. The Monsignor, then, almost in amazement, answered that it did not seem to be two fingers breadth. (Ariotti, 1973, p. 16)

Gassendi (1642): ... the Moon being nigh the Horizon and being looked at through a more Foggy Air, casts a weaker Light, and consequently forces not the Eye so much as when brighter; and therefore the Pupil does not inlarge it self, thereby transmiting a larger Projection on the Retina. (Molyneux, 1687, p. 318)

Hobbes (1655): From hence may be deduced a cause, why the moon and stars appear bigger and redder near the horizon than in the mid-heaven. For between the eye and the apparent horizon there is more impure air, such as is mingled with watery and earthy little bodies, than is between the same eye and the more elevated part of the heaven. (1839, p. 462)

Rohault (1671): Thus in the Instance of the Moon, when it is at the *highest* above the Horizon, and we look at it through the Air only in which there are no other visible Objects, we imagine it to be nearer to us, than when it *rises* or *sets*, because at those Times, there are a great many intermediate Objects upon the Earth, between us and it. (1723, p. 254)

Malebranche (1674): 'Tis for this reason that the Moon at the Rising or Setting, is seen much bigger, than when elevated a good height above the Horizon: For this elevation removes our view from off the objects lying betwixt us and her, the dimensions whereof we know; so that we cannot judge of that of the Moon by forming the comparison between them. (1700, p. 19)

Molyneux (1687): First therefore it is well known that the mean apparent Magnitude of the Moon is 30 *m*. 30*s*. . . . tis as well known also that when she is in this Posture [zenith], being looked upon by the Naked Eye she appears . . . about a foot broad. But the same Moon being Looked upon just as she rises, she appears to be three or four foot broad. . . . it would be very unreasonable to Imagine that so many Authors should rack their Brains for solving an appearance, wherein they were not certain of the matter of Fact. (pp. 314–315)

John Wallis (1616–1703) after an illustration in Wolf (1935).

Wallis (1687): But ... though as to small Distances, we may make some estimate from the known *Magnitude* of the Object: And, as to middling distances, from the Parallax (as I may call it) arising from the interval of the two Eyes: Yet even this latter will hardly reach beyond, if so far as the visible *Horizon*: and all beyond it, is lost. So that, there being nothing left to assist the fancy in estimating so great a distance, but only the intermediate Objects: Where these intermediates appear to the Eye, (as, when the Sun or Moon are near the *Horizon*:) the distance is fancied greater, than when they appear not, (as when farther from it:) and consequently (though both under the same or equal Angles) that near the *Horizon* is fancied the greater. And this I judg to be the true reason of that appearance. (pp. 328–329)

Berkeley (1709): Now, between the Eye, and the Moon, when situated in the *Horizon*, there lies a far greater Quantity of Atmosphere, than there does when

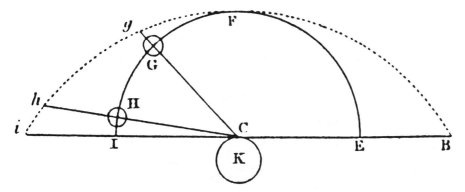

The elliptical appearance of the sky (Desaguliers, 1736a).

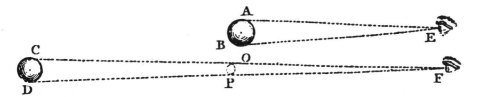

The influence of apparent brightness on apparent size (Desaguliers, 1736b).

the Moon is in the *Meridian*. Whence it comes to pass, that the Appearance of Horizontal Moon is fainter, and therefore.. it shou'd be thought bigger in that Situation, than in the *Meridian*, or any other elevation above the *Horizon*. (pp. 74–75)

Desaguliers (1736a): When we look at the Sky towards the Zenith, we imagine it to be much nearer to us, than when we look at it towards the Horizon; so that it does not appear Spherical, according to the vertical section EFGHI, but Elliptical, according to the section eFghi.... Now when the Moon is at G, we consider it is at g, not much farther than G; but when it is at H, we imagine it to be at h, almost as far again. Therefore, while it subtends the same Angle as it did before (nearly), we imagine it to be so much bigger as the Distance seems to us to be encreased. (pp. 390–391)

Desaguliers (1736b): Now in the Case of the Moon, the Deceipt is help'd, because the Vapours, thro' which we see it when low, take away its Brightness, and therefore have the same Effect as would (or does) happen in the Experiment, when the Light of the Ball *op* strikes the Eye no stronger than the light of the Ball CD. (p. 394)

Porterfield (1737): Thus also the Sun and Moon appear greater when near the Horizon, than at a greater Height, because when nigh the Horizon, they are judged at a greater Distance. (p. 192)

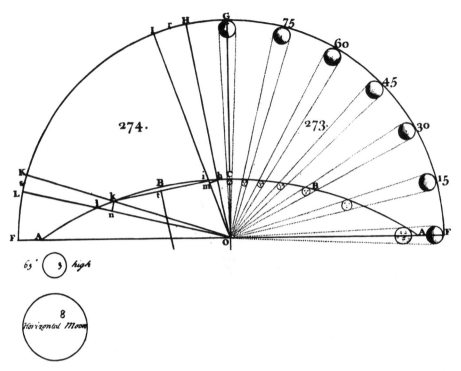

The apparent concavity of the sky, and the relative apparent sizes of the moon on the horizon and on high according to Smith.

Smith (1738): For instance, supposing the arch *ABC* to represent that apparent concavity [of the apparent figure of the sky], I find the diameter of the sun or moon will seem to be greater in the horizon than at any proposed altitude, measured by the angle *AOB*, in the proportion of its apparent distances *OA*, *OB*. The numbers that express these proportions are ... exactly represented to the eye in the 273d figure, in which the moons placed in the quadrantal arch *FG*, described about the centre, are all equal to each other ... and the unequal moons in the concavity *ABC* are terminated by the visual rays that come from the circumference of the real moon, at those heights, to the eye at *O*. (pp. 64–65)

Le Cat (1744): If we look at the Moon in the Horizon over a Wall, thro' a Paper-Tube, or with a Telescope, we see no more of these Mountains, Vallies, &c. those Indications of her Distance, and yet she ever appears larger than she is. (1750, p. 243)

Hartley (1749): The horizontal Moon appears larger than the meridional, because the Picture on the *Retina* is of nearly the same Size, and the Distance esteemed to be greater. (p. 201)

Young (1807): The sun, moon, and stars, are much less luminous when they are near the horizon, than when they are more elevated, on account of the greater

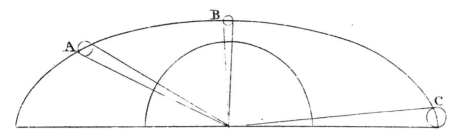

The concave vault of the heavens and its influence on the apparent size of celestial bodies (Young, 1807).

quantity of their light that is intercepted, in its longer passage through the atmosphere: we also observe a much greater variety of nearer objects almost in the same direction: we cannot, therefore, help imagining them to be more distant, when they rise or set, than at other times; and since they subtend the same angle, they appear actually larger. For similar reasons the apparent figure of the starry heavens, even when free from clouds, is that of a flattened vault, its summit appearing to be much nearer to us than its horizontal parts, and any of the constellations seems to be considerably larger when it is near the horizon than when in the zenith.... The apparent figure of the heavens being nearly like the curve ABC, the sun or moon at A or C appears to be much larger than at B. (pp. 454 and 787)

Gauss (1830): However, it appears to me that another experiment indicates a physiological explanation of the phenomenon: When I view the elevated full moon from a backwards reclining bodily position, in which the head is in the usual position with respect to the body, so that the moon shines more or less perpendicularly on the face, then I see it much larger; and conversely I see the full moon on the horizon noticeably smaller with the head inclined forwards. (Plug and Ross, 1989, p. 16)

Karl Friedrich Gauss (1777–1855) after an engraving in Augé (1898).

8.5 Motion Illusions

Several motion illusions have previously been mentioned—specifically, motion aftereffects (section 5.4), induced motion (section 5.3), and vertigo (section 5.5). The last two were mentioned by Ptolemy, but not the first. He did include motion in his categories of illusions, and described relative motions (as did Euclid: section 5.1), illusions due to objects moving too quickly (section 4.11) or too slowly (section 5.1). The illusion described here is one of motion contrast, and it was the phenomenon considered in Book II of his *Optics*.

The statements by both Aristotle and Malebranche indicate the importance of frames of reference in the perception of motion, as was discussed in the context of motion perception generally (see Wade and Swanston, 1996). Perhaps one of the most acute descriptions of frames of reference in operation is that by Locke. It was precisely the absence of well-defined spatial references that results in the ambiguous motion of distant objects (as in Smith's windmill) or the apparent motion of stars (as described by Humboldt). Smith noted that the direction of rotation of a windmill was difficult to discern from a distance, and could appear to change. Smith's windmill observation was illustrated in Porterfield (1759b) and Harris (1775) and rediscovered by Sinsteden (1860); Boring's (1942) reference to Sinsteden has resulted in the neglect of the reports by Smith, Porterfield, and Harris.

The great German naturalist Alexander von Humboldt made his observation during a scientific voyage to South America, and he genuinely believed that the stars did undergo the rapid motions that he perceived. The phenomenon became called "star wandering" because it could easily be seen, but it was only as a consequence of the lack of consistency in the reported motions that more serious study was undertaken. Observations of the same star at the same time by several astronomers yielded motions but in no coherent direction. It became clear that a perceptual rather than a physical phenomenon had been under study and Aubert (1865) renamed it the autokinetic phenomenon.

Several of Purkinje's motion illusions were described in chapter 4, but the one given here was seen following fixation on a stationary pattern. If it consisted of fine, regular, parallel lines, then prolonged observation produced an aftereffect of motion perpendicular to the lines in the adapting pattern. This aftereffect has been rediscovered many times and has become called the complementary aftereffect (see Wade, 1977b, 1982).

Poggendorff is better known for his spatial illusion (in which broken but aligned line segments appear displaced) than for a motion illusion. This is not surprising since his description was neither given prominence when it was published nor recognition thereafter. As editor of the *Annalen der Physik und Chemie* he made a short comment on the translation of Faraday's paper on a motion illusion (see section 5.2). Almost in passing, Poggendorff mentioned the motion seen in the thread of a rotating screw. This has subsequently been called the barber-pole illusion.

Aristotle (ca. 330 B.C.): ... to persons sailing past the land seems to move, when it is really the eye that is being moved by something else [the moving ship]. (Ross, 1931, p. 460b)

Ptolemy (ca. 150): A motion illusion occurs when a horse and carriage are travelling slowly, and the horse and wheels are observed at the same time. When we look at these we believe that the horse moves quickly because of the regular and repetitive rotation of the parts of the wheel. (Lejeune, 1956, pp. 84–85)

Malebranche (1674): ... what would happen to a Man that were asleep in a Vessel, who starting on a sudden, saw nothing when he wak'd, besides the top of a Mast of some Vessel that made towards him. For in case he saw not the

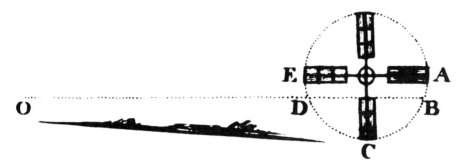

Smith's illustration of the windmill illusion.

Sails swell'd with the Wind, nor the Mariners at work, nor felt any Agitation or concussion of the Ship, or the like, he would absolutely be at a loss and in doubt, without knowing which of the two Vessels was in Motion; neither his Eyes nor his Reason could give him any Information. (1700, p. 21)

Locke (1690): Thus a Company of Chess-men, standing on the same squares of the Chess-board, where we left them, we say are all in the *same Place*, or unmoved; though, perhaps, the Chess-board hath been in the mean time carried out of one Room into another, because we compared them only to the Parts of the Chess-board, which keep the same distance one with another. The Chess-board, we also say, is in the *same Place* it was, if it remains in the same part of the Cabin, though, perhaps, the Ship it is in, sails all the while: and the Ship is said to be in the *same Place*, supposing it kept the same distance with the Parts of the neighbouring Land; though, perhaps, the Earth hath turned round; and so both Chess-man, and Board, and Ship, have every one *changed Place* in respect of remoter Bodies. (pp. 76–77)

Smith (1738): . . . by taking the nearest end of the sail of a wind-mill for the remotest, we sometimes mistake its circular motion. (p. 61)

Humboldt (1799): In the Malpays [on Tenerife], at a height of 10,700 feet above sea level, I saw with my naked eye low-lying stars in a wonderful wandering motion. Illuminated points rose upwards, moved sideways and fell back to their original positions. The phenomenon lasted 7 to 8 minutes and came to a halt long before the appearance of the sun above the horizon. The same motion could be seen through a telescope; there remains no doubt that it was the stars themselves that moved. (1850, p. 73)

Purkinje (1823a): If I fixate on the parallel lines of a sharply drawn engraving for fifteen to twenty seconds and then close the eye, there appears in the same place a scintillation of undefined light and dark zigzag lines, which run like waves through one another and perpendicular to the previously fixated lines. (pp. 119–120)

Alexander von Humboldt (1769–1859) after an engraving in *The World's Great Men*. London: The London Printing and Publishing Company, 1854.

Johann Christian Pog-
gendorff (1796–1877)
after an engraving in
Augé (1898).

Poggendorff (1831): Set a screw in rapid rotation and one believes one sees the thread move along the axis, forwards or backwards, according to the direction of rotation. (Editorial comment in Faraday, 1831b, p. 603)

8.6   Orientation Illusions

Ptolemy did include orientation in the stimulus dimensions that affect size perception, but he was referring to orientation with respect to the frontal plane. We would now probably call it bearing. The term "orientation" is used here to indicate inclination with respect to gravity, and it is an aspect of perception that received very little attention in the period under study. To be sure, disturbances of orientation of the type that occur in vertigo (section 5.5) were discussed, as were torsional eye movements (section 5.7), but no body of phenomena accrued to it. From the point of science this is perhaps understandable because the functions of the vestibular system remained a mystery until later in the nineteenth century. From the point of art, or more precisely artifice, it is surprising because illusions of orientation have been food and drink for fairground operators for centuries. However, they would have had a vested interest in keeping the tricks of their trade secret.

One of the first orientation illusions to be commented on (by Lucretius) has been exploited in the fairground: buildings constructed awry result in disorientation for anyone within them. The visual frame of reference is in conflict with that provided by posture, as is the case in a sailing ship. This last was discussed by both Melvill and Wells. Thomas Melvill's observation was recorded as one of many queries in his essay on light and colors, which was addressed to Newton's theory. He delivered two papers at meetings of the Edinburgh Philosophical Society, but died at the age of 27 before they were published. Wells applied his penetrating logic to answer the queries set by Melvill: although the head remained at rest with respect to the cabin, both it and the cabin were in motion with respect to the horizon. Montaigne, on the other hand, described a purely visual illusion—what is now called the Zöllner illusion (see Johannsen, 1971).

Lucretius (ca. 56 B.C.): Lastly, as in a building, if the original rule is warped, if the square is faulty and deviates from straight lines, if the level is a trifle wrong in any part, the whole house will necessarily be made in a faulty fashion and be falling over, warped, sloping, leaning forward, leaning back, all out of proportion, so that some parts seem about to collapse, all betrayed by false principles at the beginning. So therefore your reasoning about things must be warped and false whenever it is based upon false senses. (1975, p. 317)

Montaigne (1580): Those rings which are cut out in the form of feathers, which are called [in heraldry] *endless feathers*, no eye can discern their size, or can keep itself from the deception that on one side they enlarge, and on the other contract, and come to a point, even when the ring is being turned round the finger; yet when you feel them, they seem all of an equal size. (1853, p. 280)

Melvill (1756): I have observed, when at sea, that, though I pressed my body and head firmly to a corner of the cabin, so as to be at rest in respect of every object about me, the different irregular motions of the ship, in rolling or pitching, were still discernible by the sight: How is this fact to be reconciled to optical principles? Shall we conclude, that the eye, by the sudden motions of the vessel, is rolled out of its due position? Or, if it retains a fixed situation in the head, is the perception of the ship's motion owing to a vertigo in the brain, a deception of the imagination; or to what other cause? (pp. 80–81)

Michel de Montaigne (1533–1592) after a frontispiece engraving in Montaigne (1853).

Wells (1792): In both pitching and rolling then, the relative position of a vessel to a horizontal plane is necessarily changed. Consequently, though, in the abovementioned experiment, Mr. Melvill's body and head were at rest with respect to every object about him, still a different degree of muscular effort was required to keep them so, in every such different position of the vessel. . . . Should the necessity of supporting the body against its gravity, by the actions of its voluntary muscles, be suspended in whole, or in part, our judgments of the situation of objects, with respect to the horizon, must become irregular and uncertain. (p. 89)

8.7   Depth Reversals

Ptolemy devoted several paragraphs of his *Optics* to the appearance of concave and convex surfaces, and Smith suggested that the mistaken judgments were a consequence of the direction of the shadows. Such reversals were much more common when using microscopes (Gmelin, Haller, and Rittenhouse), and Brewster described how it could be achieved with a hollow mask of a face.

Depth reversals could also take place with outline figures. Necker had been startled by the effect when looking at engravings of crystals, like the rhomboid shown. Others, notably Wheatstone and Brewster (1844b), studied the effect further and each reproduced slightly different drawings of the rhomboid from that given by Necker. The figure is now typically drawn as a cube, without the connecting transversal, and it is called the Necker cube. Wheatstone, in fact, first described the reversals that occur with a cube, although he used a three dimensional one rather than a drawing. Not only did he demonstrate that reversals occur with solid figures but he also commented on the changes of shape that accompany the reversals in depth. Perhaps it should be renamed the Wheatstone cube, although this reversal is unlikely to take place.

Ptolemy (ca. 150): . . . we believe a concave sail [on a boat] is convex when we look at it from afar. (Lejeune, 1956, p. 76)

Smith (1738): . . . being led into the mistake [of confusing concave with convex surfaces] by an imperfect judgment of the distances of the parts of the object, and confirmed in it by a contrary position of the shadows cast by a side light. (p. 61)

Gmelin (1744): . . . if a common seal was applied to the focus of a compound microscope, or optical tube, which has two or three convex or plano-convex lenses, that part which is cut the deepest in it would appear very convex, and so on the contrary; and that sometimes, but very seldom it would appear in the same state as to the naked eye. (Brewster, 1826a, p. 100)

Haller (1767): The *convexity* or *protuberance* of a body is not seen; but is afterwards judged of by experience, after we have learned, that a body, which is convex to the feeling, causes light and shadow to be disposed in a certain manner. Hence it is, that microscopes frequently pervert the judgment, by transposing or changing the shadows. The same also happens in that phenomenon which is not yet sufficiently understood, by which the concave parts of a seal are made to seem convex, and the contrary. (1786, pp. 30–31)

David Rittenhouse (1732–1796) after an illustration in *A Catalogue of Portraits and other Works of Art in the Possession of the American Philosophical Society*. Philadephia: The American Philosophical Society, 1961.

Rittenhouse (1786): It has often been a matter of surprize to me, when viewing the moon through a good telescope, in company with persons not accustomed to such observations, that whilst the cavities and eminences of the moon's surface appeared to be marked out with the utmost certainty by their light and shades, my companions generally conceived it to be a plain surface of various degrees of brightness. The reason I suppose to be this; the astronomer knows from the moon's situation with respect to the sun, and even from the figure of its enlightened part, precisely in what direction the light falls on its surface, and therefore judges rightly of its hills and vallies, from their different degrees of light, according to those rules which are imperceptibly formed in the mind, and confirmed by long experience.... If we look through such a tube and glasses [two convex lenses] at the hearth or other object, suppose a piece of chocolate, the furrows in it appear so many ridges, on removing the tube they sink into furrows, on applying it again they rise into ridges, and the illusion might perhaps be repeated a thousand times, without the mind being able to conceive the object to appear through the tube like what it really is. But if whilst you are looking through the tube, and the object appears in its unnatural state, that is, when its furrows appear ridges, you apply your finger and feel that they really are furrows, the deception vanishes in a moment and the object appears in its natural state.... The application of a writing pen or pencil will produce the same effect. And, which is very remarkable, after the mind has been undeceived by these means once or twice, it does not readily admit of the imposition again: Though, as I observed before, if it be done by removing the glasses, the deception will return again as often as you please. The truth seems to be, that the mind chuses the least difficulty. (pp. 38 and 40–41)

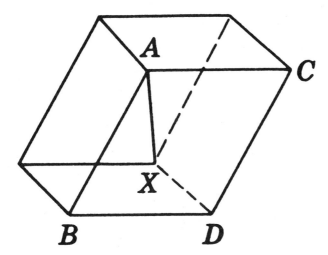

The Necker rhomboid.

Brewster (1826a): We have succeeded in carrying this deception so far, as to be able, by the eye alone, to raise a complete hollow mask of the human face into a projecting head. In order to do this, we must exclude the vision of other objects; and also the margin of thickness of the cast. (p. 108)

Necker (1832): The object I have now to call your attention to, is an observation ... which has often occurred to me while examining figures and engraved plates of crystalline forms: I mean a sudden and involuntary change in apparent position of a crystal or solid represented in an engraved figure.... What I mean will be more easily understood from the figure annexed. The rhomboid AX is drawn so that the solid angle A should be seen the nearest to the spectator, and the solid angle X the furthest from him, and that the face ACBD should be the foremost, while the face XDC is behind. But in looking repeatedly at the same figure, you will perceive that at times the apparent position of the rhomboid is so changed that the solid angle X will appear the nearest, and the solid angle A the furthest; and that the face ACDB will recede behind the face XDC, which will come forward; which effect gives the whole solid a quite contrary apparent inclination. (p. 336)

Wheatstone (1838): The inversion of relief is very striking when a skeleton cube [an outline figure of three dimensions] is looked at with one eye.... and the skeleton figure will appear to be continually undergoing a change of shape. (p. 384)

9 Conclusion

Joseph Priestley (1733–1804) after an engraving in Knight (1835b).

I have adopted the historical method *because it appears to me to have many obvious advantages over any other for my purpose, being particularly calculated to engage the attention, and to communicate knowledge with the greatest ease, certainty, and pleasure.* (*Priestley, 1772, p. vii)*

A vast range of visual phenomena had been described by the mid-nineteenth century. Most of these were based on observations of naturally occurring events, although some phenomena, like those associated with color, had been subjected to experimental manipulations. Light itself was treated as a phenomenon of physics, and it was clearly distinguished from vision. The wave theory of light prevailed over corpuscular theory, and it was analyzed by increasingly intricate mathematical methods. The appreciation of the distinction between physical and visual optics assisted in understanding many aspects of the latter. For example, knowledge of the dioptrics of the eye and the formation of an image on the retina led to a variety of hypotheses concerning the mechanism of accommodation. Young's theory, based on variations in the curvature of the crystalline lens, was supported by logic, but the means of such modulation remained a mystery. Similarly, differences between individuals in their range of accommodation could be corrected by the prescription of appropriate eyeglasses. Centuries before the dioptrics of the eye was understood, eyeglasses had been used for correcting presbyopia, although their application was not based on sound theoretical ground. Indeed, the basic anatomy of the eye was not known at those early times, and the location of the lens itself remained in considerable doubt. Eyeglasses were enlisted to correct gross errors of refraction, but relatively little attention had been given to more minor aberrations. Spherical and chromatic aberrations presented problems in manufacturing optical instruments, like microscopes and telescopes, but were not generally considered to hinder vision until Young measured the chromatic aberration in his own eyes. He also described his own astigmatism, and Airy introduced tests to detect this aberration and cylindrical lenses to correct it.

The gross anatomy of the eye, in terms of its optical structures, had been correctly illustrated by Scheiner in the seventeenth century, and the advances made thereafter were with regard to its more detailed anatomy and to methods of preparing the eye prior to dissection. Soemmerring was particularly concerned with the latter, and produced beautiful drawings of the eye and its associated structures. These included the retina, which was accepted as the receptive surface of the eye. Soemmerring identified variations in its structure and delineated the yellow spot, the central region around the fovea. The simple microscope was applied to the retina from a very early stage, and compound achromatic microscopes available in the early nineteenth century provided more detail of its cellular structure. This was achieved initially by Treviranus, who described and illustrated cylindrical cells in the retinas of a variety of animals, but the receptors were drawn facing toward rather than away from the direction of incoming light. An element of fancy might have been introduced because of the difficulties in fixing the preparations; the years following 1840 saw rapid advances in fixing, sectioning, and staining preparations (see Finger, 1994). The differentiation between rod and cone receptors and their separate functions was made in the 1860s, based principally on species differences and microanatomy (Schultze, 1866). Many of the phenomenological distinctions used subsequently to support this duplicity theory had been described much earlier.

One of the more obvious structures in the retina was the optic disc, which is devoid of receptors. It corresponds to the blind spot, which was described in the seventeenth century, but was falsely interpreted by its discoverer, Mariotte. The precise paths pursued by the two optic nerves to the brain were the object of much debate, which was not resolved until the early nineteenth century. Taylor had represented the partial decussation at the optic chiasm correctly in 1738, but this was based more on speculation than dissection. By the early nineteenth century, Wollaston was able to marshal evidence from his own hemianopia as well as from anatomy to support partial decussation; the detailed anatomy of the crossings was described later in the century by Munk (1879). Few speculated on the more central pathways of vision. This was to become a topic of considerable interest, largely as a consequence of Ferrier's (1876) studies of stimulation and ablation of the occipital cortex.

Speculations about the physiological basis of color vision abounded long before any retinal receptors were identified. The perennial preoccupation with color primaries led to the proposal (by Palmer in the eighteenth century) that there was a receptor for each primary, and that color-blind individuals were lacking one of them. However, the primaries were taken to be those used by painters: red, yellow, and blue. Apart from the specification of three rather than seven primaries, most of Newton's proposals about color were accepted. Initially, Young advanced his hypothesis that color vision was based on selective responses to the three primaries of red, yellow, and blue, but he soon amended them to red, green, and violet. Distinctions between mixing lights and pigments had been made, but awaited Helmholtz's analysis for their formalization. Helmholtz also embraced Young's trichromatic theory, but shifted the selectivity to the specific energies of nerves rather than positing them in the receptors. These approaches to color vision were based on an analogy with the rules of color mixing: if all colors could be produced by mixing three primaries in appropriate proportions, then all perceived colors could result from the appropriate combination of the actions of three color mechanisms in the visual system.

The alternative approach, which was based on color experience rather than color mixing, was championed by Goethe. Newton had stated that "the Rays to speak properly are not coloured," thus accepting the subjective dimension in color vision, but he did not subordinate the physics of light to the philosophy of sight in the manner of Goethe. One of Goethe's greatest difficulties was reconciling the purity of the perception of white light with the conception of its compound nature. However, he was able to enlist a variety of phenomena (like color contrasts, color shadows, accidental colors, and aspects of color blindness) which posed severe difficulties for the trichromatic theory. One color phenomenon he did not investigate was that of the colors seen with black-and-white patterns. Fechner, who rediscovered such subjective colors by accident, remarked that Goethe would have taken the phenomenon to be support for his theory. Despite the wealth of observations contained in his *Theory of Colors*, few students of vision saw Goethe's theory as other than evidence of the distance that separated art from science. In a lecture surveying Goethe's scientific researches, Helmholtz attempted to take a sympathetic view by stating that Goethe was primarily a poet, and that he was not disposed to support experimental inquiries into natural phenomena: "Thus, in the theory of colour, Goethe remains faithful to his principle, that Nature must reveal her secrets of her own free will; that she is but the transparent representation of the ideal world" (Helmholtz, 1895, p. 45)

Goethe sought to shift the study of vision away from physics toward phenomenology. Purkinje adopted the phenomenological methods for examining a wide variety of subjective visual phenomena, although he did try to relate them to the underlying physiology; for Purkinje, visual illusions revealed visual truths. Later in the nineteenth century the individual phenomena (like afterimages, dark adaptation, and visual persistence) continued to be studied, but not as a part of a defined subject area (like color). To be sure, chapters with this heading appeared in some books, as in Aubert's *Physiologie der Netzhaut* (Physiology of the Retina), for example, but the tenuous threads that bound the phenomena together were severed. Plateau's surveys in 1878 treated visual persistence, afterimages, and irradiation as separate topics, as did Helmholtz (1867). The generality of some of the phenomena were also called into question, notably by Helmholtz: " … there are many of the phenomena described by Purkinje which have never been seen by anyone else, although it cannot be certainly held that they depended on individual peculiarities of this acute observer's eyes" (1895, p. 200).

One subjective phenomenon was seen by many and was to provide the basis for the synthesis of motion from a sequence of static scenes: visual persistence could be harnessed to yield visual motion. This was initially exploited in philosophical toys, like the stroboscopic disc, but it was to provide an experimental means for studying motion perception, which was to become increasingly important in the nineteenth century (see Wade and Heller, 1997). Not only could motion be seen in suitably sequenced images but stationary objects could appear to move following prolonged observation of moving patterns. This motion aftereffect was rendered experimentally tractable with Plateau's spiral, described in 1849, and the phenomenon provided a vehicle for one of the first neural models of vision (Exner, 1894). Both Purkinje and Addams attributed the motion aftereffect to eye movements, although they were neither measured nor assessed following prolonged observation of visual motion. They had been assessed in the context of postrotational nystagmus: Wells used afterimages to provide an effective stabilized retinal image which could be compared with unstabilized images viewed after rotation. This remained the standard method of compensating for any eye movements for much of the nineteenth century. More objective methods of recording eye movements were introduced by Javal and by Dodge only toward the end of the century (see Heller, 1988). One of the features of eye movements that had been noted in antiquity was their binocularity—both eyes seem to move as one. The principles at the basis of equal innervation had long been known, but they were made more explicit by Hering (1868), with whose name they are still linked.

Binocularity has proved to be one of the most consistently investigated phenomena over the period here under study. The overriding concern has been to account for binocular single vision in terms of visual direction. A wide range of observations accrued from this endeavor, and in the early nineteenth century they were interpreted in terms of the Vieth-Müller circle. Wheatstone's insight was to realize that singleness of vision did not preclude depth perception on the basis of retinal disparity. Wheatstone wrote a second article on binocular vision in 1852; in it he set out the experimental approach to binocular vision and space perception, as well as describing some novel binocular instruments. The variables involved in binocular vision could be determined, and then manipulated. The methods that had proved successful in the physical sciences could be applied to visual phenomena too. He determined that retinal disparity, retinal size, accommodation, and convergence vary with the approach of a three-dimensional object; he then isolated each variable and manipulated it to determine its importance in stereoscopic vision. It was this approach that so attracted Helmholtz to the study of binocular vision and space perception, and he acknowledged Wheatstone's influence. The stereoscope could be enlisted to examine many of the constant themes of binocularity: the range of disparities over which singleness of vision was retained, and beyond which double vision occurred—binocular color and contour rivalries, and eye dominance. However, its principal purpose was to investigate the phenomenon that it disclosed—stereoscopic depth perception and its relation to retinal disparity. Wheatstone himself drew a distinction between what he called experimental and mental philosophies (which we would now refer to as physics and psychology), placing stereopsis in the latter (see Wade, 1983, 1987). His interpretation in terms of empiricist philosophy was elaborated by Helmholtz, and extended to a wider range of phenomena.

Wheatstone had reunited binocularity with space perception, and he employed Berkeleian empiricism to cement the union. Space perception had remained the province of philosophers for centuries, but it could now be addressed by the methods of science. Helmholtz adopted a similar empiricist stance, and addressed most spatial phenomena in those terms. The perception of shape, form, and position had been the subject of much inquiry. Wells, with his characteristic incisiveness, appreciated that direction and distance were the essential dimensions for locating objects in space, but his own experimental inquiries were confined mainly to direction. A range of cues to distance had been identified, notably by Descartes and Berkeley, but they were difficult to manipulate experimentally. The significance of the stereoscope was that it provided a means of manipulating a novel cue, retinal disparity, in a systematic way and with predictable results. Direction and distance could both be examined in the context of binocular vision, and in the laboratory.

What we now think of as visual illusions—the geometrical optical illusions—were rarely examined prior to the nineteenth century. Most attention was paid to those naturally occurring phenomena, like the moon illusion or induced motion. Painters had been producing allusions to the third dimension for centuries, and occasional references to ambiguous figures were made, but the systematic study of spatial distortions was rare. This is most puzzling, because the figures themselves are simple to draw, being arrangements of lines on paper. However, most of the named illusions, like the Ponzo and Müller-Lyer figures, were first produced toward the end of the nineteenth century (see Wundt, 1898). They represented the psychological dimension of seeing, and provided a means of displaying its independence from the physiology of vision. It is noteworthy that the interest in visual spatial illusions coincided with the emergence of psychology as an independent discipline. The illusions that were most studied just prior to 1840 were those associated with motion. The popularity of the philosophical toys that could synthesize motion from sequences of still pictures matched that of the stereoscope in the succeeding decades.

By 1840 the scene was set for an experimental assault on vision. Although the instruments used in this attack were generally invented in Britain and Belgium, the base from which the assault was launched was Germany. German universities were more numerous, more highly organized, and more adaptable than those in other European countries. They responded quickly to developments in science, creating new departments to adopt novel approaches. Sensory physiology was widely studied within German universities, and the greatest achievements of the nineteenth century were to be made in them. The instruments themselves were adapted and modified within the German context. Many variations on the stereoscope were presented by Dove (1851), and Helmholtz (1857) invented the telestereoscope. Wheatstone invented the electromagnetic chronoscope in 1840 (although he did not publish an account of it until 1845), but it was Hipp's modification of it that was widely used to measure reaction times (see Edgell and Symes, 1906). Wheatstone (1834) had employed electric sparks to illuminate stimuli too briefly for eye movements to occur, but it was Helmholtz's adoption of this technique that was most telling (see Wade, 1994a). Electric sparks were used by Dove (1841) and Volkmann (1859) to illuminate stereopairs, in order to determine whether depth was still seen. Appreciating the limitations of this technique, Volkmann devised the tachistoscope so that greater control of stimulus presentation time was available. Many versions of tachistoscopes were constructed in the second half of the nineteenth century (see Wade and Heller, 1997). Thus, one important avenue that emerged in mid-nineteenth century was stimulus control. The other was behavioral measurement. Virtually no attention had been paid to the ways in which perception could be measured. Weber had started his work, but it was with Fechner's (1860) *Elemente der Psychophysik* that response measurement could match stimulus manipulation.

Despite these advances, the major division within German sensory physiology in the latter part of the nineteenth century was along philosophical lines. Stimulus control had enabled vision to be studied empirically and this was allied by some, most notably Helmholtz, to philosophical empiricism. Not only was this contrary to the continental tradition, but it also met with some fierce opposition. The empiricist-nativist conflict was embodied in the disputes between Helmholtz and Hering. Hering was in the phenomenological tradition stemming from Goethe and Purkinje; in fact, he followed Purkinje in the chair of physiology at Charles university, Prague. On many of the phenomena described in this book, Helmholtz and Hering held diverging opinions. Color vision was analyzed in stimulus terms by Helmholtz, who adopted Young's trichromatic theory to account for it. Hering, on the other hand, placed more emphasis on color naming and developed an opponent process theory involving three pairs of colors. His influence was Goethe, and he studied many of the phenomena that the latter had described in his book on the *Theory of Colors*. Binocularity and space perception were other areas of combat. For Helmholtz stereoscopic depth perception was learned, and the invention of the stereoscope "made the difficulties and imperfections of the Innate Theory of sight much more obvious than before" (Helmholtz, 1873, p. 274). Hering considered that local signs for the three dimensions of space were innate, and that binocular fusion was physiological rather than psychological. The same arguments were applied to visual direction and to space perception generally. Fechner's psychophysical methods were well known to Helmholtz, but he continued to place more reliance on his own observations than on those of groups of subjects.

Helmholtz surveyed and extended the physics, physiology, and psychology of vision in the three volumes of his *Handbuch der physiologischen Optik*. They were published separately in volumes of Karsten's *General Encyclopaedia of Physics* in 1856, 1860, and 1866, and the three volumes were republished together in 1867. In addition to the experiments he described and the empiricist position he espoused, each section terminated with a brief history. Helmholtz's *Handbuch* defined the experimental approach to vision that distinguished the discipline from its observational past. It is, therefore, appropriate to conclude with a quotation by, and a portrait of, Helmholtz.

Helmholtz (1867): The fundamental thesis of the empirical theory is: *The sensations of the senses are tokens of our consciousness, it being left to our intelligence to learn how to comprehend their meaning.* The tokens which we get by the sense of sight may vary in intensity and in quality, that is, in luminosity and colour. There may be some other difference between them depending on the place on the retina that is stimulated, a so-called *local sign*. The local signs of the sensations from one eye are entirely different from those in the other eye. We feel also the *degree of innervation* which we cause to be communicated to the nerves of the ocular muscles. The apperceptions of space-relations and motion are not necessarily derived from the visual perceptions.... any other sensations, not only of sight but of the other senses also, produced by a visible object when we move our eyes or our body ... may be learned by experience. (1925, p. 533)

Hermann Ludwig Ferdinand von Helmholtz (1821–1894) embedded in a diagram of the eye taken from his *Handbuch der physiologischen Optik* (1867). The portrait is after an engraving made in 1867 from Koenigsberger (1902).

Bibliography

Addams, R. 1834. An account of a peculiar optical phænomenon seen after having looked at a moving body. *London and Edinburgh Philosophical Magazine and Journal of Science 5*:373–374.

Addams, R. 1835. Optische Täuschung nach Betrachtung eines in Bewegung begriffenen Körpers. *Annalen der Physik und Chemie 34*:348.

Aguilonius, F. 1613. *Opticorum Libri Sex. Philosophis juxta ac Mathematicis utiles.* Antwerp: Moreti.

Airy, G. B. 1827. On a peculiar defect in the eye, and a mode of correcting it. *Edinburgh Journal of Science 7*:322–325.

Alberti, L. B. 1435. *Della Pittura.* Florence.

Alberti, L. B. 1966. *On Painting.* J. R. Spencer, trans. New Haven, Conn.: Yale University Press.

Alhazen. 1572. *Opticae Thesaurus.* F. Risner, ed. Basel: Gallia.

Anon. 1797. Eyes. *Encyclopædia Britannica*, ed. 3, Vol. 7. Edinburgh: Bell and MacFarquhar.

Anon. 1972. Purkinje-Phänomen. *Brockhaus Enzyklopädie*, ed. 17, Vol. 15. Wiesbaden: Brockhaus, p. 269.

Ariotti, P. E. 1973. A little known early seventeenth century treatise on vision: Benedetto Castelli's Discorso sopra la Vista. *Annals of Science 30*:1–30.

Attinger, A., M. Godet, and H. Türler. 1932. *Dictionnaire Historique & Biographique de la Suisse*, Vol. 6. Neuchatel: Administration du Dictionnaire Historique et Biographique de la Suisse.

Aubert, H. 1865. *Physiologie der Netzhaut.* Breslau: Morgenstern.

Augé, C., ed. 1898. *Nouveau Larousse Illustré.* Paris: Libraire Larousse.

Averroes (Abu'l Walid ibn Rushd). 1961. *Averroes Epitome of Parva Naturalia.* H. Blumberg, trans. Cambridge, Mass.: Mediaeval Academy of America.

Bacon, F. 1857a. *Sylva Sylvarum: or A Natural History.* In *The Works of Francis Bacon*, Vol. 2. J. Spedding, R. L. Ellis, and D. D. Heath, eds. London: Longman, Simpkin, Hamilton, Wittaker, Bain, Hodgson, Washbourne, Bohn, Richardson, Houlston, Bickers and Bush, Willis and Sotheran, Cornish, Booth, and Snow.

Bacon, F. 1857b. *New Atlantis: A Work unfinished*. In *The Works of Francis Bacon*, Vol. 3. J. Spedding, R. L. Ellis, and D. D. Heath, eds. London: Longman, Simpkin, Hamilton, Wittaker, Bain, Hodgson, Washbourne, Bohn, Richardson, Houlston, Bickers and Bush, Willis and Sotheran, Cornish, Booth, and Snow.

Baird, J. W. 1903. The influence of accommodation and convergence upon the perception of depth. *American Journal of Psychology 14*:150–200.

Barbaro, D. 1569. *La Pratica della Perspettiva*. Venice.

Bartisch, G. 1583. *ΟΦΘΑΛΜΟΔΟΥΛΕΙΑ*. Das ist, Augendienst. Dresden: Stockel.

Baumeister, A., ed. 1885. *Denkmäler des klassischen Altertums*, Vol. 1. Leipzig: Oldenbourg.

Beare, J. I. 1906. *Greek Theories of Elementary Cognition from Alcmaeon to Aristotle*. Oxford: Clarendon.

Beattie, W. 1838. *Caledonia Illustrated in a Series of Views Taken Expressly for the Work of W. H. Bartlett, T. Allom, and Others*. London: Virtue.

Bell, C. 1803. *The Anatomy of the Human Body*, Vol. 3. London: Longman, Rees, Cadell and Davies.

Bell, C. 1823. On the motions of the eye, in illustration of the uses of the muscles and of the orbit. *Philosophical Transactions of the Royal Society 113*:166–186.

Bell, C. 1836. *The Nervous System of the Human Body*, ed. 3. Edinburgh: Adam and Charles Black.

Bell, G. J., ed. 1870. *Letters of Sir Charles Bell, Selected from his Correspondence with his Brother George Joseph Bell*. London: John Murray.

Benham, C. E. 1894. The artificial spectrum top. *Nature 51*:200.

Berkeley, G. 1709. *An Essay towards a New Theory of Vision*, ed. 2. Dublin: Pepyat.

Berkeley, G. 1732. The theory of vision vindicated and explained. *Daily Post-boy*, Sept. 9. Reprinted in A. A. Luce and T. E. Jessop, eds. *The Works of George Berkeley Bishop of Cloyne*, Vol. 1. Edinburgh: Nelson, pp. 251–276.

Bernoulli, J. J. 1882. *Römische Ikonographie*. Munich: Bruckmann.

Bernoulli, J. J. 1901. *Griechische Ikonographie*. Munich: Bruckmann.

Berry, W. 1922. The flight of colors in the after-image of a bright light. *Psychological Bulletin 19*:307–337.

Binet, J.-L., and J. Roger. 1977. *Un autre Buffon*. Paris: Hermann.

Birch, T. 1743. *The Heads of Illustrious Persons of Great Britain*, Vol. 1. London: Knapton.

Birch, T. 1757. *The History of the Royal Society of London*, Vol. 3. London: Millar.

Birch, T., ed. 1966. *Robert Boyle. The Works*, Vol. 1. Hildesheim: Olms.

Birch, T., ed. 1966. *Robert Boyle. The Works*, Vol. 6. Hildesheim: Olms.

Blagden, C. 1813. An appendix to Mr. Ware's paper on vision. *Philosophical Transactions of the Royal Society 103*:110–113.

Boerhaave, H. 1703. *Oratio de usu ratiocinii mechani in medicina*. Lugduni Batavorum: Teering.

Boerhaave, H. 1983. *Boerhaave's Orations*. E. Kegel-Brinkgreve and A. M. Luyendijk-Elshout, trans. Leiden: Leiden University Press.

Borelli, A. 1673. Observations touchant la force inégale des yeux. *Journal des Sçavans 3*:291–294.

Boring, E. G. 1929. *A History of Experimental Psychology*. New York: Appleton-Century.

Boring, E. G. 1942. *Sensation and Perception in the History of Experimental Psychology*. New York: Appleton-Century.

Bouguer, P. 1729. *Essai d'Optique, sur la Gradation de la Lumière*. Paris: Jombert.

Bouguer, P. 1760. *Traité d'Optique sur la Gradation de la Lumière*. Paris: Guerin and Delatour.

Bouguer, P. 1961. *Optical Treatise on the Gradation of Light*. W. E. Knowles Middleton, trans. Toronto: University of Toronto Press.

Bowers, B. 1975. *Sir Charles Wheatstone FRS 1802–1875*. London: Her Majesty's Stationery Office.

Boyer, C. B. 1959. *The Rainbow from Myth to Mathematics*. New York: Yoseloff.

Boyle, R. 1663. *Experiments and Considerations touching Colours*. London: Herringman. See also Birch, T., ed. 1966. *Robert Boyle, The Works*, Vol. 1. Hildesheim: Olms.

Boyle, R. 1688. *A Disquisition about Final Causes of Natural Things: Wherein it is Inquir'd, Whether, And (if at all) With what Cautions, a Naturalist should admit Them? To which is Subjoyn'd, by way of Appendix, Some Uncommon Observations about Vitiated Sight*. London: Taylor. See also Birch, T., ed. 1966. *Robert Boyle, The Works*, Vol. 5. Hildesheim: Olms.

Brewster, D. 1818. On a singular affection of the eye in the healthy state, in consequence of which it loses the power of seeing objects within the sphere of distinct vision. *Annals of Philosophy 11*:151.

Brewster, D. 1824. On the accommodation of the eye to different distances. *Edinburgh Journal of Science 1*:77–83.

Brewster, D. 1825. On some remarkable affections of the retina, as exhibited in its insensibility to indirect impressions, and to the impressions of attenuated light. *Edinburgh Journal of Science 3*:288–293.

Brewster, D. 1826a. On the optical illusion of the conversion of cameos into intaglios, and of intaglios into cameos, with an account of other analogous phenomena. *Edinburgh Journal of Science 4*:99–108.

Brewster, D. 1826b. Farther observations on the supposed optical and physiological discoveries of Mr Charles Bell. *Edinburgh Journal of Science 5*:259–268.

Brewster, D. 1827. Observations by the Editor. *Edinburgh Journal of Science* 7:325–326.

Brewster, D. 1830a. Accidental colours. In *Edinburgh Encyclopædia*, Vol. 1. Edinburgh: Blackwoods, pp. 88–93.

Brewster, D. 1830b. Optics. In *Edinburgh Encyclopædia*, Vol. 15. Edinburgh: Blackwoods, pp. 460–662.

Brewster, D. 1831a. Experiments on ocular spectra produced by the action of the sun's light on the retina. By Sir Isaac Newton. *Edinburgh Journal of Science* 4:75–77.

Brewster, D. 1831b. *A Treatise on Optics*. London: Longman, Rees, Orme, Brown & Green, and Taylor.

Brewster, D. 1832a. On the undulations excited in the retina by the action of luminous points and lines. *London and Edinburgh Philosophical Magazine and Journal of Science* 1:169–174.

Brewster, D. 1832b. *Letters on Natural Magic Addressed to Sir Walter Scott, Bart*. London: John Murray.

Brewster, D. 1833. On the anatomical and optical structure of the crystalline lenses of animals, particularly that of the cod. *Philosophical Transactions of the Royal Society* 123:323–332.

Brewster, D. 1834. On the influence of successive impulses of light upon the retina. *London and Edinburgh Philosophical Magazine and Journal of Science* 4:241–245.

Brewster, D. 1844a. Observations on colour-blindness, or insensibility to the impressions of certain colours. *London and Edinburgh Philosophical Magazine and Journal of Science* 25:134–141.

Brewster, D. 1844b. On the law of visible position in single and binocular vision, and on the representation of solid figures by the union of dissimilar plane pictures on the retina. *Transactions of the Royal Society of Edinburgh* 15:349–368.

Brewster, D. 1844c. On the knowledge of distance given by binocular vision. *Transactions of the Royal Society of Edinburgh* 15:663–674.

Brewster, D. 1855. *Memoires of the Life, Writings, and Discoveries of Sir Isaac Newton*. Edinburgh: Constable.

Brewster, D. 1856. *The Stereoscope. Its History, Theory, and Construction*. London: John Murray.

Briggs, W. 1682. A new theory of vision. *Philosophical Transactions of the Royal Society*. 167–178.

Briggs, W. 1683. A continuation of a discourse on vision, with an examination of some late objections to it. *Philosophical Transactions of the Royal Society*. 171–182.

Brown, T. 1820. *Sketch of a System of the Philosophy of the Human Mind*. Edinburgh: Bell, Bradfute, Manners & Miller, and Waugh & Innes.

Brožek, J. 1989. Contributions to the history of psychology: LII. Purkinje phenomenon: The original and a later account. *Perceptual and Motor Skills* 68:566.

Brožek, J., and V. Kuthan. 1990. Contributions to the history of psychology: LXXI. "Purkinje phenomenon": The genesis and the early uses of the term in French, German, and English. *Perceptual and Motor Skills 71*:1253–1254.

Brožek, J., V. Kuthan, and K. Arens. 1991. Contributions to the history of psychology: LXXXIII. J. E. Purkinje and Mathias Klotz: Who first described "the phenomenon"? *Perceptual and Motor Skills 73*:511–514.

Brucker, J. J. 1741–1755. *Pinacotheca Scriptorum nostra Aetate Literis Illustrium.* Augustae Vindelicorum: Haidius.

Brush, C. B., trans and ed. 1972. *The Selected Works of Pierre Gassendi.* New York: Johnson Reprint.

Buffon, (Leclerc, G. L.) Comte de. 1743a. Dissertation sur les couleurs accidentelles. *Mémoires de Mathématique et de Physique de l'Académie Royale des Sciences* 147–158.

Buffon, (Leclerc, G. L.) Comte de. 1743b. Dissertation sur la cause du strabismes ou des yeux louches. *Mémoires de Mathématique et de Physique de l'Académie Royale des Sciences* 231–248.

Buffon, (Leclerc, G. L.) Comte de. 1749. *Histoire Naturelle, générale et particulière.* Paris: Imprimerie Royale.

Buffon, (Leclerc, G. L.) Comte de. 1774. Supplément a Dissertation sur les couleurs accidentelles. *Mémoires de Mathématique et de Physique de l'Académie Royale des Sciences* [cited by Binet and Roger, 1977].

Buffon, (Leclerc, G. L.) Count of. 1866. *Natural History, General and Particular: Containing the History and Theory of the Earth, a General History of Man, the Brute Creation, Vegetables, Minerals, &c. &c.* Vol. 1. W. Smellie, trans. London: Thomas Kelly.

Bullart, I. 1682. *Academie des Sciences et des Arts, Contenant les Vies, & les Eloges Historiques des Hommes Illustres.* Amsterdam: Elzevier.

Burke, R. B. 1928. *The Opus Majus of Roger Bacon.* Philadelphia: University of Pennsylvania Press.

Burton, H. E. 1945. The optics of Euclid. *Journal of the Optical Society of America 35*:357–372.

Butter, J. 1822. Remarks on the insensibility of the eye to certain colours. *Edinburgh Philosophical Journal 6*:135–140.

Cahan, D., ed. 1994. *Hermann von Helmholtz and the Foundations of Nineteenth-Century Science.* Berkeley: University of California Press.

Campbell, F. W. 1986. In search of the spectrum's elusive yellow. *Ophthalmic and Physiological Optics 6*:129–133.

Cantor, G. 1977. Berkeley, Reid, and the mathematization of mid-eighteenth-century optics. *Journal of the History of Ideas 38*:429–448.

Cantor, G. N. 1984. Brewster on the nature of light. In *"Martyr of Science": Sir David Brewster 1781–1868.* A. D. Morrison-Low and J. R. R. Christie, eds. Edinburgh: Royal Scottish Museum, pp. 67–76.

Carmichael, L. 1926. Sir Charles Bell: A contribution to the history of physiological psychology. *Psychological Review 33*:188–217.

Caulfield, J. 1820. *Portraits, Memoirs, and Characters of Remarkable Personages*, Vol. 4. London: Whitely.

Chérubin d'Orléans. 1671. *La Dioptrique Oculaire, ou la Theorique, la Positive, et la Mechanique, de l'Oculaire Dioptrique en toutes ses Espèces*. Paris: Jolly & Bernard.

Chérubin d'Orléans. 1677. *La Vision Parfaite: ou les Concours des deux Axes de la Vision en un seul Point de l'Objet*. Paris: Marbre-Cramoisy.

Cheselden, W. 1728. An account of some observations made by a young gentleman, who was blind, or lost his sight so early, that he had no rememberance of ever having seen, and was couch'd between 13 and 14 years of age. *Philosophical Transactions of the Royal Society 35*:447–450.

Chevreul, M.-E. 1839. *De la Loi du Contraste Simultané des Couleurs, et de l'Assortiment des Objets Colorés*. Paris: Pitois-Levrault.

Chevreul, M.-E. 1854. *The Principles of Harmony and Contrast of Colours, and Their Applications to the Arts*. C. Martel, trans. London: Longman, Brown, Green, and Longmans.

Choulant, L. 1945. *History and Bibliography of Anatomic Illustration*. M. Frank, trans. New York: Schuman.

Coe, B. 1978. *Cameras. From Daguerreotypes to Instant Pictures*. London: Marshall Cavendish.

Cohen, B. 1984. Erasmus Darwin's observations on rotation and vertigo. *Human Neurobiology 3*:121–128.

Cohen, J., and D. A. Gordon. 1949. The Prevost-Fechner-Benham subjective colors. *Psychological Bulletin 46*:97–133.

Cohen, M. R., and I. E. Drabkin. 1958. *A Source Book in Greek Science*. Cambridge, Mass.: Harvard University Press.

Collins, M. 1925. *Colour-Blindness. With a Comparison of Different Methods of Testing Colour-Blindness*. London: Kegan Paul, Trench, Trubner.

Comrie, J. D. 1932. *History of Scottish Medicine*, ed. 2. London: The Wellcome Historical Medical Museum.

Condillac, Abbé de, 1754. *Traité des Sensations*. Paris.

Condillac, Abbé de, 1982. *Philosophical Writings of Etienne Bonnot, Abbé de Condillac*. F. Philip and H. Lans, trans. Hillsdale, N.J.: Erlbaum.

Coren, S., and C. P. Kaplan. 1973. Patterns of ocular dominance. *American Journal of Optometry and Physiological Optics 50*:283–292.

Coren, S., and C. Porac. 1977. Fifty centuries of right-handedness: The historical record. *Science 198*:631–632.

Crider, B. 1941. Eye-closure facility and eye dominance. *The Journal of Genetic Psychology 58*:425–426.

Crombie, A. C. 1952. *Augustine to Galileo. The History of Science A.D. 400–1650.* London: Falcon.

Crombie, A. C. 1953. *Robert Grosseteste and the Origins of Experimental Science, 1100–1700.* Oxford: Oxford University Press.

Crombie, A. C. 1964. Kepler: De Modo Visionis. In *Mélange Alexandre Koyré I. L'Aventure de la Science.* Paris: Hermann, pp. 135–172.

Crombie, A. C. 1967. The mechanistic hypothesis and the scientific study of vision: Some optical ideas as a background to the invention of the microscope. In S. Bradbury and G. L'E. Turner, eds. *Historical Aspects of Microscopy.* Cambridge: Heffer, pp. 3–112.

Crone, R. A. 1992. The history of stereoscopy. *Documenta Ophthalmologica 81*:1–16.

C.W. 1830. Contributions to the physiology of vision. No I. *Journal of the Royal Institution of Great Britain 1*:101–117.

C.W. 1831. Contributions to the physiology of vision. No II. *Journal of the Royal Institution of Great Britain 1*:534–537.

Dalton, J. 1798. Extraordinary facts relating to the vision of colours: with observations. *Memoirs of the Literary and Philosophical Society of Manchester 5*:28–45.

D'Arcy, Chevalier. 1765. Mémoire sur la durée de la sensation de la vue. *Mémoires de l'Académie des Science de Paris* 439–451.

Darwin, C. 1900. *The Origin of Species by means of Natural Selection or the Preservation of Favoured Races in the Struggle for Life*, ed. 6. London: Murray.

Darwin, E. 1778. A new case in squinting. *Philosophical Transactions of the Royal Society 68*:86–96.

Darwin, E. 1794. *Zoonomia; or, the Laws of Organic Life*, Vol. 1. London: Johnson.

Darwin, R. W. 1786. New experiments on the ocular spectra of light and colours. *Philosophical Transactions of the Royal Society 76*:313–348.

Delambre, J. B. J. 1812. Die Optik des Ptolemäus, verglichen mit der Euclid's, Alhazen's und Vitellio's. *Annalen der Physik 40*:371–388.

Delbrück, R. 1912. *Antike Porträts.* Bonn: Marcus and Weber.

Desaguliers, J. T. 1716. A plain and easy experiment to confirm Sir Isaac Newton's doctrine of the different refrangibility of the rays of light. *Philosophical Transactions of the Royal Society 34*:448–452.

Desaguliers, J. T. 1719. *Lectures of Experimental Philosophy.* London: Mears, Creake, and Sackfield.

Desaguliers, J. T. 1736a. An attempt to explain the phænomenon of the *horizontal moon* appearing bigger, than when elevated many degrees above the *horizon*: Supported by an experiment. *Philosophical Transactions of the Royal Society 39*:390–392.

Desaguliers, J. T. 1736b. An explication of an experiment made in May 1735, as a farther confirmation of what was said in the paper given in January 30, 1734–5. to account for the appearance of the *horizontal moon* seeming larger than when higher. *Philosophical Transactions of the Royal Society 39*:392–394.

Desaguliers, J. T. 1745. *A Course of Experimental Philosophy*. London.

Descartes, R. 1637/1902. *La Dioptrique*. In *Oeuvres de Descartes*, Vol. 6. C. Adam and P. Tannery, eds. Paris: Cerf, pp. 81–228.

des Cartes, Renatus. 1662. *De homine figuris et latinitate*. F. Schuyl, trans. Lugduni Batavorum: Leffen & Moyardum.

Descartes, R. 1664/1909. *Traité de l'Homme*. In *Oeuvres de Descartes*, Vol. 11. C. Adam and P. Tannery, eds. Paris: Cerf, pp. 119–215.

Descartes, R. 1677. *Tractatus de Homine et de Formatione Foetus*. L. de la Forge, illus. Amsterdam: Elsevir.

Descartes, R. 1965. *Discourse on Method, Optics, Geometry, and Meteorology*. P. J. Olscamp, trans. Indianapolis: Bobbs-Merrill.

Dewan, L. 1980. St. Albert, the sensibles, and spiritual being. In *Albertus Magnus and the Sciences. Commemorative Essays 1980*. J. A. Weisheipl, ed. Toronto: Pontifical Institute of Mediaeval Studies, pp. 291–320.

Diderot, D. 1749. *Lettre sur les Aveugles*. London.

Diderot, D. 1750. *An Essay on Blindness, in a Letter to a Person of Distinction*. Trans. from the French. London: Barker.

Digby, K. 1644. *Two Treatises*. Paris: Blaizot.

Donders, F. C. 1864. *On the Anomalies of Accommodation and Refraction of the Eye*. W. D. Moore, trans. Boston: Milford House.

Dove, H. W. 1841. Die Combination der Eindrücke beider Ohren und beider Augen zu einem Eindruck. *Monatsberichte der Berliner preussische Akademie der Wissenschaften* 41:251.

Dove, H. W. 1851. Beschreibung mehrerer Prismenstereoskope und eines einfachen Spiegelstereoskops. *Annalen der Physik und Chemie* 83:183–189.

Duke-Elder, S. 1961. *System of Ophthalmology*, Vol. 2. *The Anatomy of the Visual System*. London: Kimpton.

Duke-Elder, S. 1968. *System of Ophthalmology*, Vol. 4. *The Physiology of the Eye and of Vision*. London: Kimpton.

Duke-Elder, S., and D. Abrams. 1970. *System of Ophthalmology*, Vol. 5. *Ophthalmic Optics and Refraction*. London: Kimpton.

Duke-Elder, S., and K. Wybar. 1973. *System of Ophthalmology*, Vol. 6. *Ocular Motility and Strabismus*. London: Kimpton.

Du Laurens, A. 1613. *Toutes les oeuvres de M$^{E.}$ André du Laurens*. Paris: Mettayer.

Dumesuil, R., and F. Bonnet-Roy. 1947. *Les Medecins Celebres*. Paris: Lucien Mazenod.

Duncan, G. 1984. Brewster's contribution to the study of the lens of the eye: An experimental foundation of modern biophysics. In *"Martyr of Science": Sir David Brewster 1781–1868*. A. D. Morrison-Low and J. R. R. Christie, eds. Edinburgh: Royal Scottish Museum, pp. 101–103.

Durand, A. C., and G. M. Gould. 1910. A method of determining ocular dominance. *Journal of the American Medical Association 55:*369–370.

Du Tour, E.-F. 1760. Discussion d'une question d'optique. *Académie des Sciences. Mémoires de Mathématique et de Physique Présentés pars Divers Savants 3:*514–530.

Du Tour, E.-F. 1761. Addition au Mémoire intitulé. Discussion d'une question d'optique. *Académie des Sciences. Mémoires de Mathématique et de Physique Présentés pars Divers Savants 4:*499–511.

Eastwood, B. 1979. Al-Farabi on extramission, intromission, and the use of Platonic visual theory. *Isis 70*:423–425.

Eder, J. M. 1945. *History of Photography.* E. Epstean, trans. New York: Columbia University Press.

Edgell, B., and W. L. Symes. 1906. The Wheatstone-Hipp chronoscope: Its adjustments, accuracy, and control. *British Journal of Psychology 2*:58–87.

Edgerton, S. Y. 1975. *The Renaissance Rediscovery of Linear Perspective.* New York: Basic Books.

Elliott, J. 1780. *Philosophical Observations on the Senses of Vision and Hearing.* London: Murray.

Elliott, J. 1782. *Elements of the Branches of Natural Philosophy connected with Medicine.* London: Johnstone.

Elliott, J. 1786. *Experiments and Observations on Light and Colours: to which is prefixed, the Analogy between Heat and Motion.* London: Johnson.

Emmert, E. 1881. Grössenverhältnisse der Nachbilder. *Klinische Merkblatt der Augenheilkunde 19*:443–450.

Erb, M. B., and K. M. Dallenbach. 1939. "Subjective" colors from line patterns. *American Journal of Psychology 52*:227–241.

Euler, L. 1769. *Dioptricae Pars Prima.* Petropoli.

Exner, S. 1894. *Entwurf zu einer physiologischen Erklärung der physischen Erscheinungen.* Vienna: Deuticke.

Faraday, M. 1831a. On a peculiar class of optical deception. *Journal of the Royal Institution of Great Britain 1*:205–223.

Faraday, M. 1831b. Ueber eine besondere Klasse von optischen Täuschungen. *Annalen der Physik und Chemie 22*:601–606.

Fechner, G. T. 1838. Ueber eine Scheibe zur Erzeugung subjectiver Farben. *Annalen der Physik und Chemie 45*:227–232.

Fechner, G. T. 1840. Ueber die subjectiven Nachbilder und Nebenbilder. *Annalen der Physik und Chemie 50*:193–221.

Fechner, G. T. 1860. *Elemente der Psychophysik.* Leipzig: Breitkopf & Härtel.

Ferrier, D. 1876. *The Functions of the Brain.* London: Smith, Elder.

Finger, S. 1994. *Origins of Neuroscience. A History of Explorations into Brain Function.* New York: Oxford University Press.

Franklin, B. 1765/1970. Letter to Lord Kames. In A. H. Smyth, ed. *The Writings of Benjamin Franklin. Volume IV. 1760–1766.* New York: Haskell House.

Franklin, B. 1785/1970. Letter to George Whatley. In A. H. Smyth, ed. *The Writings of Benjamin Franklin. Volume IX. 1783–1788.* New York: Haskell House.

Franz, S. I. 1899. After-images. *Psychological Review. Monograph Supplements 3*: 1–61.

Fraser, A. C. 1871. *Life and Letters of George Berkeley, D.D. formerly Bishop of Cloyne; and an Account of his Philosophy.* Oxford: The Clarendon Press.

Freher, P. 1688. *Theatrum Virorum Eruditione Clarorum.* Noribergae: Hofmann.

Fulton, J. F., ed. 1966. *Selected Readings in the History of Physiology*, ed. 2. Springfield, Ill.: Thomas.

Galen, C. 1550. *De Usu Partium Corporis Humani, Libri XVII.* N. R. Calabro, trans. Lugduni: Rouillium.

Gall, F. J. 1822–1825. *Sur les Fonctions du Cerveau et sur Chacune de ses Parties*, 6 vols. Paris.

Gall, F. J. 1835a. *On the Functions of the Brain and Each of its Parts, with Observations on the Possibility of Determining the Instincts, Propensities and Talents, and the Moral and Intellectual Dispositions of Men and Animals by the Configuration of the Brain and Head*, Vol. 1. *On the Origin of the Moral Qualities and Intellectual Faculties of Man, and the Conditions of their Manifestation.* W. Lewis, trans. Boston: Marsh, Capen and Lyon.

Gall, F. J. 1835b. *On the Functions of the Brain and Each of its Parts, with Observations on the Possibility of Determining the Instincts, Propensities and Talents, and the Moral and Intellectual Dispositions of Men and Animals by the Configuration of the Brain and Head*, Vol. 5. *Organology; or, an Exposition of the Instincts, Propensities, Sentiments, and Talents, or of the Moral Qualities, and the Fundamental Intellectual Faculties in Man and Animals, and the Seat of their Organs.* W. Lewis, trans. Boston: Marsh, Capen and Lyon.

Gall, F. J. 1835c. *On the Functions of the Brain and Each of its Parts, with Observations on the Possibility of Determining the Instincts, Propensities and Talents, and the Moral and Intellectual Dispositions of Men and Animals by the Configuration of the Brain and Head*, Vol. 6. *Critical Review of some Anatomico-Physiological Works; with an Explanation of a New Philosophy of the Moral Qualities and Intellectual Faculties.* W. Lewis, trans. Boston: Marsh, Capen and Lyon.

Gall, F. J., and J. C. Spurzheim. 1810. *Anatomie et Physiologie du Système Nerveux en general, et du Cerveau en particulier*, Vol. 1. Paris: Schoell.

Garrison, F. H. 1914. *An Introduction to the History of Medicine.* London: Saunders.

Gauss K. F. 1830/1880. *Briefwechsel zwischen Gauss und Bessel.* Leipzig: Engelmann.

Gibson, J. J. 1966. *The Senses considered as Perceptual Systems.* Boston: Houghton Mifflin.

Gmelin, P. F. 1744. De fallaci visione per microscopia composita notata nonnulla continens. *Philosophical Transactions of the Royal Society 43*:382–391.

Goethe, J. W. 1810. *Zur Farbenlehre*. Tübingen: Cotta.

Goethe, J. W. 1840. *Theory of Colours*. C. L. Eastlake, trans. London: John Murray.

Gordon, B. L. 1949. *Medicine throughout Antiquity*. Philadelphia: Davis.

Gordon, Mrs. 1869. *The Home Life of Sir David Brewster*. Edinburgh: Edmonston and Douglas.

Grant, E. ed. 1974. *A Source Book in Medieval Science*. Cambridge, Mass.: Harvard University Press.

Gregory, R. 1996. What are illusions? *Perception 25*:503–504.

Gregory, R. L. 1981. *Mind in Science*. London: Weidenfeld and Nicolson.

Grimaldi, F. M. 1665. *Physico-mathemis de Lumine coloribis et iride*. Bononiae.

Grüsser, O.-J. 1984. J. E. Purkyně's contributions to the physiology of the visual, the vestibular and the oculomotor systems. *Human Neurobiology 3*:29–144.

Grüsser, O.-J., and M. Hagner. 1990. On the history of deformation phosphenes and the idea of internal light generated in the eye for the purpose of vision. *Documenta Ophthalmologica 74*:57–85.

Grüsser, O.-J., and T. Landis. 1991. *Vision and Visual Dysfunction*, Vol. 12. *Visual Agnosias and other Disturbances of Visual Perception and Cognition*. London: Macmillan.

Gulick, W. L., and R. B. Lawson. 1976. *Human Stereopsis: A Psychophysical Approach*. Oxford: Oxford University Press.

Gunther, R. T. 1930. *Early Science in Oxford*. Vol. 7. *The Life and Work of Robert Hooke (Part II)*. Oxford: Printed for the author.

Gunther, R. T. 1937. *Early Science in Oxford*. Vol. 11. *Oxford Colleges and their Men of Science*. Oxford: Printed for the author.

Hall, T. S. 1972. *Treatise of Man. René Descartes*. Cambridge, Mass.: Harvard University Press.

Haller, A. von. 1767. *Primae Lineae Physiologiae*, ed. 3. Goettingen: Vandenhoeck.

Haller, A. von. 1786. *First Lines of Physiology*. Vol. 2. W. Cullen, trans. Edinburgh: Elliot.

Hammond, J. H. 1981. *Tha Camera Obscura. A Chronicle*. Bristol: Hilger.

Harris, J. 1775. *A Treatise of Optics: Containing Elements of the Science; in two Books*. London: White.

Hartenstein, G., ed. 1853. *Immanuel Kant's Kritik der reinen Vernunft*. Leipzig: Voss.

Hartley, D. 1749. *Observations on Man, his Frame, his Duty, and his Expectations*. London: Richardson.

Hartley, D. 1791. *Observations on Man, his Frame, his Duty, and his Expectations*. London: Johnson.

Hayter, C. 1826. *A New Practical Treatise on the Three Primitive Colours*. London: Printed for the author.

Hekler, A. 1912. *Greek and Roman Portraits*. London: Heinemann.

Held, R. 1976. Single vision with doubled images: An historic problem. In M. Henle, ed. *Vision and Artifact*. New York: Springer, pp. 5–18.

Heller, D. 1988. On the history of eye movement recording. In G. Lüer, U. Lass, and J. Shallo-Hoffmann, eds. *Eye Movement Research. Physiological and Psychological Aspects*. Göttingen: Hogrefe, pp. 37–51.

Helmholtz, H. 1852. On Sir David Brewster's new analysis of solar light. *London, Edinburgh and Dublin Philosophical Magazine and Journal of Science 4*:401–416.

Helmholtz, H. 1857. Das Telestereoskop. *Annalen der Physik und Chemie 101*:494–496.

Helmholtz, H. 1867. *Handbuch der physiologischen Optik*. In G. Karsten, ed. *Allgemeine Encyklopädie der Physik*, Vol. 9. Leipzig: Voss.

Helmholtz, H. 1873. *Popular Lectures on Scientific Subjects*. E. Atkinson, trans. London: Longmans, Green.

Helmholtz, H. von. 1895. *Popular Lectures on Scientific Subjects*, First Series, New Impression. E. Atkinson, trans. London: Longmans, Green.

Helmholtz, H. von. 1896. *Handbuch der physiologischen Optik*, ed. 2. Hamburg: Voss.

Helmholtz, H. von. 1925. *Helmholtz's Treatise on Physiological Optics*, Vol. 3. J. P. C. Southall, trans. New York: Optical Society of America.

Hering, E. 1862. *Beiträge zur Physiologie*, Vol. 2. Leipzig: Engelmann.

Hering, E. 1865. Die Gesetze der binocularen Tiefenwahrnehmung. *Archiv für Anatomie, Physiologie und Wissenschaftliche Medicin* 152–165.

Hering, E. 1868. *Die Lehre vom binocularen Sehen*. Leipzig: Engelmann.

Herrnstein, R. J., and E. G. Boring, eds. 1965. *A Source Book in the History of Psychology*. Cambridge, Mass: Harvard University Press.

Herschel, J. F. W. 1833. *A Treatise on Astronomy*. London: Longman, Rees, Orme, Brown, Green & Longman.

Herschel, W. 1786. Investigation of the cause of that indistinctness of vision which has been ascribed to the smallness of the optic pencil. *Philosophical Transactions of the Royal Society 76*:500–507.

Hershenson, M. 1989. That most puzzling illusion. In *The Moon Illusion*. M. Hershenson, ed. Hillsdale, N.J.: Erlbaum, pp. 1–3.

Hill, E. 1915. History of eyeglasses and spectacles. In C. A. Wood, ed. *The American Encyclopedia and Dictionary of Ophthalmology*, Vol. 7. Chicago: Cleveland Press, pp. 4894–4953.

Hippocrates. 1923a. *Hippocrates*, Vol. 2. W. H. S. Jones, trans. London: Heinemann.

Hippocrates. 1923b. *Hippocrates*, Vol. 4. W. H. S. Jones, trans. London: Heinemann.

Hirsch, A. 1877. *Geschichte der Augenheilkunde*. Leipzig: Engelmann.

Hirschberg, J. 1899. Geschichte der Augenheilkunde. In *Graefe-Saemisch Handbuch der gesamten Augenheilkunde*, Vol. 12. Leipzig: Engelmann.

Hirschberg, J. 1908. Geschichte der Augenheilkunde. In *Graefe-Saemisch Handbuch der gesamten Augenheilkunde*, Vol. 13. Leipzig: Engelmann.

Hirschberg, J. 1911a. Geschichte der Augenheilkunde. In *Graefe-Saemisch Handbuch der gesamten Augenheilkunde*, Vol. 14, Pt. 1. Leipzig: Engelmann.

Hirschberg, J. 1911b. Geschichte der Augenheilkunde. In *Graefe-Saemisch Handbuch der gesamten Augenheilkunde*, Vol. 14, Pt. 2. Leipzig: Engelmann.

Hirschberg, J. 1912. Geschichte der Augenheilkunde. In *Graefe-Saemisch Handbuch der gesamten Augenheilkunde*, Vol. 14, Pt. 3. Leipzig: Engelmann.

Hobbes, T. 1640/1840. *Human Nature*. In W. Molesworth, ed. *The English Works of Thomas Hobbes. Volume IV*. London: Bohn.

Hobbes, T. 1651. *Leviathan: or, the Matter, Form, and Power of Commonwealth, Ecclesiastical and Civil*. London: Andrew Crooke.

Hobbes, T. 1651/1839. *Leviathan*. In W. Molesworth, ed. *The English Works of Thomas Hobbes. Volume III*. London: Bohn.

Hobbes, T. 1655/1839. *Elements of Philosophy*. In W. Molesworth, ed. *The English Works of Thomas Hobbes. Volume III*. London: Bohn.

Hochberg, J. E. 1963. Nativism and empiricism in perception. In L. Postman, ed. *Psychology in the Making. Histories of Selected Research Problems*. New York: Knopf, pp. 255–330.

Home, E. 1794. Some facts relating to the late Mr. John Hunter's preparation for the Croonian lecture. *Philosophical Transactions of the Royal Society 84*:21–27.

Home, E. 1795. The Croonian lecture on muscular motion. *Philosophical Transactions of the Royal Society 85*:1–23.

Home, E. 1814. *Lectures on Comparative Anatomy; in which are explained the Preparations in the Hunterian Collection*, Vol. 1. London: Nicol.

Hooke, R. 1665. *Micrographia: or some Physiological Descriptions of Minute Bodies made by Magnifying Glasses with Observations and Inquiries thereupon*. London: Martyn and Allestry.

Hooke, R. 1668. A contrivance to make the picture of any thing appear on a wall, cub-board, or within a picture-frame, &c. in the midst of a light room in the day time; or in the night-time in any room that is enlightened with a considerable number of candles. *Philosophical Transactions of the Royal Society 2*:741–743.

Hoorn, W. van. 1972. *As Images Unwind. Ancient and Modern Theories of Visual Perception*. Amsterdam: University Press.

Horner, W. G. 1834a. On the properties of the *Dædaleum*, a new instrument of optical illusion. *London and Edinburgh Philosophical Magazine and Journal of Science 4*:36–41.

Horner, W. G. 1834b. On the autoptic spectrum of certain vessels within the eye, as delineated in shadows on the retina. *London and Edinburgh Philosophical Magazine and Journal of Science 4*:262–271.

Hosack, D. 1794. Observations on vision. *Philosophical Transactions of the Royal Society 84*:196–216.

Howard, I. P. 1996. Alhazen's neglected discoveries of visual phenomena. *Perception 25*:1203–1217.

Howard, I. P., and B. J. Rogers. 1995. *Binocular Vision and Stereopsis*. New York: Oxford University Press.

Howard, I. P., and N. J. Wade. 1996. Ptolemy's contributions to the geometry of binocular vision. *Perception 25*:1189–1202.

Huddart, J. 1777. An account of persons who could not distinguish colours. *Philosophical Transactions of the Royal Society 67*:260–265.

Hueck, A. 1838. *Die Achsendrehung des Auges*. Dorpat: Kluge.

Humboldt, A. von. 1850. *Kosmos. Entwurf einer physischen Weltbeschreibung*, Vol. 3. Stuttgart: Cotta.

Hunt, D. M., K. S. Dulai, J. K. Bowmaker, and J. D. Mollon. 1995. The chemistry of John Dalton's color blindness. *Science 267*:984–988.

Hunter, J. 1786. *Observations on certain Parts of the Animal Oeconomy*. London.

Hunter, J. 1794. Letter to Sir Joseph Banks. *Philosophical Transactions of the Royal Society 84*:23–25.

Huygens, C. 1653. *Dioptrique*. La Haye: Nijhoff.

Huygens, C. 1673. An Extract of a Letter lately written by an ingenious person from Paris, containing some Considerations on Mr. Newtons doctrine of Colors, as also upon the effects of the Refractions of the Rays in Telescopical Glasses. *Philosophical Transactions of the Royal Society 8*:6086–6087.

Huygens, C. 1690. *Traité de la Lumiere*. Leiden: Vander.

Huygens, C. 1912. *Treatise on Light*. S. P. Thompson, trans. London: Macmillan.

Hykes, O. V. 1936. Jan Evangelista Purkyně [Purkinje] [1787–1869]. *Osiris 2*:463–471.

James, R. R. 1932. William Briggs, M. D. (1650–1704). *British Journal of Ophthalmology 16*:360–368.

James, R. R. 1933. *Studies in the History of Ophthalmology in England Prior to the Year 1800*. Cambridge: University Press.

Jeffries, B. J. 1883. *Color-Blindness: Its Dangers and its Detection*. Boston: Houghton Mifflin.

Johannsen, D. E. 1971. Early history of perceptual illusions. *Journal of the History of the Behavioral Sciences 7*:127–140.

John, H. J. 1959. *Jan Evangelista Purkyně. Czech Scientist and Patriot 1787–1869*. Philadelphia: American Philosophical Society.

Julesz, B. 1971. *Foundations of Cyclopean Perception*. Chicago: University of Chicago Press.

Jurin, J. 1738. An essay on distinct and indistinct vision. In R. Smith *A Compleat System of Opticks in Four Books*. Cambridge, pp. 115–171.

Kant, I. 1781. *Critik der reinen Vernunft*. Riga: Hartknoch.

Kant, I. 1929. *Immanuel Kant's Critique of Pure Reason*. N. K. Smith, trans. London: Macmillan.

Keele, K. D. 1955. Leonardo da Vinci on vision. *Proceedings of the Royal Society of Medicine 48*:384–390.

Kemp, M., ed. 1989. *Leonardo on Painting*. New Haven, Conn.: Yale University Press.

Kemp, M. 1990. *The Science of Art. Optical Themes in Western Art from Brunelleschi to Seurat*. New Haven, Conn.: Yale University Press.

Kepler, J. 1604. *Ad Vitellionem Paralipomena*. Frankfurt: Marinium and Aubrii.

Kepler, J. 1611. *Dioptrice*. Augsburg: Franci.

Kirby, J. 1755. *Dr. Brook Taylor's Method of Perspective made easy, both in Theory and Practice. In Two Books*, ed. 2. Ipswich.

Kircher, A. 1646. *Ars Magna Lucis et Umbrae*. Rome: Scheus.

Klotz, M. 1806. *Meldung einer Farbenlehre und eines Farben-Systems*. Munich: Scherer.

Klotz, M. 1816. *Gründliche Farbenlehre*. Munich: Lindauer'schen Buchhandlung.

Knight, C. 1833a. *The Gallery of Portraits: with Memoirs*, Vol. 1. London: Charles Knight.

Knight, C. 1833b. *The Gallery of Portraits: with Memoirs*, Vol. 2. London: Charles Knight.

Knight, C. 1834. *The Gallery of Portraits: with Memoirs*, Vol. 3. London: Charles Knight.

Knight, C. 1835a. *The Gallery of Portraits: with Memoirs*, Vol. 4. London: Charles Knight.

Knight, C. 1835b. *The Gallery of Portraits: with Memoirs*, Vol. 5. London: Charles Knight.

Knight, C. 1837. *The Gallery of Portraits: with Memoirs*, Vol. 7. London: Charles Knight.

Koelbing, H. M. 1967. *Renaissance der Augenheilkunde. 1540–1630*. Bern: Huber.

Koelbing, H. M. 1968. Ocular physiology in the seventeenth century and its acceptance by the medical profession. In G. Scheiz, ed. *Analecta Medico-Historica 3*. Oxford: Pergamon, pp. 219–225.

Koenigsberger, L. 1902. *Hermann von Helmholtz*, Vol. 1. Brauschweig: Vieweg.

Kuntze, J. E. 1892. *Gustav Theodor Fechner (Dr. Mises). Ein deutsches Gelehrtenleben*. Leipzig: Breitkopf and Härtel.

La Hire, P. de. 1685. Dissertation sur la conformation de l'oeil. *Journal des Sçavans* *13*:335–363, 398–405.

La Hire, P. de. 1694. Dissertation sur les differens accidens de la vue. In *Mémoires de Mathématique et de Physique*. Paris: Anisson, pp. 233–302.

La Hire, P. de. 1709/1742. An explanation of some facts in opticks, and the manner in which vision is performed. *The Philosophical History and Memoirs of the Royal Academy of Sciences at Paris*, Vol. 3. J. Martyn and E. Chambers, trans. London: Knapton and Nourse, pp. 190–200.

La Hire, P. de. 1711/1742. Remarks on some colours. *The Philosophical History and Memoirs of the Royal Academy of Sciences at Paris*, Vol. 4. J. Martyn and E. Chambers, trans. London: Knapton and Nourse, pp. 133–135.

Lambert, J. H. 1760. *Photometria, sive de Mensura et Gradibus Luminis Colorum, et Umbrae.* Augsburg.

Landon, C. P. 1805. *Galerie Historique des Hommes le plus Célèbres de tous les Nations.* Paris.

Langguth, G. A. 1742. *De Luce ex Pressione Oculi.* Wittenberg: Eichsfeld.

Laurentius, A. 1599/1938. *A Discourse of the Preservation of the Sight: of Melancholic Diseases; of Rheumes, and of Old Age.* R. Surphlet, trans. London: The Shakespeare Association.

le Blon, J. C. 1720. *Il Coloritto, or the Harmony of Colouring in Painting.* London.

Le Cat, C. N. 1744. *Traité des Sens.* Amsterdam: Wetstein.

Le Cat, C. N. 1750. *A Physical Essay on the Senses*, trans. from the French. London: Griffiths.

Le Cat, C. N. 1767. *Traité des Sensations et des Passions en général*, Vol. 2. Paris: Vallat-la-Chapelle.

Le Clerc, G., Count of Buffon. 1866. *Natural History, General and Particular: Containing the History and Theory of the Earth, a General History of Man, the Brute Creation, Vegetables, Minerals, &c. &c.*, Vol. 1. W. Smellie, trans. London: Thomas Kelly.

Le Clerc, S. 1679. *Discours Touchant de Point de Veue, dans lequel il es prouvi que les chose qu'on voit distinctement, ne sont veues que d'un oeil.* Paris: Jolly.

Le Clerc, S. 1712. *Système de la Vision.* Paris: Delaulne.

Leeuwenhoek, A. van. 1674. More observations from Mr. Leewenhook ... *Philosophical Transactions of the Royal Society 9*:178–182.

Leeuwenhoek, A. van. 1675. Microscopical observations from Mr. Leewenhoeck, concerning the optick nerve ... *Philosophical Transactions of the Royal Society 9*:378–380.

Lejeune, A. 1948. *Euclide et Ptolémée. Deux Stades de L'Optique Geométrique Grecque.* Louvain: Université de Louvain.

Lejeune, A. ed., 1956. *L'Optique de Claude Ptolémée dans la version Latine d'après l'Arabe de l'Émir Eugène de Sicile.* Louvain: Université de Louvain.

Lejeune, A. ed. and trans., 1989. *L'Optique de Claude Ptolémée dans la version Latine d'après l'Arabe de l'Émir Eugène de Sicile.* Leiden: Brill.

Lenard, P. 1933. *Great Men of Science.* H. S. Hatfield, trans. London: Bell.

Leonardo da Vinci, 1721. *A Treatise of Painting.* London: Senex and Taylor.

Lindberg, D. C. 1967. Alhazen's theory of vision and its reception in the West. *Isis* *58*:321–341.

Lindberg, D. C., ed. and trans. 1970. *John Pecham and the Science of Optics.* Madison: University of Wisconsin Press.

Lindberg, D. C. 1974. Late thirteenth-century synthesis in optics. In E. Grant, ed. *A Source Book in Medieval Science.* Cambridge, Mass.: Harvard University Press, pp. 392–435.

Lindberg, D. C. 1976. *Theories of Vision from Al-Kindi to Kepler.* Chicago: University of Chicago Press.

Lindberg, D. C. 1978. The science of optics. In D. C. Lindberg, ed. *Science in the Middle Ages.* Chicago: University of Chicago Press, pp. 338–368.

Lindberg, D. C. 1983a. *Studies in the History of Medieval Optics.* London: Variorum Reprints.

Lindberg, D. C. 1983b. *Roger Bacon's Philosophy of Nature.* Oxford: Clarendon.

Lindberg, D. C., and N. H. Steneck. 1972. The sense of vision and the origins of modern science. In A. G. Debus, ed. *Science, Medicine and Society in the Renaissance. Essays to Honor Walter Pagel*, Vol. 1. New York: Science History Publications, pp. 29–45.

Locke, J. 1690. *An Essay Concerning Humane Understanding.* London: Basset.

Locke, J. 1694. *An Essay Concerning Humane Understanding*, ed. 2. London: Awnsham, Churchill, and Manship.

Lohne, J. A. 1972. Thomas Harriot. In C. C. Gillespie, ed. *Dictionary of Scientific Biography*, Vol. 6, New York: Scribner, pp. 121–129.

Lomonosow, M. 1757. *Oratio de Origine Lucis, sistens Novam Theorium Colorum.* Petropoli.

Lomonosov, M. 1970. Oration on the origin of light. A new theory of color. In *Mikhail Vasil'evich Lomonosov on the Corpuscular Theory.* H. M. Leicester, trans. Cambridge, Mass.: Harvard University Press, pp. 247–269.

Lucretius. 1975. *De Rerum Natura.* W. H. D. Rouse, trans. Cambridge, Mass: Harvard University Press.

MacCurdy, E. 1938a. *The Notebooks of Leonardo da Vinci*, Vol. 1. London: Cape.

MacCurdy, E. 1938b. *The Notebooks of Leonardo da Vinci*, Vol. 2. London: Cape.

Mach, E. 1926. *The Principles of Physical Optics. An Historical and Philosophical Treatment.* J. S. Anderson and A. F. A. Young, trans. London: Methuen.

Mackenzie, W. 1830. *A Practical Treatise on the Diseases of the Eye*, London: Longman, Rees, Orme, Brown, Green, & Longman.

Mackenzie, W. 1835. *A Practical Treatise on the Diseases of the Eye*, ed. 2. London: Longman, Rees, Orme, Brown, Green, & Longman.

McMurrich, J. P. 1930. *Leonardo da Vinci the Anatomist (1452–1519)*. Baltimore: Williams & Wilkins.

Magnus, H. 1901. *Die Augenheilkunde der Alten*. Breslau: Kern.

Malebranche, N. 1674. *De la Recherche de la Vérité*. Paris: Pralard.

Malebranche, N. 1699/1742. Reflections on light and colours, and on the generation of fire. *The Philosophical History and Memoirs of the Royal Academy of Sciences at Paris*, Vol. 1. J. Martyn and E. Chambers, trans, London: Knapton and Nourse, pp. 32–46.

Malebranche, N. 1700. *Treatise concerning the Search after Truth*. T. Taylor, trans. London: Bennet, Leigh and Midwinter.

Malebranche, N. 1712. *De la Recherche de la Vérité*, ed. 6. Paris: Pralard.

Malebranche, N. 1980. *The Search after Truth*. T. M. Lennon and P. J. Olscamp, trans. Columbus: Ohio State University Press.

Malus, E. L. 1810. *Théorie de la Double Réfraction de la Lumière dans les Substances Cristallisées*. Paris: Baudouin.

Mariotte, E. 1668. A new discovery touching vision. *Philosophical Transactions of the Royal Society 3*:668–669.

Mariotte, E. 1670. The answer of Monsieur Marriotte to Monsieur Pecquet, concerning the principal organ of vision; where occurr divers considerable experiments. *Philosophical Transactions of the Royal Society 5*:1023–1042.

Mariotte, E. 1681. *De la Nature des Couleurs*. Paris: Michallet.

Mariotte, E. 1683. An account of two letters of Mr. Perrault and Mr. Mariotte, concerning vision. *Philosophical Transactions of the Royal Society 13*:266–267.

Mariotte, E. 1717. *Oeuvres de Mr. Mariotte*. Leiden: Vander.

Matousek, O. 1961. J. E. Purkyněs Leben und Tätigkeit im Lichte der Berliner und Prager Archive. *Nova Acta Leopoldina 24*:109–129.

Maurolico, F. 1611. *Photismi de Lumine et Umbrae*. Naples.

Maurolico, F. 1940. *The Photismi de Lumine of Maurolycus: A Chapter in Late Medieval Optics*. H. Crew, trans. New York: Macmillan.

May, M. T. 1968. *Galen. On the Usefulness of the Parts of the Body*. Ithaca, N.Y.: Cornell University Press.

Mayor, A. H. 1946. The photographic eye. *Bulletin of the Metropolitan Museum of Art 5*:15–26.

Melvill, T. 1756. *Edinburgh Philosophical Society. Essays and Observations, Physical and Literary 2*:12–90.

Meyerhof, M. 1928. *The Book of the Ten Treatises on the Eye, Ascribed to Hunain ibn Is-Hâq (809–877 A.D.)*. Cairo: Government Press.

Meyering, T. C. 1989. *Historical Roots of Cognitive Science*. Dordrecht: Kluwer.

Minnaert M, 1940. *Light and Colour in the Open Air*. H. M. Kremer-Priest, trans. London: Bell.

Molesworth, W., ed. 1839. *The English Works of Thomas Hobbes*, Vol. 1. London: Bohn.

Mollon, J. 1985. Colourful notions. Studies in scarlet. *The Listener* 10 January, pp. 6–7.

Mollon, J. D. 1987. John Elliot MD (1747–1787). *Nature 329*:19–20.

Mollon, J. D. 1989. "Tho' she kneel'd in that place where they grew … " The uses and origins of primate colour vision. *Journal of Experimental Biology 146*:21–38.

Mollon, J. D., and A. J. Perkins. 1996. Errors of judgement at Greenwich in 1796. *Nature 380*:101–102.

Molyneux, W. 1687. Concerning the apparent magnitude of the sun and moon, or the apparent distance of two stars, when nigh the horizon, and when higher elevated. *Philosophical Transactions of the Royal Society 16*:314–323.

Molyneux, W. 1692. *Dioptrica Nova. A Treatise of Dioptricks in two Parts*. London: Tooke.

Montaigne, M. 1580. *Les Essais de Messire Michel, Seigneur de Montaigne*. Bordeaux: Millanges.

Montaigne, M. 1853. *The Works of Michael de Montaigne*, ed. 3. W. Hazlitt, trans. London: Templeman.

Montucla, J. E. 1758. *Histoire des Mathematiques*. Vol. 2. Paris: Jombert.

Morgan, M. J. 1977. *Molyneux's Question. Vision, Touch and the Philosophy of Perception*. Cambridge: Cambridge University Press.

Morgagni, J. B. 1719. *Adversaria Anatomica Omnia*. Padua: Cominus.

Morgagni, J. B. 1761. *De sedibus, et causis morborum per anatomen indagatis. Libri quinque*, Vol. 1. Venice: Remondinianis.

Morgagni, J. B. 1769. *The Seats and Causes of Diseases Investigated by Anatomy; in Five Books*, Vol. 1. B. Alexander, trans. London: Millar, Cadell, Johnson, and Payne.

Müller, J. 1826a. *Zur vergleichenden Physiologie des Gesichtssinnes des Menschen und der Thiere, nebst einen Versuch über die Bewegung der Augen und über den menschlichen Blick*. Leipzig: Cnobloch.

Müller, J. 1826b. *Über die phantastischen Gesichtserscheinungen*. Coblenz: Hölscher.

Müller, J. 1834. *Handbuch der Physiologie des Menschen*, Vol. 1. Coblenz: Hölscher.

Müller, J. 1838. *Handbuch der Physiologie des Menschen*, Vol. 2. Coblenz: Hölscher.

Müller, J. 1840. *Elements of Physiology*, Vol. 1. W. Baly, trans. London: Taylor and Walton.

Müller, J. 1843. *Elements of Physiology*, Vol. 2. W. Baly, trans. London: Taylor and Walton.

Munk, H. 1879. Physiologie der Sehsphäre der Grosshirnrinde. *Centralblatt für praktische Augenheilkunde 3*: 255–266.

Muspratt, S. 1853. *Chemistry Theoretical, Practical & Analytic, as Applied and Relating to the Arts and Manufactures*. Glasgow: Mackenzie.

Nagel, A. 1861. *Das Sehen mit zwei Augen und die Lehre von den identischen Netzhautstellen*. Leipzig: Winter.

Necker, L. A. 1832. Observations on some remarkable phenomena seen in Switzerland; and an optical phenomenon which occurs on viewing a figure of a crystal or geometrical solid. *London and Edinburgh Philosophical Magazine and Journal of Science 1*:329–337.

Needham, J. 1954. *Science and Civilization in China*, Vol. 3. London: Cambridge University Press.

Needham, J., and L. Gwei-Djen. 1967. The optick artists of Chiangsu. In S. Bradbury and G. L'E. Turner, eds. *Historical Aspects of Microscopy*. Cambridge: Heffer, pp. 113–138.

Newton, I. 1672. A Letter of Mr. Isaac Newton. . . . containing his New Theory about Light and Colours. *Philosophical Transactions of the Royal Society 7*:3075–3087.

Newton, I. 1673. Mr. Newtons Answer to the foregoing Letter further explaining his Theory of Light and Colors, and particularly that of Whiteness; together with his continued hopes of perfecting Telescopes by Reflections rather than Refractions. *Philosophical Transactions of the Royal Society 8*:6087–6092.

Newton, I. 1691/1829. Letter to John Locke. In Lord King *The Life of John Locke, with Extracts from his Correspondence, Journals and Common-Place Books*. London: Henry Colburn.

Newton, I. 1704. *Opticks: or, a Treatise of the Reflections, Refractions, Inflections and Colours of Light*. London: Smith and Walford.

Newton, I. 1717. *Opticks: or, a Treatise of the Reflections, Refractions, Inflections and Colours of Light*, ed. 2. London: Innys.

Newton, I. 1730. *Opticks: or, a Treatise of the Reflections, Refractions, Inflections and Colours of Light*, ed. 4. London: Innys.

Nuttin, J. 1961. *Psychology in Belgium*. Louvain: Studia Psychologica.

O'Leary, De L. 1949. *How Greek Science Passed to the Arabs*. London: Routledge and Kegan Paul.

Ono, H. 1981. On Wells' 1792 law of visual direction. *Perception and Psychophysics 30*:403–406.

Ono, H., and N. J. Wade. 1985. Resolving discrepant results of the Wheatstone Experiment. *Psychological Research 47*:135–142.

Oppel, J. J. 1855. Über geometrisch-optische Täuschungen. *Jahresbericht des physikalischen Vereins zu Frankfurt am Main 37*–47.

Palmer, G. 1777a. *Theory of Colours and Vision*. London: Leacroft.

Palmer, G. 1777b. *Theory of Light and Colours*. Paris.

Palmer, G. 1786. *Théorie de la Lumière, Applicable aux Arts, et principalement à la Peinture*. Paris: Hardouin and Gattey.

Panum, P. L. 1858. *Physiologische Untersuchungen über das Sehen mit zei Augen*. Kiel: Schwerssche Buchhandlung.

Paré, A. 1564. *Dix Livres de la Chirurgie*. Paris: Le Royer.

Paré, A. 1649. *The Workes of the famous Chirurgion Ambrose Parey*. T. Johnson, trans. London: Cotes and Dugard.

Paris, J. A. 1827. *Philosophy in Sport made Science in Earnest! Being an attempt to illustrate the first Principles of Natural Philosophy*, Vol. 3. London: Longman, Rees, Orme, Brown, and Green.

Pastore, N. 1971. *Selective History of Theories of Visual Perception: 1650–1950*. New York: Oxford University Press.

Pastore, N. 1972. Sebastien Le Clerc on retinal disparity. *Journal of the History of the Behavioral Sciences 8*:336–339.

Paulus Ægineta. 1844. *The Seven Books of Paulus Ægineta*, Vol. 1. F. Adams, trans. London: The Sydenham Society.

Pecquet, A. 1668. Answer to Mr. Mariotte.... *Philosophical Transactions of the Royal Society 3*:669–671.

Perrault, Charles. 1696. *Les Hommes Illustres qui ont paru en France pendent ce Siecle: avec leurs Portraits naturel*, Vol. 1. Paris: Dezallier.

Perrault, Claude. 1683. An account of two letters of Mr. Perrault and Mr. Mariotte, concerning vision. *Philosophical Transactions of the Royal Society 13*:265–266.

Petit de Julleville, L. 1898. *Histoire de la Langue et de la Littérature française*, Vol. 6. Paris: Colin.

Pettigrew, T. J. 1838. *Medical Portrait Gallery*, Vol. 2. London: Fisher.

Pettigrew, T. J. 1840a. *Medical Portrait Gallery*, Vol. 3. London: Whittaker.

Pettigrew, T. J. 1840b. *Medical Portrait Gallery*, Vol. 4. London: Whittaker.

Picard, E. ed. 1924. *Histoire de la Nation Française*, Vol. 14. Paris: Plon-Nourrit.

Plateau, J. 1829. Dissertation sur quelques propriétés des impressions produites par la lumière sur l'organe de la vue. Liége.

Plateau, J. 1833. Des illusions sur lesquelles se fonde le petit appareil appelé récemment Phénakisticope. *Annales de Chimie et de Physique de Paris 53*:304–308.

Plateau, J. 1835. Sur un principe de photométrie. *Bulletins de l'Académie Royale des Sciences et de Belles-Lettres de Bruxelles 2*:52–59.

Plateau, J. 1836. Notice sur l'anorthoscope. *Bulletins de l'Académie Royale des Sciences et de Belles-Lettres de Bruxelles 3*:7–10.

Plateau, J. 1839. Answer to the objections published against a general theory of the visual appearances which arise from the contemplation of coloured objects. *London and Edinburgh Philosophical Magazine and Journal of Science 14*:330–340; 439–446.

Plateau, J. 1849. Quatrième note sur de nouvelles applications curieuses de la persist-ance des impressions de la rétine. *Bulletin de l'Académie Royale des Sciences, des Lettres et des Beaux-Arts de Belgique 16*:254–260.

Plateau, J. 1878a. Bibliographie analytique des principaux phénomènes de la vision, depuis les temps anciens jusqu'a la fin du XVIIIe siècle. Première section. Persistance des impressions sur la rétine. *Mémoires de l'Académie Royale de Belgique 42*:1–59.

Plateau, J. 1878b. Bibliographie analytique des principaux phénomènes de la vision, depuis les temps anciens jusqu'a la fin du XVIIIe siècle, suivie d'une bibliographie simple pour la partie écoulée du siècle actuel. Deuxième section. Couleurs accidentelles ordi-naires de succession. *Mémoires de l'Académie Royale de Belgique 42*:1–59.

Plateau, J. 1878c. Bibliographie analytique des principaux phénomènes de la vision, depuis les temps anciens jusqu'a la fin du XVIIIe siècle, suivie d'une bibliographie simple pour la partie écoulée du siècle actuel. Troisième section. Images qui succédent a la contemplation d'objets d'un grand éclat ou mème d'objets blanc bien éclairés. *Mémoires de l'Académie Royale de Belgique 42*:1–26.

Plateau, J. 1878d. Bibliographie analytique des principaux phénomènes de la vision, depuis les temps anciens jusqu'a la fin du XVIIIe siècle, suivie d'une bibliographie simple pour la partie écoulée du siècle actuel. Quatrième section. Irradiation. *Mémoires de l'Académie Royale de Belgique 42*:1–44.

Plater, F. 1583. *De corporis humani structura et usu*. Basel: König.

Plato. 1896. *The Works of Plato*, Vol. 1. H. Cary, trans. London: Bell.

Plato. 1935. *The Republic*. P. Shorey, trans. London: Heinemann.

Plato. 1946. *Timaeus*. R. G. Bury, trans. London: Heinemann.

Plehn, F. 1920. J. Keplers Behandlung des Sehens. *Zeitschrift für ophthalmologische Optik 8*:154–157.

Plehn, F. 1921. J. Keplers Behandlung des Sehens. *Zeitschrift für ophthalmologische Optik 9*:13–26, 40–54, 73–87, 103–109, 143–152, 177–182.

Pliny. 1940. *Natural History*, Vol. 3. Books VIII–XI. H. Rackham, trans. London: Heinemann.

Plug, C. 1989. Annotated bibliography. In *The Moon Illusion*. M. Hershenson, ed. Hillsdale, N.J.: Erlbaum, pp. 385–407.

Plug, C., and H. Ross. 1989. Historical review. In *The Moon Illusion*. M. Hershenson, ed. Hillsdale, N.J.: Erlbaum, pp. 5–27.

Polyak, S. 1957. *The Vertebrate Visual System*. Chicago: University of Chicago Press.

Polyak, S. L. 1942. *The Retina*. Chicago: University of Chicago Press.

Porac, C., and S. Coren. 1976. The dominant eye. *Psychological Bulletin 83*:880–897.

Porta, G. B. 1589. *Magiae Naturalis Libri XX*. Naples: Salviani.

Porta, J. B. 1593. *De Refractione. Optices Parte. Libri Novem*. Naples: Carlinum and Pacem.

Porta, J. B. 1669. *Natural Magick*. J. Wright, trans. London.

Porterfield, W. 1737. An essay concerning the motions of our eyes. Part I. Of their external motions. *Edinburgh Medical Essays and Observations 3*:160–263.

Porterfield, W. 1738. An essay concerning the motions of our eyes. Part II. Of their internal motions. *Edinburgh Medical Essays and Observations 4*:124–294.

Porterfield, W. 1759a. *A Treatise on the Eye, the Manner and Phænomena of Vision*, Vol. 1. Edinburgh: Hamilton and Balfour.

Porterfield, W. 1759b. *A Treatise on the Eye, the Manner and Phænomena of Vision*, Vol. 2. Edinburgh: Hamilton and Balfour.

Prévost, B. 1826. Sur une apparence de décomposition de la lumière blanche par la mouvement du corps qui la réfléchit. *Mémoires de la de la Société de Physique et d'Histoire Naturelle 3*:121.

Prevost, P. 1804. *Essais de Philosophie ou Étude de l'Esprit Humain*. Geneva: Paschoud.

Priestley, J. 1772. *The History and Present State of Discoveries relating to Vision, Light, and Colours*. London: Johnson.

Ptolemy, C. 1952. *The Almagest*, R. C. Taliaferro, trans. In *The Encyclopaedia Britannica Great Books of the Western World*, Vol. 16, pp. 1–478.

Purkinje, J. 1820. Beyträge zur näheren Kenntniss des Schwindels aus heautognostischen Daten. *Medicinische Jahrbücher des kaiserlich-königlichen österreichischen Staates 6*:79–125.

Purkinje, J. 1823a. *Beobachtungen und Versuche zur Physiologie der Sinne. Beiträge zur Kenntniss des Sehens in subjectiver Hinsicht*. Prague: Calve.

Purkinje, J. 1823b. *De Examine Physiologico Organi visus et Systematis Cutanei*. Breslau.

Purkinje, J. 1825. *Beobachtungen und Versuche zur Physiologie der Sinne. Neue Beiträge zur Kenntniss des Sehens in subjectiver Hinsicht*. Berlin: Reimer.

Reid, T. 1764. *An Inquiry into the Human Mind, On the principles of common sense*. Edinburgh: Millar, Kincaid & Bell.

Reisch, G. 1503. *Margarita Philosophica*. Freiburg: Schott.

Reusner, N. 1589. *Icones sives imagines vivae, literis*. Basel: Valdkirch.

Richter, G. M. A. 1965. *The Portraits of the Greeks*, 3 vols. London: Phaidon.

Rittenhouse, D. 1786. Explanation of an optical deception. *Transactions of the American Philosophical Society 2*:37–42.

Robinson, N. H. 1980. *The Royal Society Catalogue of Portraits*. London: The Royal Society.

Roget, P. M. 1825. Explanation of an optical deception in the appearance of the spokes of a wheel seen through vertical apertures. *Philosophical Transactions of the Royal Society 115*:131–140.

Roget, P. M. 1834. *Animal and Vegetable Physiology Considered with Reference to Natural Theology. Bridgewater Treatise V*, Vol. 2. London: Pickering.

Rohault, J. 1671. *Traité de Physique*. Paris: Savreux.

Rohault, J. 1723. *Rohault's System of Natural Philosophy*. J. Clarke, trans. London: Knapton.

Rohr, M. von. 1919. Ausgewählte Stücke aus Christoph Scheiners Augenbuch. *Zeitschrift für ophthalmologische Optik* 7:35–44, 53–64, 76–91, 101–113, 121–133.

Ronchi, V. 1968. The general influence of the development of optics in the seventeenth century on science and technology. In A. Beer, ed. *Vistas in Astronomy. IX*. Oxford: pp. 123–133.

Ronchi, V. 1970. *The Nature of Light. An Historical Survey*. V. Barocas, trans. London: Heinemann.

Rosen, E. 1956. The invention of eyeglasses. *Journal of the History of Medicine 11*:13–46, 183–218.

Ross, H. E., and D. J. Murray. 1978. *E. H. Weber: The Sense of Touch*. London: Academic Press.

Ross, H. E., and C. Plug. 1998. The history of size constancy and size illusions. In V. Walsh and J. J. Kulikowski, eds. *Visual Constancies: Why Things Look as They Do*. Cambridge: Cambridge University Press, pp. 499–528.

Ross, H. E., and G. W. Ross. 1976. Did Ptolemy understand the moon illusion? *Perception 5*:377–385.

Ross, W. D., ed. 1913. *The Works of Aristotle*, Vol. 6. Oxford: Clarendon.

Ross, W. D., ed. 1927. *The Works of Aristotle*, Vol. 7. Oxford: Clarendon.

Ross, W. D., ed. 1931. *The Works of Aristotle*, Vol. 3. Oxford: Clarendon.

Rumford, (Thompson, B.) Count. 1794. An account of some experiments upon coloured shadows. *Philosophical Transactions of the Royal Society 84*:107–118.

Runes, D. D. 1959. *Pictorial History of Philosophy*. New York: Philosophical Library.

Russell, B. 1959. *Wisdom of the West*. New York: Doubleday.

Ryff, W. H. 1541. *Das aller fürtrefflichsten . . . geschöpffs . . . wahrhafftige Beschreibung der Anatomie*. Strassburg: Beck.

Sabra, A. I. 1966. Ibn al-Haytham's criticisms of Ptolemy's *Optics*. *Journal of the History of Philosophy 4*:145–149.

Sabra, A. I. 1967. *Theories of Light from Descartes to Newton*. London: Oldbourne.

Sabra, A. I. 1987. Psychology versus mathematics: Ptolemy and Alhazen on the moon illusion. In E. Grant and J. E. Murdoch, eds. *Mathematics and its Application to Natural Philosophy in the Middle Ages*. Cambridge: Cambridge University Press, pp. 217–247.

Sabra, A. I., trans. and ed. 1989. *The Optics of Ibn Al-Haytham. Books I–III. On Direct Vision*. London: The Warburg Institute.

Saunders, J. B. de C. M., and C. D. O'Malley. 1950. *The Illustrations from the Works of Andreas Vesalius of Brussels*. New York: World Publishing.

Saunders, J. C. 1816. *A Treatise on some practical Points relating to the Diseases of the Eye*. New ed. with additions. London: Longman, Hurst, Rees, Orme, and Brown.

Scheerer, E. 1984. Nachbild. In *Historisches Wörterbuch der Philosophie*, Vol. 6. J. Ritter and K. Gründer, eds. Basel: Schwabe, pp. 341–348.

Scheerer, E. 1987. Tobias Mayer—Experiments on visual acuity (1755). *Spatial Vision* 2:81–97.

Scheiner, C. 1619. *Oculus, hoc est fundamentum opticum....* Innsbruck: Agricola.

Scheiner, C. 1630. *Rosa Ursina....* Bracciani: Phaeum.

Scherffer, K. 1761. *De coloribus accidentalibus*. Vindobone: Trattner.

Schmitz, E.-H. 1981. *Handbuch zur Geschichte der Optik*, Vol. 1. Bonn: Wayerborgh.

Schmitz, E.-H. 1982. *Handbuch zur Geschichte der Optik*, Vol. 2. Bonn: Wayerborgh.

Schmitz, E.-H. 1983. *Handbuch zur Geschichte der Optik*, Vol. 3. Bonn: Wayerborgh.

Schultze, M. 1866. Zur Anatomie und Physiologie der Retina. *Archiv für mikroskopische Anatomie 2*:175–286.

Seebeck, A. 1837. Ueber den bei manchen Personen verkommenden Mangel an Farbensinn. *Annalen der Physik und Chemie 42*:177–233.

Senden, M von. 1960. *Space and Sight*. P. Heath, trans. London: Methuen.

Seneca, 1971. *Naturales Quaestiones*. T. H. Corcoran, trans. London: Heinemann.

Shapiro, A. E. 1980. The evolving structure of Newton's theory of white light and color. *Isis 71*:211–235.

Shastid, T. H. 1917. History of ophthalmology. In C. A. Wood, ed. *The American Encyclopedia and Dictionary of Ophthalmology*, Vol. 11. Chicago: Cleveland Press, pp. 8524–8904.

Sherman, P. D. 1981. *Colour Vision in the Nineteenth Century*. Bristol: Hilger.

Shipley, T., and S. C. Rawlings. 1970. The nonius horopter—I. History and theory. *Vision Research 10*:1225–1262.

Shryock, R. H. 1944. The strange case of Wells' theory of natural selection. In M. F. A. Montagu, ed. *Studies and Essays in the History of Science and Learning offered in Homage to George Sarton*. New York: Schuman, pp. 197–207.

Siegel, R. E. 1959. Theories of vision and color perception of Empedocles and Democritus: Some similarities to the modern approach. *Bulletin of the History of Medicine 33*:145–159.

Siegel, R. E. 1970. *Galen on Sense Perception*. Basel: Karger.

Singer, C. 1921. Steps leading to the invention of the first optical apparatus. In C. Singer, ed. *Studies in the History and Method of Science*. Oxford: Clarendon, pp. 385–413.

Sinsteden, W. J. 1860. Ueber ein neues pseudoskopisches Bewegungsphänomen. *Annalen der Physik und Chemie 187*:336–339.

Smith, A. M. 1996. *Ptolemy's Theory of Visual Perception: An English Translation of the Optics with Introduction and Commentary.* Philadelphia: The American Philosophical Society.

Smith, J. A., and W. D. Ross, eds. 1910. *The Works of Aristotle*, Vol. 4. Oxford: Clarendon.

Smith, R. 1738. *A Compleat System of Opticks in Four Books.* Cambridge.

Smythies, J. R. 1957. A preliminary analysis of the stroboscopic patterns. *Nature* *179*:523–524.

Soemmerring, S. T. von. 1801. *Abbildungen des menschlichen Auges.* Frankfurt: Varrentrapp and Wenner.

Stampfer, S. 1833. *Die stroboskopischen Scheiben; oder, die optischen Zauberscheiben, deren Theorie und wissenschaftliche Anwendung, erklärt von dem Erfinder S. Stampfer.* Vienna: Trentsensky & Vieweg.

Steinbuch, J. G. 1811. *Beytrag zur Physiologie der Sinne.* Nuremberg: Schrag.

Stewart, D. 1800. *Account of the Life and Writings of Thomas Reid, D.D. F.R.S.Edin.* Edinburgh: Bell & Bradfute.

Stirling, W. 1902. *Some Apostles of Physiology.* London: Waterlow.

Stratton, G. M. 1917. *Theophrastus and the Greek Physiological Psychology before Aristotle.* New York: Macmillan.

Strong, D. S. 1979. *Leonardo on the Eye.* New York: Garland.

Struik, D. J. 1975. Willebrord Snel (Snellius or Snel van Royen). In C. C. Gillespie, ed. *Dictionary of Scientific Biography*, Vol. 12. New York: Scribner's, pp. 499–502.

Sudhoff, K. 1907. Tradition und Naturbeobachtung in den Illustrationen medizinischer Handschriften und Frühdrucke vornehmlich des 15. Jahrhunderts. *Studien zur Geschichte der Medizin.* Leipzig: Barth, pp. 1–92.

T. J. 1794. Letter to the Editor. *The Gentleman's Magazine and Historical Chronicle* *64*:1093.

Talbot, H. F. 1834. Experiments on light. *The London and Edinburgh Philosophical Magazine and Journal of Science 5*:321–334.

Taylor, Brook. 1719. *New Principles of Linear Perspective*, ed. 2. London.

Taylor, J. 1727. *An Account of the Mechanism of the Eye.* Norwich.

Taylor, J. 1738. *Le Mechanisme ou le Nouveau Traité de l'Anatomie du globe de l'oeil, avec l'usage de ses différentes parties, & de celles qui lui sont contigues.* Paris: David.

Taylor, W. C. 1846. *The National Portrait Gallery of Illustrious and Eminent Personages*, Vol. 2. London: Jackson.

Thompson, S. P. 1880. Optical illusions of motion. *Brain 3*:289–298.

Thorwald, J. 1962. *Science and Secrets of Early Medicine.* R. and C. Winston, trans. London: Thames & Hudson.

Treviranus, G. R. 1828. *Beiträge zur Anatomie und Physiologie der Sinneswerkzeuge des Menschen und der Thiere.* Bremen: Heyse.

Treviranus, G. R. 1837. *Beiträge zur Aufklärung der Erscheinungen und Gesetze des organischen Lebens*, Vol. 1, Issue 3. *Resultate neuer Untersuchungen über die Theorie des Sehens und über den innern Bau der Netzhaut des Auges.* Bremen: Heyse.

Troxler, D. 1804. Ueber das Verschwinden gegebener Gegenstände innerhalb unseres Gesichtskreises. *Ophthalmologische Bibliothek 2*:1–53.

Turbervile, D. 1684. Two Letters.... containing several remarkable cases in Physick, relating chiefly to the Eyes. *Philosophical Transactions of the Royal Society 14*:736–738.

Türler, H., V. Attinger, and M. Godet. 1929. *Historisch-Biographisches Lexikon der Schweiz*, Vol. 5. Neurenburg: Administration des historisch-biographischen Lexikons der Schweiz.

Turnbull, H. W., ed. 1960. *The Correspondence of Isaac Newton*, Vol. 2. *1676–1687.* Cambridge: Cambridge University Press.

Urmson, J. O. 1960. *The Concise Encyclopædia of Western Philosophy and Philosophers.* London: Hutchinson.

Vaizey, M. 1982. *The Artist as Photographer.* London: Sidgwick and Jackson.

Veltman, K. H. 1986. *Studies on Leonardo da Vinci. l. Linear Perspective and the Visual Dimensions of Science and Art.* Munich: Deutscher Kunstverlag.

Verriest, G. 1990. Life, eye disease and work of Joseph Plateau. *Documenta Ophthalmologica 74*:9–20.

Verstraten, F. A. J. 1996. On the ancient history of the direction of the motion after-effect. *Perception 25*:1177–1187.

Vesalius, A. 1543. *De humani corporis fabrica Libri septem.* Basel.

Vieth, G. U. A. 1818. Ueber die Richtung der Augen. *Annalen der Physik 28*:233–253.

Vitellonis (Witelo). 1572. *Perspectiva. Thuringopoloni Opticae Libri Decem.* F. Risner, ed. Basel: Gallia.

Vitruvius. 1962. *On Architecture.* F. Granger, trans. London: Heinemann.

Volkmann, A. W. 1836. *Neue Beiträge zur Physiologie des Gesichtssinnes.* Leipzig: Breitkopf and Härtel.

Volkmann, A. W. 1859. Das Tachistoscop, ein Instrument, welches bei Untersuchung des momentanen Sehens den Gebrauch des elektrischen Funkens ersetz. *Berichte über die Verhandlungen der königlichen sächsischen Akademie der Wissenschaften mathematische-naturwissenschaftliche Klasse 2*:90–98.

Vollgraff, J. A. 1936. Snellius' notes on the reflection and refraction of rays. *Osiris 1*:718–725.

Voltaire (François Marie Arouet). 1738. *Elémens de la Philosophi de Neuton.* Amsterdam: Desbordes.

Wade, N. J. 1977a. A note on the discovery of subjective colours. *Vision Research 17*:671–672.

Wade, N. J. 1977b. Distortions and disappearances of geometrical patterns. *Perception* 6:407–433.

Wade, N. J. 1978a. Why do patterned afterimages fluctuate in visibility? *Psychological Bulletin 85*:338–352.

Wade, N. J. 1978b. Sir Charles Bell on visual direction. *Perception 7*:359–362.

Wade, N. J. 1981. A note on the history of binocular microscopes. *Perception 10*:591–592.

Wade, N. 1982. *The Art and Science of Visual Illusions*. London: Routledge & Kegan Paul.

Wade, N. J., ed. 1983. *Brewster and Wheatstone on Vision*. London: Academic Press.

Wade, N. J. 1987. On the late invention of the stereoscope. *Perception 16*:785–818.

Wade, N. 1990. *Visual Allusions: Pictures of Perception*. Hove, East Sussex: Erlbaum.

Wade, N. J. 1994a. Hermann von Helmholtz (1821–1894). *Perception 23*:981–989.

Wade, N. J. 1994b. A selective history of the study of visual motion aftereffects. *Perception 23*:1111–1134.

Wade, N. J. 1998. Early studies of eye dominances. *Laterality 3*:97–109.

Wade, N. J., and D. Heller. 1997. Scopes of perception: The experimental manipulation of space and time. *Psychological Research 60*:227–237.

Wade, N. J., and M. T. Swanston. 1996. A general model for the perception of space and motion. *Perception 25*:187–194.

Wallis, J. 1687. The sentiments of the Reverend and Learned Dr. Wallis R. S. Soc. upon the aforesaid appearance. *Philosophical Transactions of the Royal Society 16*:323–329.

Walls, G. L. 1956. The G. Palmer story. (Or, what it's like, sometimes, to be a scientist). *Journal of the History of Medicine and Allied Sciences 11*:66–96.

Walsh, J. 1935. Galen's writings and influences inspiring them. *Annals of Medical History 7*:570–589.

Walsh, J. W. T. 1958. *Photometry*. London: Constable.

Wardrop, J. 1808. *Essays on the Morbid Anatomy of the Human Eye*, Vol. 1. Edinburgh: Archibald Constable.

Wardrop, J. 1818. *Essays on the Morbid Anatomy of the Human Eye*, Vol. 2. Edinburgh: Archibald Constable.

Wardrop, J. 1826. Case of a lady born blind, who received sight at an advanced age by the formation of an artificial pupil. *Philosophical Transactions of the Royal Society 116*:529–540.

Wardrop, J. 1834. *Essays on the Morbid Anatomy of the Human Eye*, ed. 2, Vol. 2. London: Churchill.

Ware, J. 1801. Case of a young gentleman, who recovered his sight when seven years of age, after having been deprived of it by cataracts, before he was a year old; with remarks. *Philosophical Transactions of the Royal Society 91*:382–396.

Ware, J. 1813. Observations relative to the near and distant sight of different persons. *Philosophical Transactions of the Royal Society 103*:31–50.

Wartmann, É. F. 1844. *Mémoire sur la Daltonisme.* Geneva.

Wasserman, G. S. 1978. *Color Vision: An Historical Introduction.* New York: Wiley.

Waterhouse, J. 1902. Camera obscura. *Encyclopædia Britannica,* ed. 10, Vol. 26, Edinburgh and London: Adams and Charles.

Weale, R. A. 1957. Trichromatic ideas in the seventeenth and eighteenth centuries. *Nature 179*:648–651.

Wells, W. C. 1792. *An Essay upon Single Vision with two Eyes: together with Experiments and Observations on several other Subjects in Optics.* London: Cadell.

Wells, W. C. 1794a. Reply to Dr. Darwin on vision. *The Gentleman's Magazine and Historical Chronicle 64*:794–797.

Wells, W. C. 1794b. Reply to Dr. Darwin on vision. *The Gentleman's Magazine and Historical Chronicle 64*:905–907.

Wells, W. C. 1811. Observations and experiments on vision. *Philosophical Transactions of the Royal Society 101*:378–391.

Welsh, D. 1825. *Account of the Life and Writings of Thomas Brown, M.D.* Edinburgh: Tait.

Westfall, R. S. 1980. *Never at Rest. A Biography of Isaac Newton.* Cambridge: Cambridge University Press.

Westheimer, G. 1976. Oculomotor control: The vergence system. In R. A. Monty and J. W. Senders, eds. *Eye Movements and Psychological Processes.* Hillsdale, N.J.: Erlbaum, pp. 55–64.

Wheatstone, C. 1827. Description of the kaleidophone, or phonic kaleidoscope; A new philosophical toy, for the illustration of several interesting and amusing acoustical and optical phenomena. *Quarterly Journal of Science, Literature and Art 23*:344–351.

Wheatstone, C. 1834. An account of some experiments to measure the velocity of electricity and the duration of electric light. *Philosophical Transactions of the Royal Society 124*:583–591.

Wheatstone, C. 1838. Contributions to the physiology of vision—Part the first. On some remarkable, and hitherto unobserved, phenomena of binocular vision. *Philosophical Transactions of the Royal Society 128*:371–394.

Wheatstone, C. 1852. Contributions to the physiology of vision—Part the second. On some remarkable, and hitherto unobserved, phenomena of binocular vision. *Philosophical Transactions of the Royal Society 142*:1–17.

Wheatstone, C. 1853. On the binocular microscope, and on stereoscopic pictures of microscopic objects. *Transactions of the Microscopical Society of London 1*:99–102.

Whewell, W. 1837. *History of the Inductive Sciences, from the Earliest to the Present Times,* Vol. 2. London: Parker.

Whisson, Rev. 1778. An account of a remarkable imperfection of sight. In a letter from J. Scott to the Rev. Mr. Whisson. *Philosophical Transactions of the Royal Society* *68*:611–614.

Wilde, E. 1838. *Geschichte der Optik, vom Ursprunge dieser Wissenschaft bis auf die gegenwärtige Zeit*. Berlin: Rücken and Püchler.

Wilkes, A., and N. J. Wade. 1997. Bain on neural networks. *Brain and Cognition* *33*:295–305.

Williams, S. W. 1845. *American Medical Biography*. Greenfield, Mass.: Merriam.

Willis, T. 1664. *Cerebri Anatome: cui accessit Nervorum Descriptio et Usus*. London: Martyn and Allestry.

Willis, T. 1681. *The Anatomy of the Brain*. S. Pordage, trans. In *The Remaining Medical Works of that Famous and Renowned Physician Dr Thomas Willis*. London: Dring, Harper, Leigh, and Martyn.

Wilson, G. 1855. *Researches on Colour-Blindness*. Edinburgh: Sutherland & Knox.

Wohlgemuth, A. 1911. On the after-effect of seen movement. *British Journal of Psychology. Monograph Supplement* *1*:1–117.

Wolf, A. 1935. *A History of Science, Technology, and Philosophy in the 16th & 17th Centuries*. London: Allen & Unwin.

Wolf, A. 1938. *A History of Science, Technology, and Philosophy in the Eighteenth Century*. London: Allen & Unwin.

Wollaston, W. H. 1824a. On the semi decussation of the optic nerves. *Philosophical Transactions of the Royal Society* *114*:222–231.

Wollaston, W. H. 1824b. On the apparent direction of eyes in a portrait. *Philosophical Transactions of the Royal Society* *114*:247–256.

Wood, C. A. 1929. *Benevenutus Grassus of Jerusalem De Oculis*. Stanford: Stanford University Press.

Wood, W. ed. 1885. *Portraits of the One Hundred Greatest Men in History*. London: Sampson Low, Marston, Searle, & Rivington.

Wundt, W. 1898. Die geometrisch-optischen Täuschungen. *Abhandlungen der mathematisch-physischen Classe der königlichen sächsischen Gesellschaft der Wissenschaften* *42*:55–178.

Young, T. 1793. Observations on vision. *Philosophical Transactions of the Royal Society* *83*:169–181.

Young, T. 1800. Outlines of experiments and enquiries respecting sound and light. *Philosophical Transactions of the Royal Society* *90*:106–150.

Young, T. 1801. On the mechanism of the eye. *Philosophical Transactions of the Royal Society* *91*:23–88.

Young, T. 1802a. On the theory of lights and colours. *Philosophical Transactions of the Royal Society* *92*:12–48.

Young, T. 1802b. An account of some cases of the production of colours, not hitherto described. *Philosophical Transactions of the Royal Society 92*:387–397.

Young, T. 1807. *A Course of Lectures on Natural Philosophy and the Mechanical Arts*, 2 vols. London: Johnson.

Zaunick, R. 1961. ed. Purkyně-symposium. *Nova Acta Leopoldina*, Vol. 24. Leipzig: Barth.

Ziggelaar, A. 1983. *François de Aguilón S. J. (1567–1617) Scientist and Architect*. Rome: Institutum Historicum S. I.

Ziggelaar, A. 1993. The early debate concerning wave-theory. In M. J. Petry, ed. *Hegel and Newtonianism*. Amsterdam: Kluwer, pp. 517–529.

Zöllner, F. 1862. Ueber eine neue Art anorthoscopische Zerrbilder. *Annalen der Physik und Chemie 117*:477–484.

Zusne, L. 1968. Optical illusions: Output of publications. *Perceptual and Motor Skills 27*:175–177.

Name Index

Numbers in boldface indicate the pages on which quotations are given. The italicized numbers refer to pages that contain portraits, and those in roman are for entries in the commentary, text, and legends.

Subject Index

Entries in boldface denote chapter sections.